Virus-Based Nanomaterials and Nanostructures

Virus-Based Nanomaterials and Nanostructures

Special Issue Editors

Dong-Wook Han
Jin-Woo Oh

MDPI • Basel • Beijing • Wuhan • Barcelona • Belgrade • Manchester • Tokyo • Cluj • Tianjin

Special Issue Editors

Dong-Wook Han
Pusan National University (PNU)
Korea

Jin-Woo Oh
Pusan National University (PNU)
Korea

Editorial Office
MDPI
St. Alban-Anlage 66
4052 Basel, Switzerland

This is a reprint of articles from the Special Issue published online in the open access journal *Nanomaterials* (ISSN 2079-4991) (available at: https://www.mdpi.com/journal/nanomaterials/special_issues/Virus-Based_Nano).

For citation purposes, cite each article independently as indicated on the article page online and as indicated below:

LastName, A.A.; LastName, B.B.; LastName, C.C. Article Title. *Journal Name* **Year**, *Article Number*, Page Range.

ISBN 978-3-03928-694-2 (Pbk)
ISBN 978-3-03928-695-9 (PDF)

Contents

About the Special Issue Editors

Dong-Wook Han obtained his B.S. from the Department of Biochemistry at Yonsei University, Seoul, Korea, in 1998. He completed his M.S. and Ph.D. in the graduate program of biomedical engineering from Yonsei University in 2000 and 2004, respectively. With two and a half years of experience as a postdoctoral fellow (under the supervision of Prof. Suong-Hyu Hyon) at the Institute for Frontier Medical Sciences, Kyoto University, Japan, Dr. Han returned to Korea. In 2008, he joined the faculty of Pusan National University (PNU) where he began his academic carrier as an assistant professor and is currently a full professor and chair in the Department of Optics and Mechatronics Engineering at PNU. Dr. Han serves as a board member of several biomedical societies and an editorial board member for many scientific journals including BioMed Research International, Nanomaterials, World Journal of Stem Cells, Journal of Nanotheranostics, Biomaterials Research, etc. Since 2008, he has authored or co-authored over 120 scientific publications, possesses over 10 international and national patents, and joined several book chapters. His research interest concerns BT-NT convergence; especially tissue engineering and regenerative/translational medicine using smart nanobiomaterials; development of artificial tissues/organs, medical devices, and organs-on-chips; 3D bioprinting; cell imaging; and evaluation of biocompatibility and nanotoxicity.

Jin-Woo Oh, For more than five years, Dr. Jin-Woo Oh has honed his knowledge of M13 bacteriophage, bio-photonics, the synthesis of nano-biomaterials, biometric self-assembly, and bio-photoelectronics through extensive research. He was inspired to enter his profession due to his desire to contribute to the advancement of life. He has worked as director of the Future Materials Discovery Business: Center for Phage-Meta Materials since 2018, director of the Institute of BIT Convergence Technology since 2017, and associate professor in the department of nanoenergy engineering at Pusan National University in Busan, Republic of Korea, since 2016. He commenced his career as an assistant professor in the department on nanoenergy engineering at Pusan National University in 2012, holding the role for four years. Throughout his career, he has contributed to 30 patents and has helped to develop a virus-based electronic nose and phage meta-materials, which he cites as a highlight of his career. An expert in his field, Dr. Jin-Woo Oh holds a Bachelor of Science in chemistry, Master of Science in physical chemistry, and Doctor of Philosophy in physical chemistry from Hanyang University in Seoul, Korea. Subsequently, he completed two postdoctoral fellowships: one in chemical engineering at Hanyang University and one in bioengineering at the University of California, Berkeley. He concluded his academic pursuits in 2012.With a vast amount of knowledge at his disposal, Dr. Oh has contributed more than 100 papers to prestigious professional journals, including Biomimetic Virus-Based Colourimetric Sensors, Bio-Inspired Piezoelectric Nanogenerators-Based on Vertically Aligned Phage Nanopillars, and Micro Heater-Based Virus Templating Full Colour Pixels, among several others. In recognition of his success, he earned an Excellent Doctorate Thesis from Hanyang University in 2009 and a Young Academic Award from The Polymer Society of Korea in 2016. Looking to the future, Dr. Jin-Woo Oh intends to become an expert in the bio-photonic field.

Editorial

Virus-Based Nanomaterials and Nanostructures

Jin-Woo Oh [1],* and Dong-Wook Han [2],*

[1] Department of Nanofusion Technology, College of Nanoscience & Nanotechnology, Pusan National University, Busan 46241, Korea
[2] Department of Cogno-Mechatronics Engineering, College of Nanoscience & Nanotechnology, Pusan National University, Busan 46241, Korea
* Correspondence: ojw@pusan.ac.kr (J.-W.O.); nanohan@pusan.ac.kr (D.-W.H.)

Received: 28 February 2020; Accepted: 19 March 2020; Published: 20 March 2020

Abstract: This Special Issue highlights the recent developments and future directions of virus-based nanomaterials and nanostructures in energy and biomedical applications. The virus-based biomimetic materials formulated using innovative ideas presented herein are characterized for the applications of biosensors and nanocarriers. The research contributions and trends based on virus-based materials, covering energy-harvesting devices to tissue regeneration over the last two decades, are described and discussed.

Keywords: virus-based nanomaterials; energy devices; biomedical applications; self-assembly; piezoelectric biomaterials

---

Virus-based biomimetic materials derived from plant viruses and bacteriophages rarely generate harmful side effects in human beings since phages do not consist of mammalian promoter sequences in their genomes [1]. The structures of viruses consist of two or three parts: (1) the genetic material composed from either DNA or RNA, which carries genetic information; (2) a protein coat, called the capsid, which surrounds and protects the genetic material; and in some cases (3) an envelope of lipids that surrounds the protein coat [2]. Advantageously, the monodispersed phages can self-assemble to develop rope-like bundles or liquid crystals and their surface can be modified either by chemical or genetic modifications [1,3]. Due to these unique properties, many research findings based on phage-based biomaterials have been developed for applications in drug delivery, biosensors, biomedical imaging, tissue regeneration, energy, and catalysis [4,5].

The current Special Issue, including 11 original research works, focuses on highlighting the progress, challenges, and future directions in the area of virus-based nanomaterials and nanostructures with multiple applications in biomedicine and energy. Researchers have studied mineralization, magnetization, bioconjugation, and drug delivery for the development of biosensors and vaccines [6–12]. Research findings of virus-based biomimetic materials in energy, biosensors, and tissue regeneration over the last two decades are comprehensively discussed in reviews [13–16].

Kim et al. propose an M13-bacteriophage-based colorimetric multi-array biosensor that can classify four different antibiotics (duricef, citopcin, amoxicillin, and rifampin) and hormone (estrogen) drugs including mercilon, gestodene, estrone, and estradiol by analyzing the color change [6]. The sensor can be fabricated using self-assembly of genetically engineered M13 bacteriophages, which incorporates peptide libraries on its surface. The fabricated sensor platform consists of a sensor chip that is 1 cm^2 wide, a chamber of about 30 cc capacity, and a small webcam. This biosensor system is inexpensive and easy to apply in monitoring the environment and health care. The color change in the biosensor is caused by a reaction between the sensor array and external substances, detected by a complementary metal–oxide–semiconductor detector, and followed by employing hierarchical cluster analysis.

Yuste-Calvo et al. developed turnip mosaic virus (TuMV)-based nanoparticles to detect antibodies with high sensitivity [7]. TuMV is a virion with an elongated and flexuous structure. It is 700 nm long and 12 nm wide. Nearly 2000 copies of coat protein are present in each particle. The research group modified TuMV virus-like particles (VLPs) with a peptide from the chaperonin Hsp60, which is known to be involved in inflammation processes and autoimmune diseases. The quantitative detection of anti-Hsp60 autoantibodies is demonstrated through the multimeric presentation of the epitopes on TuMV VNPs through an in vivo murine (adult C57BL/6J) model. In particular, the high sensitivity of the developed Hsp60-VLPs provides a novel effective tool for diagnosis, progression, and prognosis in inflammation-mediated disorders. The model is suggested to reproduce the clinical, histopathological, and immune characteristics observed in humans by the induction of chronic colitis associated with diarrhea and weight loss.

Damm et al. functionalized the surface of calcium phosphate (CaP) nanoparticles with stabilized trimers of the HIV-1 envelope (Env), resulting in Env-CaP-p30 nanoparticles, to demonstrate improvement of Env antibody responses by intrastructural help (ISH) [8]. The Env trimers (MW = 140 kDa) are coupled to the nanoparticle surface using sulfosuccinimidyl-trans-4-(N-maleimidomethyl)cyclohexane-1-carboxylate (sulfo-SMCC) cross-linker, which reacts primary amines in Env with the thiol groups on the nanoparticle surface. The in vitro studies explored the Env-CaP-p30 nanoparticles' induction of the activation of naive Env-specific B-cells in contrast to soluble Env trimers. The authors applied the nanoparticles to study the effects of ISH in mice immunized with a licensed vaccine against tetanus toxoid.

Atanasova et al. determined whether the surface hydrophilicity of tobacco mosaic virus (TMV) particles can be manipulated through covalent attachment of polymer molecules [9]. Three different polymers, namely perfluorinated poly(pentafluorostyrene) (PFS), the thermo-responsive poly(propylene glycol) acrylate (PPGA), and the block copolymer polyethylene-block-poly(ethylene glycol), were examined. In this study, wild-type tobacco mosaic virus (wt-TMV) and a TMV-cysteine (Cys) mutant presenting thiol groups help with the covalent binding of the organic molecules. The covalent attachment makes the virus surface hydrophobic without affecting the integrity of the capsid and suppresses the virus mineralization by inorganic deposits. The growth mechanism of the inorganic material on the virus surface was analyzed in terms of the optical properties, bandgap (Eg), and particle size of solution-grown zinc sulfide (ZnS) nanoparticles. The authors concluded from the ZnS mineralization test that the degree of the virus hydrophobicity can be tuned by the polymer properties.

Suhaimi et al. developed a vaccine by expressing *Neospora caninum* profilin (NcPROF) in silkworm larvae by recombinant *Bombyx mori* nucleopolyhedrovirus (BmNPV) bacmid [10]. They investigated three NcPROF-based constructs for the recombination viz. native NcPROF fused with an N-terminal protective antigen (PA)tag (PA-NcPROF), PA-NcPROF with the signal sequence of bombyxin from *B. mori* (bx-PA-NcPROF), and bx-PA-NcPROF with additional C-terminal transmembrane and cytoplasmic domains of GP64 from BmNPV (bx-PA-NcPROF-GP64TM). Only bx-PA-NcPROF-GP64TM was found to be secreted as nanoparticles with binding affinity to its receptors and pelleted, whereas the remaining two were not.

Rybka et al. characterized the assembly of hepatitis B virus capsid protein into virus-like particles (HBc VLPs) with the magnetic core of superparamagnetic iron oxide nanoparticles (SPIONs) [11]. The synthesized SPIONs were functionalized with two different ligands—1,2-,istearoyl-sn-glycero-3-phosphoethanolamine-N-[carboxy-(polyethyleneglycol)] (PL-PEG-COOH) or dihexadecyl phosphate (DHP)—to further assess crucial parameters guiding SPION-HBc VLP assembly. They evaluated the mechanism of self-assembly as well as the antigenicity of SPION-HBc VLPs.

Finbloom et al. investigated three morphologically distinct VLPs, namely a 27 nm MS2 sphere, an 18 nm tobacco mosaic virus (TMV) disk, and a 50 nm nanophage filamentous rod conjugated to doxorubicin (DOX), for their drug delivery potential to glioblastoma [12]. Although all VLPs exhibited adequate drug delivery and cell uptake in vitro, the authors found that the survival rates of glioma-bearing mice treated with TMV showed the best responses in vivo. The results of physicochemical and biological characterizations suggested that these VLPs can be promising nanocarriers for the convection-enhanced delivery (CED) of traditional chemotherapeutics such as DOX in glioblastoma treatment.

Shin et al. provide a comprehensive review of the principles, properties, and role of organic piezoelectric biomaterials in energy and biomedical applications [13]. The biomedical devices featuring the biocompatible piezoelectric materials are useful in energy-harvesting devices, sensors, and scaffolds for cell and tissue engineering. The review addresses how to tackle issues related to the better integration of organic piezoelectric biomaterials into biomedical devices. When mechanically agitated, piezoelectric materials accumulate an electric charge. Although organic piezoelectric biomaterials possess weak piezoelectricity compared with their inorganic counterparts, they can serve as the functional materials in the field of medically mountable and implantable applications when they are well processed. Piezoelectric mechanisms and the properties of the materials, including collagen, glycine, M13 bacteriophage, silk, cellulose, diphenylalanine, and poly(vinylidene fluoride), are discussed throughout the review.

Park et al. describe methods to fabricate M13-bacteriophage-based piezoelectric energy-harvesting devices to develop high-performance and biocompatible energy devices for a wide range of practical applications [14]. Due to surface modification, M13 bacteriophages overcome other natural biomaterials with limitations in mass production and low piezoelectric properties. M13 bacteriophages exhibit unique features such as similar structures with collagens, mass amplification, genetic modification, liquid–crystalline phase transition, and excellent piezoelectric properties, which distinguish them from other materials. Among the M13-phage-based piezoelectric energy-harvesting devices, vertically aligned phage films exhibited the highest performance with a peak voltage of 2.8 V and a peak current of 120 nA. The review suggests some strategies, such as fabricating triboelectric devices based on M13 phages and developing composite structures composed of organic and inorganic biomaterials, to enhance the power of devices.

In another review, Moon et al. highlight the recent progress made in the application of M13-bacteriophage-based sensor systems and discuss future M13 bacteriophage technology [15]. Genetic engineering provides many possibilities to use M13 bacteriophages as the core material of sensors. Due to this engineering, these bacteriophages exhibit specific binding affinity to target materials, including chemicals and biological materials. The modification mainly occurs on the pVIII protein of the M13 bacteriophage. M13 bacteriophages display similar structure and behavior to liquid crystals during fabrication. The crystal structure of self-assembled M13 bacteriophages exhibits different phases, i.e., nematic, cholesteric, and smectic at low, medium, and high concentrations, respectively. Using a natural M13 bacteriophage without any genetic engineering also provides the opportunity to interact with ligands due to an abundant negative charge on the surface protein. The charge distribution of the C-terminus (positive) and N-terminus (negative) induces a strong dipole to M13 bacteriophages providing a natural negative charge, which enables them to interact with positively charged materials including carbon nanofibers. The sensing potential of M13 bacteriophages toward proteins, microorganisms, chemicals, and color is discussed throughout this review.

Raja et al. describe the emerging trend of virus-based biomimetic materials in tissue regeneration [16]. The review outlines the integration of the virus into biomaterials with different morphologies such as hydrogels, two-dimensional substrates, and nanofibers in the field of tissue regeneration. It discusses remarkable properties of medicinally valued viruses including fd, M13, TMV, T7, and Potato virus X (PVX). Apart from tissue regeneration, the review covers other biomedical applications such as drug delivery, bioimaging, and biosensing. In conclusion, the authors recommend a systematic study using virus-based biomimetic materials to explain the different phases of tissue regeneration. Sophisticated techniques and methodologies are required to estimate the number of peptides expressed on each phage particle. Nonetheless, many viral nanocomposites in the form of polymeric micelles, vesicles, and dendrimers remain to be developed.

In conclusion, as the Editors of this Special Issue, we would like to thank all the authors and reviewers, who contributed to this Special Issue with innovative ideas and constructive reviewers' comments. We would like to express appreciation for the consistent support from the Editorial Office. We are sure that this Special Issue will provide our readers with a platform to understand the advantageous properties of benevolent virus-based nanomaterials and nanostructures with their pivotal roles in energy and biomedical applications.

Funding: This work was supported by a 2-Year Research Grant of Pusan National University, Busan, Korea.

Acknowledgments: The Guest Editors are deeply thankful to all authors for submitting their studies to the present Special Issue and for its successful completion. We sincerely acknowledge the reviewers and editorial members for enhancing the quality and impact of all submitted papers. Special thanks to Susie Sun and Mirabelle Wang, the editorial assistants, for the smooth and efficient process.

Conflicts of Interest: The authors declare no conflict of interest.

References

1. Mohan, K.; Weiss, G.A. Dual genetically encoded phage-displayed ligands. *Anal. Biochem.* **2014**, *453*, 1–3. [CrossRef] [PubMed]

2. Mohan, K.; Weiss, G.A. Chemically modifying viruses for diverse applications. *ACS Chem. Biol.* **2016**, *11*, 1167–1179. [CrossRef] [PubMed]

3. Steinmetz, N.F. Viral nanoparticles as platforms for next-generation therapeutics and imaging devices. *Nanomedicine* **2010**, *6*, 634–641. [CrossRef] [PubMed]

4. Cao, B.; Li, Y.; Yang, T.; Bao, Q.; Yang, M.; Mao, C. Bacteriophage-based biomaterials for tissue regeneration. *Adv. Drug Deliv. Rev.* **2019**, *145*, 73–95. [CrossRef] [PubMed]

5. Wan, J.; Shu, H.; Huang, S.; Fiebor, B.; Chen, I.; Petrenko, V.A.; Chin, B.A. Phage-based magnetoelastic wireless biosensors for detecting *Bacillus anthracis* spores. *IEEE Sens. J.* **2007**, *7*, 470–477. [CrossRef]

6. Kim, C.; Lee, H.; Devaraj, V.; Kim, W.-G.; Lee, Y.; Kim, Y.; Jeong, N.-N.; Choi, E.J.; Baek, S.H.; Han, D.-W.; et al. Hierarchical cluster analysis of medical chemicals detected by a bacteriophage-based colorimetric sensor array. *Nanomaterials* **2020**, *10*, 121. [CrossRef] [PubMed]

7. Yuste-Calvo, C.; López-Santalla, M.; Zurita, L.; Cruz-Fernández, C.F.; Sánchez, F.; Garín, M.I.; Ponz, F. Elongated flexuous plant virus-derived nanoparticles functionalized for autoantibody detection. *Nanomaterials* **2019**, *9*, 1438. [CrossRef] [PubMed]

8. Damm, D.; Rojas-Sánchez, L.; Theobald, H.; Sokolova, V.; Wyatt, R.T.; Überla, K.; Epple, M.; Temchura, V. Calcium phosphate nanoparticle-based vaccines as a platform for improvement of HIV-1 Env antibody responses by intrastructural help. *Nanomaterials* **2019**, *9*, 1389. [CrossRef] [PubMed]

9. Atanasova, P.; Atanasov, V.; Wittum, L.; Southan, A.; Choi, E.; Wege, C.; Kerres, J.; Eiben, S.; Bill, J. Hydrophobization of tobacco mosaic virus to control the mineralization of organic templates. *Nanomaterials* **2019**, *9*, 800. [CrossRef] [PubMed]

10. Suhaimi, H.; Hiramatsu, R.; Xu, J.; Kato, T.; Park, E.Y. Secretory nanoparticles of neospora caninum profilin-fused with the transmembrane domain of GP64 from silkworm hemolymph. *Nanomaterials* **2019**, *9*, 593. [CrossRef] [PubMed]

11. Rybka, J.D.; Mieloch, A.A.; Plis, A.; Pyrski, M.; Pniewski, T.; Giersig, M. Assembly and characterization of HBc derived virus-like particles with magnetic core. *Nanomaterials* **2019**, *9*, 155. [CrossRef] [PubMed]

12. Finbloom, J.A.; Aanei, I.L.; Bernard, J.M.; Klass, S.H.; Elledge, S.K.; Han, K.; Ozawa, T.; Nicolaides, T.P.; Berger, M.S.; Francis, M.B. Evaluation of three morphologically distinct virus-like particles as nanocarriers for convection-enhanced drug delivery to glioblastoma. *Nanomaterials* **2018**, *8*, 1007. [CrossRef] [PubMed]

13. Shin, D.-M.; Hong, S.W.; Hwang, Y.-H. Recent advances in organic piezoelectric biomaterials for energy and biomedical applications. *Nanomaterials* **2020**, *10*, 123. [CrossRef] [PubMed]

14. Park, I.W.; Kim, K.W.; Hong, Y.; Yoon, H.J.; Lee, Y.; Gwak, D.; Heo, K. Recent developments and prospects of M13- bacteriophage based piezoelectric energy harvesting devices. *Nanomaterials* **2020**, *10*, 93. [CrossRef] [PubMed]

15. Moon, J.-S.; Choi, E.J.; Jeong, N.-N.; Sohn, J.-R.; Han, D.-W.; Oh, J.-W. Research progress of M13 bacteriophage-based biosensors. *Nanomaterials* **2019**, *9*, 1448. [CrossRef] [PubMed]
16. Raja, I.S.; Kim, C.; Song, S.-J.; Shin, Y.C.; Kang, M.S.; Hyon, S.-H.; Oh, J.-W.; Han, D.-W. Virus-incorporated biomimetic nanocomposites for tissue regeneration. *Nanomaterials* **2019**, *9*, 1014. [CrossRef] [PubMed]

 *nanomaterials*

Review

Virus-Incorporated Biomimetic Nanocomposites for Tissue Regeneration

Iruthayapandi Selestin Raja [1,†], Chuntae Kim [2,†], Su-Jin Song [3], Yong Cheol Shin [4], Moon Sung Kang [3], Suong-Hyu Hyon [5], Jin-Woo Oh [2,*] and Dong-Wook Han [3,*]

[1] Monocrystalline Bank Research Institute, Pusan National University, Busan 46241, Korea
[2] Department of Nanofusion Technology, College of Nanoscience & Nanotechnology, Pusan National University, Busan 46241, Korea
[3] Department of Cogno-Mechatronics Engineering, College of Nanoscience & Nanotechnology, Pusan National University, Busan 46241, Korea
[4] Department of Medical Engineering, Yonsei University, College of Medicine, Seoul 03722, Korea
[5] Joint Faculty of Veterinary Medicine, Kagoshima University, Kagoshima 890-8580, Japan
* Correspondence: ojw@pusan.ac.kr (J.-W.O.); nanohan@pusan.ac.kr (D.-W.H.)
† These authors contributed equally to this work.

Received: 31 May 2019; Accepted: 8 July 2019; Published: 15 July 2019

Abstract: Owing to the astonishing properties of non-harmful viruses, tissue regeneration using virus-based biomimetic materials has been an emerging trend recently. The selective peptide expression and enrichment of the desired peptide on the surface, monodispersion, self-assembly, and ease of genetic and chemical modification properties have allowed viruses to take a long stride in biomedical applications. Researchers have published many reviews so far describing unusual properties of virus-based nanoparticles, phage display, modification, and possible biomedical applications, including biosensors, bioimaging, tissue regeneration, and drug delivery, however the integration of the virus into different biomaterials for the application of tissue regeneration is not yet discussed in detail. This review will focus on various morphologies of virus-incorporated biomimetic nanocomposites in tissue regeneration and highlight the progress, challenges, and future directions in this area.

Keywords: virus; tissue regeneration; biomimetic nanocomposites; phage display

1. Introduction

1.1. Emerging Trends in Tissue Regeneration

Tissue engineering is a part of the regenerative medicine field, which emphasizes the fabrication of various functional biological constructs to reduce the increased demand for donor organs [1]. The shortage of organ donors and the increased number of people undergoing transplantation have necessitated the development of effective biomimetic materials adopting advanced technologies [2]. The aim of the field is to harness nature's ability to treat the damaged tissues., ensuring biocompatibility and supporting cellular biological events. When muscles are damaged by incidents such as illness, accidents, and microbial invasion, they lose integrity for healthy functioning at the cellular level and subsequently follow a cascade of biochemical events, including hemostasis, inflammation, proliferation, and maturation, to restore integrity [3]. However, the first ever immediate response after an injury is the production of reactive oxygen species by NADPH oxidase before inflammatory reaction [4]. Moreover, the time consumption to retain normal function in the dysfunctional organ is dependent on one's age and the seriousness of the damage. When tissue fails in the ability for self-regeneration, especially in a pathological condition, the external application of a scaffold becomes inevitable [5].

Stainless steel was first implanted as an artificial hip in 1929, which paved the way for the researchers to design biomimetic materials to replace body organs [6]. However, tissue and organ failure were considered as major medical complications until 1954, as there were not enough remedies to treat dysfunctional organs. The Nobel Laureate in medicine, Joseph Murray, transplanted a healthy kidney to a genetically identical twin brother successfully [1]. Following this historical incident, the researchers began to utilize cell structures to treat damaged tissues. These structures are autografts, isografts, allografts, and xenografts, which are biological tissues harvested from a patient's organ, a genetically identical twin, a genetically non-identical individual of the same species, and a donor of a different species, respectively [7]. The autograft has some advantages over other forms of grafts, with no rejection concerns. However, it requires a secondary procedure for the reconstructive surgery and faces donor site complication.

Meanwhile, isografts, allografts, and xenografts express slight or significant degrees of rejection as the Major Histocompatibility Complex (MHC) recognizes them as foreign antigens causing cytotoxic T-cell-mediated demolition. Unlike autografts, other forms of grafts eliminate the need for the secondary procedure without the donor site complications [8]. Though researchers are involved in finding therapeutic techniques, such as immunosuppressive medications, infectious prophylaxis, and DNA-based tissue typing, graft-versus-host disease (GVHD) remains a hurdle limiting the use of grafts in tissue regeneration [9].

The extracellular matrix (ECM) is an important component in tissues, which has the intrinsic cues to provide sites with all their cellular activities, including migration, orientation, shape, plasticity, and cell–matrix and cell–cell interactions [10]. The ECM is a complex three-dimensional network comprising proteins, such as collagen, elastin, and laminin, proteoglycans, and glycoproteins [11]. It acts as a glue and platform to bind cells together with the connective tissues. The decellularized extracellular matrix (DECM), a water-insoluble matrix, can be prepared from any organ or tissue by removing cellular components from ECM. In recent years, DECM has been used as a scaffold to hold the cells together and maintain integrity during wound healing processes [12]. Over the past two decades, the researchers have been inclined towards the preparation of artificial biomimetic grafts from the naturally available biomacromolecules, synthetic polymers, and ceramic materials, to replace tissue grafts and DECM. The remarkable features of artificial grafts are that they are economically inexpensive, their ease of preparation and storage, biocompatibility, and the tuning of the physicochemical and biological properties [13]. The most widely studied natural polymers include collagen, chitosan, hyaluronic acid, carboxymethyl cellulose, alginate, and silk fibroin, which have been reported to trigger a less immunogenic reaction in the human body, while the synthetic polymers, such as polycaprolactone, polyvinyl alcohol, nylon, and polyphosphazene, provide longer shelf life. To obtain combinatorial properties, nowadays researchers utilize a blend of polymers and their derivatives [14,15]. Ceramic and bioglass materials exhibit osteoconductive properties, showing remarkable mechanical strength and resistance to deformation, and hence, they have been widely used for bone tissue regeneration [16]. While designing a synthetic biomimetic scaffold, it is a requisite to create a benign microenvironment in a way that mimics the ECM niche for the application of effective tissue regeneration.

In the past ten years, researchers have explored a variety of biomimetic nanocomposites by incorporating bioactive molecules, such as growth factors, cytokines, genes, antibiotics, and anti-inflammatory drugs within the scaffold to increase tissue regeneration potential [17,18]. As the direct administration of therapeutic medications is not effective, leading to overdose toxicity and short time exposure to biological tissue, these biomimetic nanocomposites were much appreciated by chemists, biologists, and nanotechnologists [19]. Mostly, the bioactive molecules are cross-linked with the primary polymeric component by physical, chemical, and enzymatical treatment with or without the aid of cross-linkers and their release from the matrix is adjusted for either immediate, sustained, or slow release, according to the desired site of the target tissue. The drug release rate from the biomaterial majorly involves simple diffusion, erosion, or degradation of the matrix and swelling of the polymeric scaffold [5]. After the advent of nanotechnology, the nanoparticles were entrapped into the scaffold to exploit their medicinal properties. A composite of collagen and silver nanoparticles has been used to augment the burn tissue repair process [20]. Mieszawska et al. studied a composite film composed of silk and nanoclay to serve as a supportive biomaterial to improve bone tissue regeneration [21]. The in vivo effects of reduced graphene oxide and hydroxyapatite nanocomposite powders were investigated on critical-sized calvarial defects in a rabbit model and it was reported that the nanocomposite stimulated osteogenesis and enhanced bone formation without inflammatory responses [22]. Our research group also studied the influence of graphene oxide dispersed into a polylactic-co-glycolic acid (PLGA) electrospun nanofiber towards stimulation of myogenesis and enhanced vascular tissue regeneration [23,24]. The nanoparticles with diameters in the range of 50–700 nm acted as therapeutic drug carriers to pass through the capillaries into cells, facilitating the regeneration of new tissues [5,25].

Cell-laden scaffolds have also received attention among researchers aiming to achieve a tissue-imitating engineered graft. Kizilel et al. encapsulated pancreatic islets into nano-thin polyethylene glycol coating for enhanced insulin secretion [26]. Yoon et al. fabricated a three-dimensional layered structure using the blend of collagen epidermal keratinocytes and dermal fibroblasts to progress migration and proliferation of keratinocytes and fibroblasts during the skin repair process [27]. Within this context, microbe-based biomimetic materials have appeared as an emerging trend in tissue regeneration in recent times. Virus-based biomaterials have many biomedical applications, including cancer markers, antibacterial materials, drug carriers, and tissue regeneration [28]. Not only do they encapsulate and release the therapeutic agents to the target site, but the morphology of biomaterials also plays a pivotal role in altering physicochemical and biological properties. In this review, we have focused on summarizing the impacts of various virus-incorporated biomimetic nanocomposites with different morphologies, such as nanoparticles, nanofibers, hydrogels, and organic-inorganic hybrids, in the field of tissue regeneration. The same has been schematically shown in Figure 1. The nanoparticles that have a large surface area to volume have proven their effectiveness with long-term functionality and stability in the biological milieu. The nanoparticles can diffuse across the cell membrane and interact with cellular biomacromolecules residing inside the cell [20]. Remarkably, the hydrogel provides wettability and cell migration, while the nanofibrous matrix ensures air permeability and mechanical properties in tissue regeneration [29,30].

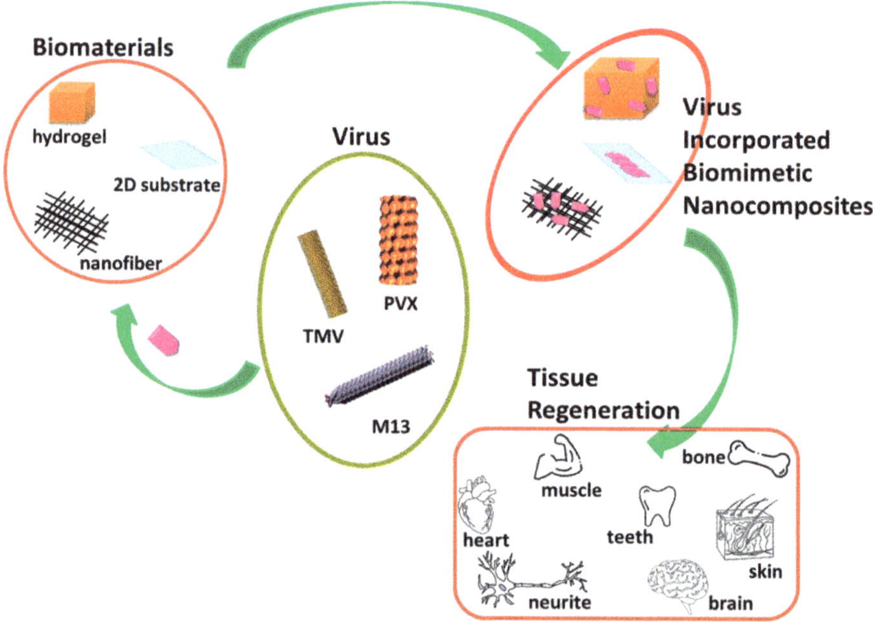

Figure 1. A compendium of tissue regeneration functionality of various virus incorporated biomimetic nanocomposites. Plant-based viruses Tobacco Mosaic Virus (TMV) and Potato Virus X (PVX) and bacteriophage M13 were shown. Reproduced with permission from [31]. Copyright Elsevier, 2010.

1.2. Remarkable Properties of Medicinally Valuable Viruses

Not all viruses cause infectious diseases in the human body. Viruses can be classified as lytic, temperate, or lysogenic based on the level of adverse effects produced in its host [32]. During infection, lytic phages kill the host bacteria, triggering the release of progeny. Lysogenic phages do not affect the host cell and infection occurs with replication of the phage genome but not the host bacterial genome. Temperate phages reside in host bacteria for amplification with no lysis, however some phages, including the λ phage, exceptionally, have both categories and thrive following either lytic or lysogenic cycles. The lytic phages, including T1–T7, contain a head and flexible tail but lack filaments. The T7 phage belongs to the Podoviridae family and structurally has an icosahedral head and a short tail. They were reported to lyse the host cell within a minute by secreting the lysozyme enzyme [33]. Professionally, T4 phages have found applications in food preservation, antibiotics, detection of bacteria, DNA and protein packing systems, and DNA-based vaccines. The literature reports revealed that Podovirus P22 assisted the assembly of cadmium sulfide nanocrystals to improve photosensitization in tissue imaging [34]. The filamentous phages, including Ff, f1, M13, N1, and Ike, are the examples of temperate phages. As they can act as a template for the synthesis of nanomaterials, the general applications of temperate phages are huge compared to lytic phages [33].

Virus-based biomimetic materials are generally derived from plant viruses and bacteriophages, as they rarely generate harmful side effects in human beings. Generally, bacteriophages (excluding fd and M13) are categorized into filamentous type of viruses, follow a non-lytic mode to infect and thrive in bacteria. It has been reported that these phages do not consist of mammalian promoter sequences in their genome, and hence, do not instigate dreadful human diseases [35]. In the human body, bacteriophages are present abundantly in the gut, bladder, and oral cavity, functioning to shape bacterial metabolisms and populations of microbial communities. It has been described that the potential role of phages increases from childhood to adulthood [36]. The monodispersed phages can self-assemble themselves into hierarchically ordered structures, such as rope-like bundles and liquid crystals. The protein surface can be modified either by covalent and non-covalent interactions or genetic alterations [35]. These unique properties have prompted the researchers to take a long stride in utilizing the phage-based biomaterials towards a wide range of biomedical applications, including biomedical imaging, drug delivery, biosensors, tissue regeneration, energy, and catalysis [37].

Owing to economically inexpensive, large scale production, ease of manipulation, and stability against a wide range of pH and temperature, a variety of phage-based biomimetic nanocomposites have been constructed for the application of effective tissue regeneration [38]. As far as morphology is concerned, a typical phage has a diameter of 68 A° and a length in the range of 800–2000 nm. The circular single-stranded DNA (ssDNA) of the phage encodes 10 genes containing 5000–8000 nucleotides, which encode a highly ordered major coat protein (p8) located around the center of phage, two minor coat proteins (p7 and p9) at one end, and two others (p3 and p6) at another terminal portion of the phage. The helical arrays of major coat proteins assemble to form the capsid shell. Generally, minor coat proteins display larger sized peptides than the major coat protein (p8) [39]. The major coat protein, p8, of M13 phage, has different segments, such as the N-terminal amphipathic, hydrophobic transmembrane (TM), and DNA binding segments. The small residues (Gly, Ala, and Ser) present on these segments have been reported to be involved in helix–helix axial and lateral interactions, which facilitate extrusion of the virion from the membrane during assembly, and hence have been known as conserved regions in the DNA sequence. Fiber diffraction and spectroscopic data show that M13 differs from fd at the 12th residue, where M13 replaces Asp of fd with Asn [40]. Filamentous phages are defined as non-enveloped bacterial viruses, having some properties in common, namely, life cycles, organization, and morphology.

The ssDNA has a left-handed helix structure possessing strong interactions with the positively charged inner surface of the capsid shell. The diffraction pattern studies classified filamentous phages into two distinct groups. Class I symmetry group consists of fd, M13, If1, and IKe, which are consistent with 5-fold symmetry. Class II symmetry group includes Pf3, Pf1, and Xf, wherein the helices are arranged with a rising per monomer of about 3.0 A° [39]. The aligned solid-state NMR studies proved that fd has O-P-O phosphate linkages in an ordered manner, whereas Pf1 did not possess such linkages [41]. According to NMR studies, phage fd has strong electrostatic interactions between the negatively charged phosphate backbone of the ssDNA nucleotide and two of four positively charged amino acid residues present at the C-terminal portion of the major coat protein, which is attributed to stabilization of the DNA core structure. The literature reports revealed that M13 and IKe showed similarity in π–π interactions between the residues of Tyr9 of one p8 and Tyr29 of an adjacent p8 [42].

Infection of E. coli by phage is initiated by the attachment of N-terminal amino acids of p3, which is present in the specialized threadlike appendage, F pilus. Subsequently, the coat protein of the phage dissolves onto the envelope of the host, which allows the only ssDNA into the cytoplasm. The host machinery synthesizes a complementary DNA strand with the involvement of two virally encoded proteins, p2 and p10, which leads to the formation of a double-stranded replicative form. The replicative form acts as a template to transcript phage genes for the synthesis of progeny ssDNAs. These progeny phage particles discharge from the bacterial cell envelope through the membrane pore complex, acquire coat proteins from the membrane, and appear as mature virions. The fact is that the infected cells undergo division at a slower rate than the uninfected cells [39].

In recent times, the researchers have sought to explore multifunctional phage-based biomaterials by precisely adjusting the surface chemistry of phage nanofibers. Covalent, non-covalent, and genetic modifications of phage coat proteins have been described comprehensively by the researchers. The genetic modification of phage coat proteins would display various foreign peptides with different functional groups at the side wall and the two termini of the phage. The endogenous amino acids of phage coat proteins are genetically combined with the foreign amino acid sequence in order to form a hybrid fusion protein, which is incorporated into phage particles and released from the cell subsequently. As a result, the foreign peptide is displayed on the surface of the phage coat protein [35]. The phage display is generally specified after N-terminal modification in its respective coat proteins. For example, if the N-terminus of p3 of the phage undergoes modification, the resulting phage is designated as a p3 display. When two or more coat proteins are controlled for modification in the same phage, then they can be known as double display, and so on [43]. In the phage coat protein, the carboxylates of aspartic and glutamic acid residues, the amine of lysine, and the phenol of tyrosine are the majorly available functional entities for the chemical modification. Introducing aldehyde into the reactive amine group has been involved in a wide range of bio-conjugation reactions, whereas the cross-linkage of p-azidophenylalanine has provided an azide handle on the phage surface, which can be easily modified for further reactions [44,45]. The EDC treatment has been helpful in cross-linking the reactive carboxylate groups with amine-functionalized moieties in phage proteins [46]. Strong nucleophile selenocysteine has been successfully genetically incorporated into phage protein using an opal stop, codon suppressing mRNA [47].

The phage-display library, with a heterogeneous mixture of phages carrying different foreign DNA insert, was created for selective binding of phage proteins with the target ligands, such as polymers, proteins, organic and inorganic crystals, small molecules, such as trinitrotoluene, and cells [48–51]. Among the phage-display libraries, the reports of p3 and p6 libraries have been well documented in research publications. Conventionally, research studies have adopted the biopanning method to find extensive use of phage particles in tissue regeneration. Biopanning is a typical technique to form a population of enriched phage-displayed peptides and specifically identify a target binding peptide [52]. According to this selection procedure, initially, a phage-display random library is incubated with the targets. Subsequently, the non-bound phage particles are eliminated with the help of detergent solubilized buffer. The target-bound phage particles are then eluted using a specialized buffer maintaining acidic pH around 2.2, and the amplification process is followed by infection of host bacteria. The resulting amplified phages form a newly enriched sub-library with more specificity to interact with the targets. The procedure is repeated several times until a only few desired peptides are predominantly available in the sub-library [53]. In the subsequent section, we will investigate the contribution of plant virus and phage-based biomimetic nanocomposites in the field of tissue regeneration.

2. Different Morphologies of Virus-Incorporated Biomimetic Nanocomposites in Tissue Regeneration

2.1. Virus-Based Nanoparticles

Many plants and phage-based viral nanoparticles have been employed so far for tissue regeneration. Plant viral nanoparticles are mono-dispersed, meta-stable, and structurally uniformed [54]. Li revealed that when the virus-based nanoparticle is more robust, the functional nanostructure is more stable, but at the same time they might be harmful to the encapsulated cargo [55].

Though the unmodified TMV nanoparticles have the potential to accelerate osteogenic differentiation in adult stem cells, the lack of affinity to the mammalian cell surface diminishes the cell adhesion property. Hence, the researchers opt for either genetic or chemical modification in viral nanoparticles in order to increase the cell binding capacity and find versatile biomedical applications. Sitasuwan et al. [62] modified the surface of a TMV nanoparticle by coupling azide-derivatized Arg-Gly-Asp-(RGD) tripeptide with tyrosine residues through Cu (I) catalyzed azide-alkyne cycloaddition reaction. When incorporated into the artificial scaffold, the RGD peptides overexpressed on ECM increase initial cell attachment by binding integrin receptors. The spacing between RGD motifs alter biological events, such as fibroblast adhesion and spreading (<440 nm), focal adhesion assembly (<140 nm), and induction of stress fiber formation (<60 nm). Owing to lack of mammalian cell infectivity, cost-effectiveness, and highly uniform size, plant viral nanoparticles have gained attention among nano and biomedical researchers. The Tobacco Mosaic Virus (TMV) constitutes a rod-like shaped nanoparticle with a diameter of 18 nm and length of 300 nm. The TMV nanoparticle consists of a capsid with 2130 identical coat protein subunits, which are responsible for assembling into a helical structure around the ssRNA. When each subunit is modified, the resulting

TMV is a polyvalent nanoparticle. The TMV can withstand temperatures up to 60 °C and can be stable in a pH range of 2–10. The TEM micrograph of wild type TMV is shown in Figure 2a [56].

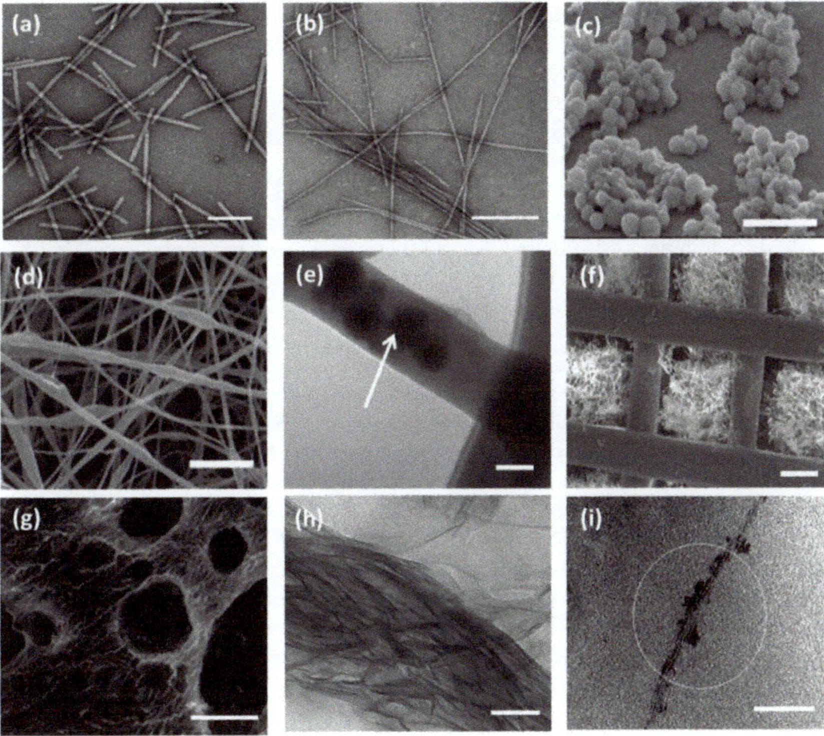

Figure 2. TEM micrographs of wild type Tobacco Mosaic Virus (TMV) nanoparticles, scale bar 200 nm (**a**) and TMV/PANI/PSS nanofiber (**b**) generated by flow assembly method, scale bar 500 nm. Reproduced with permission from [56]. Copyright American Chemical Society, 2015. (**c**) SEM micrograph of freeze-dried capsules of T4 bacteriophage/alginate water in oil emulsion in chloroform, scale bar 5 μm. (**d**) SEM and TEM micrographs (**e**) of PEO electrospun nanofiber containing T4 bacteriophage/alginate have been shown scale bars of 10 μm and 100 nm, respectively. The arrow indicates the presence of T4/alginate into the nanofiber. Reproduced with permission from [57]. Copyright John Wiley and Sons, 2013. (**f**) SEM image of the 3D printed bioceramic bone scaffold consisting of biphasic calcium phosphate with pores filled with a matrix of chitosan and RGD phage, scale bar 200 μm. Reproduced with permission from [58]. Copyright John Wiley and Sons, 2014. (**g**) SEM micrograph of porous alginate hydrogel containing TMV particles displaying interconnected channels and macropores, scale bar 100 μm. Reproduced with permission from [59]. Copyright American Chemical Society, 2012. (**h**) The TEM image of a mineralized E8-displaying phage bundle formed after 90 h incubation in the solution containing calcium and phosphate ions, scale bar 100 nm. Reproduced with permission from [60]. Copyright John Wiley and Sons, 2010. (**i**) The TEM image of the HAP-fd phage bundle with scale bar 100 nm. The circle indicates the presence of hydroxyapatite nanoparticles in the fibrous structure. Reproduced with permission from [61]. Copyright American Chemical Society, 2010.

In the human body, bone tissue regenerates to a greater extent when compared to other types of tissues. However, the regeneration process is complicated in the case of tumor resection, hip implant revision, and major fractures [63]. Pi et al. constructed a cartilage targeting gene delivery nanocomposite system by conjugating polyethylenimine (PEI) with M13 phage-displayed chondrocyte-affinity peptide

(CAP), DWRVIIPPRPSA, which was isolated after two rounds of biopanning. During incubation, the phages expressing CAP showed higher affinity towards rabbit chondrocytes at 265.5-fold when compared to unmodified phages. They reported that the CAP-conjugated PEI particles had no species specificity in binding chondrocytes of rabbit and humans. Furthermore, most of the particles were found to enter chondrocytes without being trapped in ECM, which acknowledges their larger transfection efficiency [64].

T7 viral nanoparticles were explored to display two different functional peptides CARSKNKDC (CAR) and CRKDKC (CRK) to target the microvasculature of regenerating wound tissue, including skin and tendon [25]. Skin disintegration may occur in many ways, such as bruising, abrasion, hacking, burning, stabbing, and laceration. It was observed that CAR was similar to heparin-binding sites, whereas CRK was homologous to a segment of thrombospondin type I repeat. Interestingly, CAR displayed a dominant function in the early stages of skin wound healing, while CRK showed preferences in the later stages of the same process. As the terminal residues contain cysteine, the screened peptides had more feasibility to be involved in disulfide bond formation to form a molecular cycle structure. The CAR-expressing T7 phage nanoparticles had been found to appear in wound sites 100–140-fold more efficiently than the non-recombinant phage nanoparticle [65]. The biomedical application of siRNAs is minimal owing to their low absorption across the stratum corneum, a horny outer layer of skin. Hsu et al. [66] explored M13 phage (from Ph.D-C7C library) viral nanoparticle-expressing skin penetrating and cell entering (SPACE) peptide with the sequence of AC-KTGSHNQ-CG in order to reach therapeutic macromolecules, including siRNAs, into the skin-associated cells. The in vitro physicochemical studies explored that the various macromolecules, including siRNA, penetrated across the stratum corneum into the epidermis layer of skin through the macropinocytosis pathway when the molecules were conjugated with SPACE.

A muscle binding M13 phage nanoparticle with peptide sequence ASSLNIA was identified to possess more excellent selectivity (at least five-times more) compared to the control phage nanoparticle. While investigating overall muscle selectivity on different organs, the muscle binding affinity was found to be 9–20-fold for the skeletal and 5–9-fold for cardiac muscle [67]. Sun et al. synthesized functional multivalent M13 phage (Ph.D.-7^{TM} display library) nanoparticles to express RIYKGVIQA and SEEL sequences, which are found in Nogo-66, a neurite outgrowth inhibitory protein. They selectively bound negative growth regulatory protein 1 (NgR1) with electrostatic forces of repeated leucine residues, enhancing neural differentiation of pc12 cells. Hence, this specific engineered viral nanoparticle has been appreciated for its potential use in neurite tissue regeneration, including spinal cord injury, optic nerve injury, ischemic stroke, and neurodegenerative diseases [68]. Collett et al. suggested that hepatitis C virus-based nanoparticles could act as a quadrivalent vaccine to trigger humoral and cellular immune responses. They explored biophysical, biochemical, and biomechanical properties of nanoparticles using Atomic Force Microscopy and observed that glycosylation occurred on the surface of the nanoparticle with ordered packing of the core [69]. The literature reports revealed that Sendai virus vectors displaying cardiac transcription factors could efficiently reprogram both mouse and human fibroblasts into induced cardiomyocyte-like cells in vitro. In addition, they could reduce scar formation, maintaining cardiac function in myocardial infarction affected animals [70].

The phosphate tailored TMV nanoparticle was demonstrated to induce expression of osteospecific genes of rat bone marrow stem cells (BMSCs), including osteocalcin and osteopontin, when compared to unmodified TMV nanoparticles. As shown in Figure 3d–f, the enhanced cell attachment and spreading of BMSCs were observed in phosphate grafted TMV (TMV-Phos) coated Ti substrates more than TMV coated substrates after 14 days of incubation in cell culture [73].

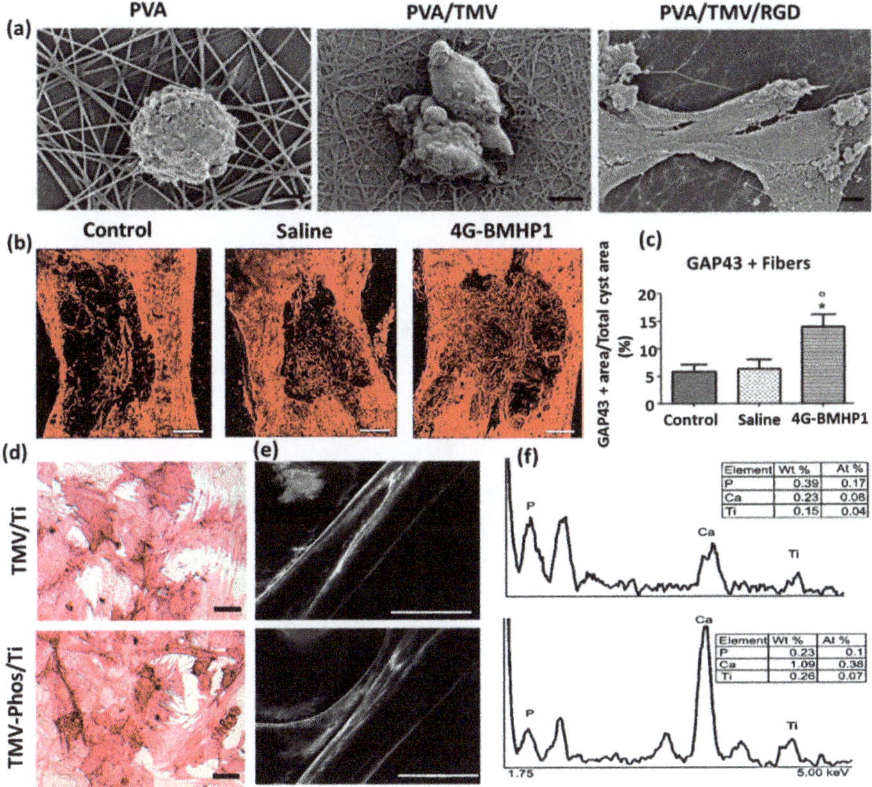

Figure 3. (**a**) Field emission scanning electron micrographs of Baby Hamster Kidney cells after 1 h incubation on electrospun nanofibrous substrates PVA, PVA-TMV, and PVA-TMV/RGD. Scale bar 2 μm. Reproduced with permission from [71]. Copyright John Wiley and Sons, 2014. (**b**) Immunofluorescence staining of longitudinal sections of total cyst area in spinal cord injury (SCI)-treated female Sprague-Dawley rats during the chronic phase. The synthesis of GAP-43 immunopositive fibers was significantly greater in bone marrow homing peptide-expressed phages (4G-BMHP1) treated group when compared to control groups (SCI control and saline). Scale bar 400 mm. The percentage of immunopositive fibers into the cyst area per total cyst area is represented in a bar graph (**c**) from six independent experiment results, and the values are reported with ± SEM. Significant factor * $p < 0.05$, 4G-BMHP1 vs. SCI control; 4G-BMHP1 vs. saline. Reproduced with permission from [72]. Copyright PLOS, 2011. (**d**) Bright-field optical microscope images of histochemical staining of alkaline phosphatase (ALPL) in bone marrow-derived stem cells (BMSCs) in TMV and phosphate grafted TMV (TMV-Phos)-coated Ti substrates after 14 days of culture. BMSCs have been shown to form neighboring well-spread cells under osteogenic conditions, which are stained highly positive for ALPL. Scale bar is 100 μm. (**e,f**) SEM micrographs and EDX analyses of TMV and TMV-Phos coating on Ti substrates. Scale bar for SEM is 100 μm. Reproduced with permission from [73]. Copyright Elsevier, 2010.

2.2. Virus-Incorporated 2D Films and Nanofibers

A combinatorial biomaterial consisting of PVX-based cyclic RGD, containing filament (RGD-PVX) and polyethylene glycol conjugated stealth filament (PEG-PVX), was developed to analyze biodistribution in mice xenograft models. The comparative studies demonstrated that PEG-PVX was preferentially accumulated into tumor cells, while RGD-PVX was trapped into the lung site in a large quantity. It has been reported that the filamentous and elongated nanoparticles are more advantageous in drug targeting than the spherical counterparts. Non-spherical nanoparticles present more ligands on their surfaces and show significant accumulation towards the vessel wall, improving the efficiency of tumor homing. Owing to the flexible nature of viral capsid, PVX-based nanoparticles could pass through restrictions in the complex biological environments and permeate into tissue cells without difficulty [74]. Wu et al. [56] successfully synthesized TMV-based electroactive nanofibers for neural tissue regeneration from the blend of polyaniline (PANI) and sodium polystyrene sulfonate (PSS). The morphology of the TMV/PANI/PSS nanofiber has been shown through TEM micrograph in Figure 2b.

An electrospun nanofiber of blends of polyvinyl alcohol (PVA) and TMV/RGD afforded higher cell density of baby hamster kidney (BHK) cells in culture. The enhanced cell adhesion and spreading and the formation of F-actin filaments were observed more on PVA/TMV/RGD nanofiber than PVA and PVA/TMV nanofibrous substrates, which were noticed in SEM micrographs (Figure 3a). The resulting nanofiber provided electroactivity and topographical cues to the neural cells and was reported to augment the length of neurites, increase the population of cells, and lead the cellular bipolar morphology more than TMV-based non-conductive nanofibers [71]. Korehei et al. [57] produced a virus incorporated nanofiber by electrospinning the blends of polyethylene oxide (PEO) and T4 bacteriophage suspension. The SEM measurement showed that T4 bacteriophages were protected from severe electrospinning conditions, as they were concentrated within the alginate capsule, as can be seen in Figure 2c. The alginate beads containing phages were found to exhibit smooth rounded surfaces. The size of the electrospun nanofiber of PEO/alginate/T4 bacteriophages had an average diameter 500 ± 100 nm (Figure 2d). According to TEM measurement, the capsules of T4 bacteriophages were distributed without uniformity throughout the fiber matrix (Figure 2e).

The induced pluripotent stem cells (iPSCs) are a promising cell source, which can rise to different cell lineages and construct a well-developed functional bone substitute. However, there is a challenge in osteoblastic differentiation of iPSCs by a conventional biomaterial, as it may form teratoma, raising health risks. Wang et al. [75] demonstrated that a phage (M13)-based nanofiber with four different signal peptides aiming to influence stem cell fate could be potentially utilized for bone tissue regeneration. The aligned nanofibrous matrix provided biochemical and biophysical cues to the cells promoting differentiation of iPSCs into osteoblasts. Among the signal peptides investigated by them were two adhesive-directing peptides RGD and RGD/PHSRN from fibronectin, and the remaining two included ALKRQGRTLYGFGG and KIPKASSVPTELSAISTLYL sequences, which are the growth regulating peptides from osteogenic growth factor and bone morphogenetic protein 2 (BMP2), respectively. The layer-by-layer technique produced a phage-assembled nanofiber assuming nanotopography of the ridge-groove structure, wherein the phage strands were parallel to each other but separated by grooves. Due to this specialized nanotopography of material, the occurrence of controlled osteoblastic differentiation was observed, even in the absence of osteogenic supplements. The research group reported that the phages displaying growth factor signal peptides could express a higher level of alkaline phosphatase (ALP) than the phages having adhesive signal peptides on the surface. The in vivo animal studies disclosed that iPSCs alone caused teratoma after one month of cells injection into nude mice, whereas the group of iPSC-derived osteoblasts did not. Cigognini and co-workers engineered an electrospun nanofibrous scaffold dispersing phage-displayed bone marrow homing peptide (BMHP (1) with sequence PFSSTKT and investigated its potential use in a chronically damaged spinal cord, which was caused by the degeneration of the central nervous system [72]. The clinical data showed that the biomimetic material enhanced nervous tissue regeneration, owing to porosity and nanostructure at the microscopic level, and improved the locomotor recovery of experimental rats. From Figure 3b,c, the histological analyses revealed that the scaffold affected increased cellular infiltration and axonal regeneration after eight weeks of experimental investigation in rats. They found a higher synthesis of growth-associated protein 43 (GAP-43) in engineered scaffold-treated animal when compared to saline and control groups with spinal cord defects. Our research group has explored electrospun nanofibrous matrices of PLGA containing self-assembled M13 bacteriophages along with additives RGD and graphene oxide to show enhanced differentiation of fibroblasts, smooth muscle cells, and myoblasts [76–80].

2.3. Virus-Incorporated 3D Hydrogel Scaffolds

Cell-laden-agarose hydrogel was prepared by dispersing genetically engineered rod-shaped PVX nanoparticles, which present functional RGD peptides and mineralization inducing peptides (MIP) on its surface, into agarose polymeric components [81]. Luckanagul et al. [59] prepared freeze-dried solid foam of a porous alginate hydrogel (PAH) comprising TMV. The incorporation of TMV nanoparticles resulted in large sized and well-defined spherical pores (100–500 μm) in TMV/PAH, analyzed by Field Emission Scanning Electron Microscope image (Figure 2g).

The PVX nanoparticles adopted a nano-filamentous structural network on coated surfaces. Exploiting the synergistic effect of both peptides, the PVX nanoparticles in hydrogel expressed significant cell adhesion as well as hydroxyapatite nucleation. Confirmed by SEM and immunostaining characterizations, it was further reported that the viral nanoparticles could be preserved over 14 days in hydrogel and the whole biomaterial could act as a promising bone substitute. Maturavongsadit et al. [82] developed an injectable TMV based hydrogel under physiological conditions to imitate a cartilage microenvironment. The hydrogel was prepared by cross-linking methacrylate hyaluronic acid polymers by cysteine inserted TMV mutants involving in situ Michael addition reaction. The hydrogel was reported to influence enhancement of cartilage tissue regeneration by promoting chondrogenesis via up-regulation of BMP-2. The interaction of TMV nanoparticles with the cells assisted the high-level expression of BMP-2, an effective inducer of differentiation of mesenchymal stem cells into chondrocytes.

Luckanagul et al. [58] investigated the performance of functional TMV-RGD-blended alginate hydrogel nanocomposites to treat in vivo cranial bone defects in Sprague-Dawley rats. The TMV-functionalized sponge-like hydrogel supported cell localization without triggering any systemic toxicity in the defect area, and hence was envisaged as an active bone replacement biomimetic material in the future direction of reconstructive orthopedic surgery. Shah et al. [83] studied an integrated co-assembled hydrogel system of peptide amphiphiles, in which M13 phage coat protein was modified to express a high density of binding peptide HSNGLPL to combine with transforming growth factor β1 (TGF-β1). The research group found an enhancement in articular cartilage tissue regeneration in a rabbit model with a full-thickness chondral defect because of the slower release of growth factor from the hydrogel, with approximately 60% of cumulative drug release at 72 h, which supported the viability and chondrogenic differentiation of mesenchymal stem cells in the defective site. The in vivo evaluation of the rabbit model showed that the hydrogel treated animal group had no apparent symptoms of chronic inflammatory responses after four weeks. All of the rabbits appeared with a full range of motion in their knees at the end of the investigation. Caprini et al. [84] isolated M13 phage-displayed peptide, KLPGWSG, which could adhere on the surface of murine neural stem cells. Subsequently, the research group designed a self-assembled KLPGWSG-based biomimetic hydrogel with tunable visco-elastic properties for the regeneration of the degenerated nervous system. It was discovered that the phage-based hydrogel favored cell adherence and differentiation in the range of 100–1000 Pa, suggesting that the elastic property of the matrix is a crucial factor in tissue regeneration.

2.4. Virus-Incorporated Organic-Inorganic Hybrid Nanocomposites

The interaction of organic and inorganic biocompatible materials in scaffolds bring about significant impacts in biomedical applications. Cementum, classified as a hard mineralized tissue, surrounds tooth root and has been a part of periodontal tissue that connects the tooth to the bone. When an infectious biofilm adheres to tooth root, triggering periodontal disease, the tooth loss is more enhanced. Gungormus et al. [85] demonstrated amelogenin-derived M13 phage-displayed peptide controlled hydroxyapatite biomineralization for dental tissue regeneration. It was reported that Amelogenin directed hydroxyapatite to form a protein matrix during the formation of enamel. Hence, the research group synthesized the cementomimetic material by applying an aqueous solution of the amelogenin-displayed peptide on the human demineralized root surface to form a layer, which was subsequently immersed into the solution of calcium and phosphate ions. Ramaraju et al. [86] isolated M13 phage-displayed peptides to design a dual functional apatite-coated film for effective bone tissue regeneration. They reported that one peptide sequence of the phage, VTKHLNQISQSY, had mineral (apatite) binding affinity with 25% hydrophobicity, whereas another peptide, DPIYALSWSGMA, had cell binding affinity with 50% hydrophobicity. Also, they discovered that the dual functional apatite-based biomaterial could stimulate the adhesion strength of human bone marrow stromal cells (hMSC) and subsequently increase cell proliferation and differentiation. Due to the mineral binding affinity, the film provided a platform for the adherence of osteogenic cells with osteoconductive and osteoinductive signals. Further, the biomimetic nanocomposite showed a greater extent of proliferation of hMSCs with an elevated level of Runx2 expression when compared to biomimetic apatite without functional peptides.

Wang et al. [58] prepared a 3D-printed biomimetic nanofiber with M13 phage-displayed RGD peptides residing in the pores of the scaffold to enhance bone tissue regeneration. The nanocomposite consisted of hydroxyapatite and tri-calcium phosphate showing an ordered pattern with interconnected micro and macro scale pores, which are shown in the TEM micrograph (Figure 2f). The research group implanted a MSC-seeded biomimetic scaffold into a rat radial bone defect and discovered that the order of regeneration was found as follows: scaffold filled with modified phages > scaffolds filled with wild-type phages > pure scaffold. He et al. [60] carried out a similar kind of research work, genetically modifying M13 phage to express oligonucleotide encoding E8 and inducing self-assembly followed by oriental mineralization to synthesize nanofibrous biomimetic materials under the influence of divalent calcium ions. The resulting mineralized phage bundle has been shown in TEM micrography (Figure 2h). Wang et al. [61] used Ca^{2+} ions to prompt self-assembly of fd phage-based anionic nanofibers and transform them into a bundle sheet (Figure 2i), which provided insights into biomineralization and fabrication of organic–inorganic hybrid nanocomposites. The divalent ion-triggered bundle not only acted as a biotemplate but also served as a Ca source to initiate the ordered nucleation and growth of crystalline hydroxyapatite in the biological fluid.

3. Other Formulations of Virus-Based Nanocomposites with Different Biomedical Applications

Apart from tissue regeneration, virus-based biomimetic nanocomposites have traced their steps in different biomedical applications, such as drug delivery, bioimaging, and biosensing. Wang et al. [87] studied f8/8 phage-based polymeric micelles from the self-assembly of polymeric PEG- diacyl lipid conjugates. These polymeric micelles were reported to have cell-targeting ability to release less water-soluble drugs with more specificity towards breast cancer Michigan Cancer Foundation-7 (MCF-7) cells. The non-toxic filamentous f88.4 bacteriophage viral nanoparticle, which was designed to display a single chain antibody, delivered the vectors to the different regions of the brain in albino, laboratory-bred nude mice (BALB/c), and hence was proposed for treating Alzheimer's disease with early diagnosis [88,89]. Wang et al. [90] studied a M13 phage-displayed peptide with the sequence HSQAAVP to target fibroblast growth factor 8b (FGF8b) to treat prostate cancer. The genetic level disturbances in homeostasis between prostate epithelial and stromal cells cause prostate cancer. The major isotherm of fibroblast growth factor 8 is FGF8b, which is associated with the stages of prostate cancer and has been a potential target for appropriate therapies. In this study, the research group revealed that the biomimetic material interrupted FGF8b binding to its receptors, and thereby prevented FGF8b-induced cell proliferation. Furthermore, they reported that the biomaterial had the potential to arrest the cell cycle at the phase G0/G1 by suppressing cyclin D1 and proliferating cell nuclear antigens (PCNA).

Carrico et al. [43] chemically modified the amino acid residues present on the surface of filamentous fd phage coat protein following a two-step transamination/oxime reaction for its potential use in characterizing breast cancer cells. The research group discovered that the chemical reaction selectively targets N-terminal groups but is not involved in transamination of lysine ε-amines. They conjugated PEG polymeric chains to the phage protein in order to reduce immunogenicity, decrease non-specific binding, and increase solubility in the aqueous environment. They observed that there were no significant differences in either absorption or emission properties after fluorophores were labeled with polymer conjugated phages. Fan et al. [91] isolated cyclic peptide CAGALCY from T7 phage nanoparticles in order to target the pial microvasculature of the brain and inhibit platelet adhesion. The presence of the bulky hydrophobic core, two cysteine residues at each end, and the tyrosine residue at the carboxy terminus are considered as remarkable features for selectively binding the brain microvasculature. When pharmacokinetic properties were assessed, the non-filamentous phage, T7, showed a fast clearance rate from the blood with a half-life of 12 min, whereas the filamentous phages M13 and fUSE5 had longer half-lives of 7 h and 9 h, respectively. To identify the specificity of the T7 phage-displayed peptide, they determined selectivity indices using plaque assay for various organs of mice, including lung, liver, brain, kidney, colon, small intestine, and large intestine. The characterization results exposed that T7 displayed peptide resided (accumulated) in the brain, with a selectivity index of 1000, whereas other organs possessed low specificity for the peptide, with selectivity indices less than 50.

Bean et al. [92] prepared a bacteriophage K (ΦK) by incorporating the virus into a photo cross-linked hyaluronic acid methacrylate (HAMA)-based hydrogel that resulted in a material with antimicrobial properties. The presence of two zinc finger genomes ($CX_2CX_{22}CX_2C$ and $CX_2CX_{23}CX_2C$) in the virus caused it to be virulent against a wide range of infective Staphylococci. The secretion of hyaluronidase enzyme-mediated *S. aureus* sensitizes HAMA and triggered degradation of the hydrogel, facilitating the release of ΦK at a sustained level to inhibit bacterial growth effectively. This stimuli-responsive hydrogel was shown to reduce pain, promote cell migration and tissue hydration in the wound site, and was suggested for the application of dermal tissue regeneration. Schmidt et al. [93] identified two different adenovirus phage-displayed peptides QTRFLLH and VPTQSSG to target neural precursor cells (NPC) in the hippocampal dentate gyrus of adult mice through adenovirus-mediated gene transfer. The peptides were found to be strongly internalized into NPCs when the investigated material was added to neurosphere culture containing clusters of neural stem cells.

Kelly KA et al. [94] isolated high-throughput fluorochrome-labeled M13 phage particles (Ph.D. C7C library) to rapidly identify ligands of biological interest in vivo using secreted protein acidic and rich in cysteine (SPARC) molecules and vascular cell adhesion molecules-1 (VCAM-1). The engineered phage particles led to higher sensitivity with an attachment of 800 fluorophores per phage. Wan et al. [37] developed an f8/8 phage-based biosensor exploiting magnetoelastic wireless detection system. The genetically modified phage-expressed peptide sequence EPRLSPHS on the surface of the target biological agent, *Bacillus anthracis* spore. The resonance frequency of the sensor decreased gradually depending on the binding agent on the surface. They reported that this affinity-based phage-displayed biosensor exhibited more longevity activity as a diagnostic probe to target numerous agents with more efficiency than antibody-based biosensors.

4. Conclusions and Perspectives

The potential application of virus-incorporated biomimetic nanocomposites in the form of self-assembled nanoparticles, nanofibers, hydrogels, and organic–inorganic hybrids in the field of tissue regeneration has been elucidated in this review. Though virus-based biomaterial has displayed many beneficial properties, there are some issues to be addressed. (1) Many research groups have expressed desired peptides on the surface of phage-based viral nanoparticles exploiting phage libraries. However, whether the number of peptides exhibited by each nanoparticle is the same is questionable. (2) Biodistribution of viral nanoparticles in different organs of animal tissues has been studied by some researchers. Still, a comprehensive study to describe bioavailability must be demonstrated. (3) It has been well documented that viral nanoparticles contribute to the enhancement in tissue regeneration. However, a systematic study is required to explain the phases of tissue regeneration, in which viral nanocomposites contribute more. (4) The viral nanocomposites in the form of polymeric micelles, vesicles, and dendrimers are less formulated and have not been explored enough for the application of tissue regeneration. The following are suggestions for the future of this field. (1) Sophisticated techniques and methodology to quantify the number of peptides expressed on each phage particle. (2) Pharmacokinetic and pharmacodynamics studies to determine the required dosage of viral nanoparticles in each organ type of tissue regeneration. (3) An extensive in vivo animal study to show the influence of viral-based nanocomposites in each phase of tissue regeneration. (4) Successful bioconjugation of viral nanoparticles with amphiphilic polymers or surfactants to design various oil-in-water-type emulsions. We hope that the researchers with interdisciplinary backgrounds will advance the field of tissue regeneration using viral-based biomimetic nanocomposites by considering the problems and the concerned suggestions.

Author Contributions: I.S.R. and D.-W.H. developed the idea and structure of the review article. I.S.R. and C.K. wrote the paper using material supplied by S.-J.S., Y.C.S., and M.S.K. S.-H.H. and J.-W.O. revised and improved the manuscript. J.-W.O. and D.-W.H. supervised the manuscript. All authors have given approval to the final version of the manuscript.

Funding: This research was supported by a grant of the Korea Health Technology R&D Project through the Korea Health Industry Development Institute (KHIDI), funded by the Ministry of Health and Welfare, Republic of Korea (HI17C1662), and by the Bio and Medical Technology Development Program of the National Research Foundation (NRF) and funded by the Korean government (MSIP and MOHW, No. 2017M3A9E4048170 and MEST, No. 2015M3A9E2028643).

Conflicts of Interest: The authors declare no conflict of interest.

References

1. Shafiee, A.; Atala, A. Tissue engineering: Toward a new era of medicine. *Ann. Rev. Med.* **2017**, *68*, 29–40. [CrossRef] [PubMed]
2. Atala, A. Regenerative medicine strategies. *J. Pediatr. Surg.* **2012**, *47*, 17–28. [CrossRef] [PubMed]
3. Raja, I.S.; Fathima, N.N. Gelatin–cerium oxide nanocomposite for enhanced excisional wound healing. *ACS Appl. Biol. Mater.* **2018**, *1*, 487–495. [CrossRef]
4. Galliot, B.; Crescenzi, M.; Jacinto, A. Trends in tissue repair and regeneration. *Development* **2017**, *144*, 357–364. [CrossRef] [PubMed]
5. Dorati, R.; DeTrizio, A.; Modena, T.; Conti, B.; Benazzo, F.; Gastaldi, G.; Genta, I. Biodegradable scaffolds for bone regeneration combined with drug-delivery systems in osteomyelitis therapy. *Pharmaceuticals* **2017**, *10*, 96. [CrossRef] [PubMed]
6. Ahadian, S.; Khademhosseini, A. Smart scaffolds in tissue regeneration. *Regener. Biomater.* **2018**, *5*, 125–128. [CrossRef] [PubMed]
7. Lu, H.H.; El-Amin, S.F.; Scott, K.D.; Laurencin, C.T. Three-dimensional, bioactive, biodegradable, polymer-bioactive glass composite scaffolds with improved mechanical properties support collagen synthesis and mineralization of human osteoblast-like cells in vitro. *J. Biomed. Mater. Res. A* **2003**, *64*, 465–474. [CrossRef] [PubMed]
8. Shibuya, N.; Jupiter, D.C. Bone graft substitute: Allograft and xenograft. *Clin. Podiatr. Med. Surg.* **2015**, *32*, 21–34. [CrossRef] [PubMed]
9. Ferrara, J.L.M.; Levine, J.E.; Reddy, P.; Holler, E. Graft-versus-host disease. *Lancet* **2009**, *373*, 1550–1561. [CrossRef]
10. Higuchi, A.; Ling, Q.-D.; Chang, Y.; Hsu, S.-T.; Umezawa, A. Physical cues of biomaterials guide stem cell differentiation fate. *Chem. Rev.* **2013**, *113*, 3297–3328. [CrossRef]
11. Kular, J.K.; Basu, S.; Sharma, R.I. The extracellular matrix: Structure, composition, age-related differences, tools for analysis and applications for tissue engineering. *J. Tissue Eng.* **2014**, *5*, 2041731414557112. [CrossRef] [PubMed]
12. Taylor, D.A.; Sampaio, L.C.; Ferdous, Z.; Gobin, A.S.; Taite, L.J. Decellularized matrices in regenerative medicine. *Acta Biomater.* **2018**, *74*, 74–89. [CrossRef] [PubMed]
13. Zhao, P.; Gu, H.; Mi, H.; Rao, C.; Fu, J.; Turng, L.S. Fabrication of scaffolds in tissue engineering: A review. *Front. Mech. Eng.* **2018**, *13*, 107–119. [CrossRef]
14. Sultana, N. Natural-synthetic polymer blend composite scaffold for bone tissue engineering: Study of in vitro degradation and protein adsorption. *Appl. Mech. Mater.* **2014**, *554*, 42–46. [CrossRef]

15. Doulabi, A.H.; Mequanint, K.; Mohammadi, H. Blends and nanocomposite biomaterials for articular cartilage tissue engineering. *Materials* **2014**, *7*, 5327–5355. [CrossRef] [PubMed]
16. Gerhardt, L.C.; Boccaccini, A.R. Bioactive glass and glass-ceramic scaffolds for bone tissue engineering. *Materials* **2010**, *3*, 3867–3910. [CrossRef] [PubMed]
17. Zeng, Y.; Hoque, J.; Varghese, S. Biomaterial-assisted local and systemic delivery of bioactive agents for bone repair. *Acta Biomater.* **2019**, *93*, 152–168. [CrossRef]
18. Monteiro, N.; Martins, A.; Pires, R.; Faria, S.; Fonseca, N.A.; Moreira, J.N.; Reis, R.L.; Neves, N.M. Immobilization of bioactive factor-loaded liposomes on the surface of electrospun nanofibers targeting tissue engineering. *Biomater. Sci.* **2014**, *2*, 1195–1209. [CrossRef]
19. Raja, I.S.; Fathima, N.N. A gelatin based antioxidant enriched biomaterial by grafting and saturation: Towards sustained drug delivery from antioxidant matrix. *Colloids Surf. B Biointerfaces* **2015**, *128*, 537–543. [CrossRef]
20. Hasan, A.; Morshed, M.; Memic, A.; Hassan, S.; Webster, T.J.; Marei, H.E. Nanoparticles in tissue engineering: Applications, challenges and prospects. *Int. J. Nanomed.* **2018**, *13*, 5637–5655. [CrossRef]
21. Mieszawska, A.J.; Llamas, J.G.; Vaiana, C.A.; Kadakia, M.P.; Naik, R.R.; Kaplan, D.L. Clay enriched silk biomaterials for bone formation. *Acta Biomater.* **2011**, *7*, 3036–3041. [CrossRef] [PubMed]
22. Shin, Y.C.; Song, S.J.; Jeong, S.J.; Kim, B.; Kwon, I.K.; Hong, S.W.; Oh, J.W.; Han, D.W. Graphene-based nanocomposites as promising options for hard tissue regeneration. *Adv. Exp. Med. Biol.* **2018**, *1078*, 103–117. [PubMed]
23. Shin, Y.C.; Lee, J.H.; Kim, M.J.; Hong, S.W.; Kim, B.; Hyun, J.K.; Choi, Y.S.; Park, J.C.; Han, D.W. Stimulating effect of graphene oxide on myogenesis of C2C12 myoblasts on RGD peptide-decorated PLGA nanofiber matrices. *J. Biol. Eng.* **2015**, *9*, 22. [CrossRef] [PubMed]
24. Shin, Y.C.; Kim, J.; Kim, S.E.; Song, S.J.; Hong, S.W.; Oh, J.W.; Lee, J.; Park, J.C.; Hyon, S.H.; Han, D.W. RGD peptide and graphene oxide co-functionalized PLGA nanofiber scaffolds for vascular tissue engineering. *Regen. Biomater.* **2017**, *4*, 159–166. [CrossRef] [PubMed]
25. Saidykhan, L.; Abu Bakar, M.Z.; Rukayadi, Y.; Kura, A.U.; Latifah, S.Y. Development of nanoantibiotic delivery system using cockle shell-derived aragonite nanoparticles for treatment of osteomyelitis. *Int. J. Nanomed.* **2016**, *11*, 661–673. [CrossRef] [PubMed]
26. Kizilel, S.; Scavone, A.; Liu, X.; Nothias, J.M.; Ostrega, D.; Witkowski, P.; Millis, M. Encapsulation of pancreatic islets within nano-thin functional polyethylene glycol coatings for enhanced insulin secretion. *Tissue Eng. A* **2010**, *16*, 2217–2228. [CrossRef]
27. Yoon, H.; Lee, J.-S.; Yim, H.; Kim, G.; Chun, W. Development of cell-laden 3D scaffolds for efficient engineered skin substitutes by collagen gelation. *RSC Adv.* **2016**, *6*, 21439–21447. [CrossRef]
28. Mohan, K.; Weiss, G.A. Chemically modifying viruses for diverse applications. *ACS Chem. Biol.* **2016**, *11*, 1167–1179. [CrossRef]
29. Manoukian, O.S.; Matta, R.; Letendre, J.; Collins, P.; Mazzocca, A.D.; Kumbar, S.G. Electrospun nanofiber scaffolds and their hydrogel composites for the engineering and regeneration of soft tissues. *Methods Mol. Biol.* **2017**, *1570*, 261–278.
30. Shin, Y.C.; Lee, J.H.; Kim, M.J.; Park, J.H.; Kim, S.E.; Kim, J.S.; Oh, J.W.; Han, D.W. Biomimetic hybrid nanofiber sheets composed of RGD peptide-decorated PLGA as cell-adhesive substrates. *J. Funct. Biomater.* **2015**, *6*, 367–378. [CrossRef]
31. Steinmetz, N.F. Viral nanoparticles as platforms for next-generation therapeutics and imaging devices. *Nanomedicine* **2010**, *6*, 634–641. [CrossRef] [PubMed]
32. Howard-Varona, C.; Hargreaves, K.R.; Abedon, S.T.; Sullivan, M.B. Lysogeny in nature: Mechanisms, impact and ecology of temperate phages. *ISME J.* **2017**, *11*, 1511. [CrossRef] [PubMed]
33. Sunderland, K.S.; Yang, M.; Mao, C. Phage-enabled nanomedicine: From probes to therapeutics in precision medicine. *Angew. Chem. Int. Ed.* **2017**, *56*, 1964–1992. [CrossRef] [PubMed]
34. Kale, A.; Bao, Y.; Zhou, Z.; Prevelige, P.E.; Gupta, A. Directed self-assembly of CdS quantum dots on bacteriophage P22 coat protein templates. *Nanotechnology* **2013**, *24*, 045603. [CrossRef] [PubMed]
35. Mohan, K.; Weiss, G.A. Dual genetically encoded phage-displayed ligands. *Anal. Biochem.* **2014**, *453*, 1–3. [CrossRef] [PubMed]
36. Miller-Ensminger, T.; Garretto, A.; Brenner, J.; Thomas-White, K.; Zambom, A.; Wolfe, A.J.; Putonti, C. Bacteriophages of the urinary microbiome. *J. Bacteriol.* **2018**, *200*, e00738–17. [CrossRef] [PubMed]

37. Cao, B.; Li, Y.; Yang, T.; Bao, Q.; Yang, M.; Mao, C. Bacteriophage-based biomaterials for tissue regeneration. *Adv. Drug Deliv. Rev.* **2018**. [CrossRef]
38. Wan, J.; Shu, H.; Huang, S.; Fiebor, B.; Chen, I.; Petrenko, V.A.; Chin, B.A. Phage-based magnetoelastic wireless biosensors for detecting bacillus Anthracis spores. *IEEE Sens. J.* **2007**, *7*, 470–477. [CrossRef]
39. Smith, G.P. Filamentous fusion phage: Novel expression vectors that display cloned antigens on the virion surface. *Science* **1985**, *228*, 1315–1317. [CrossRef]
40. Williams, K.A.; Glibowicka, M.; Li, Z.; Li, H.; Khan, A.R.; Chen, Y.M.; Wang, J.; Marvin, D.A.; Deber, C.M. Packing of coat protein amphipathic and transmembrane helices in filamentous bacteriophage M13: Role of small residues in protein oligomerization. *J. Mol. Biol.* **1995**, *252*, 6–14. [CrossRef]
41. Cross, T.A.; Tsang, P.; Opella, S.J. Comparison of protein and deoxyribonucleic acid backbone structures in fd and Pf1 bacteriophages. *Biochemistry* **1983**, *22*, 721–726. [CrossRef] [PubMed]
42. Xu, J.; Dayan, N.; Goldbourt, A.; Xiang, Y. Cryo-electron microscopy structure of the filamentous bacteriophage IKe. *Proc. Natl. Acad. Sci. USA* **2019**, *116*, 5493–5498. [CrossRef] [PubMed]
43. Wang, Y.; Ju, Z.; Cao, B.; Gao, X.; Zhu, Y.; Qiu, P.; Xu, H.; Pan, P.; Bao, H.; Wang, L.; et al. Ultrasensitive rapid detection of human serum antibody biomarkers by biomarker-capturing viral nanofibers. *ACS Nano* **2015**, *9*, 4475–4483. [CrossRef] [PubMed]
44. Carrico, Z.M.; Farkas, M.E.; Zhou, Y.; Hsiao, S.C.; Marks, J.D.; Chokhawala, H.; Clark, D.S.; Francis, M.B. N-Terminal labeling of filamentous phage to create cancer marker imaging agents. *ACS Nano* **2012**, *6*, 6675–6680. [CrossRef] [PubMed]
45. Tian, F.; Tsao, M.-L.; Schultz, P.G. A phage display system with unnatural amino acids. *J. Am. Chem. Soc.* **2004**, *126*, 15962–15963. [CrossRef]
46. Li, K.; Chen, Y.; Li, S.; Nguyen, H.G.; Niu, Z.; You, S.; Mello, C.M.; Lu, X.; Wang, Q. Chemical modification of M13 bacteriophage and its application in cancer cell imaging. *Bioconjug. Chem.* **2010**, *21*, 1369–1377. [CrossRef]
47. Sandman, K.E.; Benner, J.S.; Noren, C.J. Phage display of selenopeptides. *J. Am. Chem. Soc.* **2000**, *122*, 960–961. [CrossRef]
48. Jaworski, J.W.; Raorane, D.; Huh, J.H.; Majumdar, A.; Lee, S.-W. Evolutionary screening of biomimetic coatings for selective detection of explosives. *Langmuir* **2008**, *24*, 4938–4943. [CrossRef]
49. Cho, W.; Fowler, J.D.; Furst, E.M. Targeted binding of the M13 bacteriophage to thiamethoxam organic crystals. *Langmuir* **2012**, *28*, 6013–6020. [CrossRef]
50. Li, Y.; Cao, B.; Yang, M.; Zhu, Y.; Suh, J.; Mao, C. Identification of novel short $BaTiO_3$-binding/nucleating peptides for phage-templated in situ synthesis of $BaTiO_3$ polycrystalline nanowires at room temperature. *ACS Appl. Mater. Interfaces* **2016**, *8*, 30714–30721. [CrossRef]
51. Sanghvi, A.B.; Miller, K.P.; Belcher, A.M.; Schmidt, C.E. Biomaterials functionalization using a novel peptide that selectively binds to a conducting polymer. *Nat. Mater.* **2005**, *4*, 496–502. [CrossRef] [PubMed]
52. McGuire, M.J.; Li, S.; Brown, K.C. Biopanning of phage displayed peptide libraries for the isolation of cell-specific ligands. *Methods Mol. Biol.* **2009**, *504*, 291–321. [PubMed]
53. Cao, B.; Mao, C. Identification of microtubule-binding domains on microtubule-associated proteins by major coat phage display technique. *Biomacromolecules* **2009**, *10*, 555–564. [CrossRef] [PubMed]
54. Lee, K.L.; Uhde-Holzem, K.; Fischer, R.; Commandeur, U.; Steinmetz, N.F. Genetic engineering and chemical conjugation of potato virus X. *Methods Mol. Biol.* **2014**, *1108*, 3–21. [PubMed]
55. Li, L.; Xu, C.; Zhang, W.; Secundo, F.; Li, C.; Zhang, Z.-P.; Zhang, X.-E.; Li, F. Cargo-compatible encapsulation in virus-based nanoparticles. *Nano Lett.* **2019**, *19*, 2700–2706. [CrossRef] [PubMed]
56. Wu, Y.; Feng, S.; Zan, X.; Lin, Y.; Wang, Q. Aligned electroactive TMV nanofibers as enabling scaffold for neural tissue engineering. *Biomacromolecules* **2015**, *16*, 3466–3472. [CrossRef] [PubMed]
57. Korehei, R.; Kadla, J. Incorporation of T4 bacteriophage in electrospun fibres. *J. Appl. Microbiol.* **2013**, *114*, 1425–1434. [CrossRef] [PubMed]
58. Wang, J.; Yang, M.; Zhu, Y.; Wang, L.; Tomsia, A.P.; Mao, C. Phage nanofibers induce vascularized osteogenesis in 3D printed bone scaffolds. *Adv. Mater.* **2014**, *26*, 4961–4966. [CrossRef] [PubMed]
59. Luckanagul, J.; Lee, L.A.; Nguyen, Q.L.; Sitasuwan, P.; Yang, X.; Shazly, T.; Wang, Q. Porous alginate hydrogel functionalized with virus as three-dimensional scaffolds for bone differentiation. *Biomacromolecules* **2012**, *13*, 3949–3958. [CrossRef]

60. He, T.; Abbineni, G.; Cao, B.; Mao, C. Nanofibrous bio-inorganic hybrid structures formed through self-assembly and oriented mineralization of genetically engineered phage nanofibers. *Small* **2010**, *6*, 2230–2235. [CrossRef]

61. Wang, F.; Cao, B.; Mao, C. Bacteriophage bundles with prealigned Ca^{2+} initiate the oriented nucleation and growth of hydroxylapatite. *Chem. Mater.* **2010**, *22*, 3630–3636. [CrossRef] [PubMed]

62. Sitasuwan, P.; Lee, L.A.; Li, K.; Nguyen, H.G.; Wang, Q. RGD-conjugated rod-like viral nanoparticles on 2D scaffold improve bone differentiation of mesenchymal stem cells. *Front. Chem.* **2014**, *2*, 31. [CrossRef] [PubMed]

63. Luckanagul, J.A.; Metavarayuth, K.; Feng, S.; Maneesaay, P.; Clark, A.Y.; Yang, X.; García, A.J.; Wang, Q. Tobacco mosaic virus functionalized alginate hydrogel scaffolds for bone regeneration in rats with cranial defect. *ACS Biomater. Sci. Eng.* **2016**, *2*, 606–615. [CrossRef]

64. Pi, Y.; Zhang, X.; Shi, J.; Zhu, J.; Chen, W.; Zhang, C.; Gao, W.; Zhou, C.; Ao, Y. Targeted delivery of non-viral vectors to cartilage in vivo using a chondrocyte-homing peptide identified by phage display. *Biomaterials* **2011**, *32*, 6324–6332. [CrossRef] [PubMed]

65. Jarvinen, T.A.; Ruoslahti, E. Molecular changes in the vasculature of injured tissues. *Am. J. Pathol.* **2007**, *171*, 702–711. [CrossRef] [PubMed]

66. Hsu, T.; Mitragotri, S. Delivery of siRNA and other macromolecules into skin and cells using a peptide enhancer. *Proc. Natl. Acad. Sci. USA* **2011**, *108*, 15816–15821. [CrossRef] [PubMed]

67. Samoylova, T.I.; Smith, B.F. Elucidation of muscle-binding peptides by phage display screening. *Muscle Nerve* **1999**, *22*, 460–466. [CrossRef]

68. Sun, Z.; Dai, X.; Li, Y.; Jiang, S.; Lou, G.; Cao, Q.; Hu, R.; Huang, Y.; Su, Z.; Chen, M.; et al. A novel Nogo-66 receptor antagonist peptide promotes neurite regeneration in vitro. *Mol. Cell. Neurosci.* **2016**, *71*, 80–91. [CrossRef]

69. Collett, S.; Torresi, J.; Earnest-Silveira, L.; Christiansen, D.; Elbourne, A.; Ramsland, P.A. Probing and pressing surfaces of hepatitis C virus-like particles. *J. Colloid Interface Sci.* **2019**, *545*, 259–268. [CrossRef]

70. Engel, J.L.; Ardehali, R. Sendai virus based direct cardiac reprogramming: What lies ahead? *Stem Cell Investig.* **2018**, *5*, 37. [CrossRef]

71. Zhao, X.; Lin, Y.; Wang, Q. Virus-based scaffolds for tissue engineering applications. *Wiley Interdiscip. Rev. Nanomed. Nanobiotechnol.* **2015**, *7*, 534–547. [CrossRef] [PubMed]

72. Cigognini, D.; Satta, A.; Colleoni, B.; Silva, D.; Donega, M.; Antonini, S.; Gelain, F. Evaluation of early and late effects into the acute spinal cord injury of an injectable functionalized self-assembling scaffold. *PLoS ONE* **2011**, *6*, e19782. [CrossRef] [PubMed]

73. Kaur, G.; Wang, C.; Sun, J.; Wang, Q. The synergistic effects of multivalent ligand display and nanotopography on osteogenic differentiation of rat bone marrow stem cells. *Biomaterials* **2010**, *31*, 5813–5824. [CrossRef] [PubMed]

74. Shukla, S.; DiFranco, N.A.; Wen, A.M.; Commandeur, U.; Steinmetz, N.F. To target or not to target: Active Vs. passive tumor homing of filamentous nanoparticles based on Potato virus X. *Cell. Mol. Bioeng.* **2015**, *8*, 433–444. [CrossRef] [PubMed]

75. Wang, J.; Wang, L.; Yang, M.; Zhu, Y.; Tomsia, A.; Mao, C. Untangling the effects of peptide sequences and nanotopographies in a biomimetic niche for directed differentiation of iPSCs by assemblies of genetically engineered viral nanofibers. *Nano Lett.* **2014**, *14*, 6850–6856. [CrossRef] [PubMed]

76. Han, J.; Devaraj, V.; Kim, C.; Kim, W.G.; Han, D.W.; Hong, S.W.; Kang, Y.C.; Oh, J.W. Fabrication of self-assembled nanoporous structures from a self-templating M13 bacteriophage. *ACS Appl. Nano Mater.* **2018**, *1*, 2851–2857. [CrossRef]

77. Shin, Y.C.; Lee, J.H.; Jin, L.; Kim, M.J.; Oh, J.W.; Kim, T.W.; Han, D.W. Cell-adhesive RGD peptide-displaying M13 bacteriophage/PLGA nanofiber matrices for growth of fibroblasts. *Biomater. Res.* **2014**, *18*, 14. [CrossRef]

78. Shin, Y.C.; Lee, J.H.; Jin, O.S.; Lee, E.J.; Jin, L.H.; Kim, C.S.; Hong, S.W.; Han, D.W.; Kim, C.; Oh, J.-W. RGD peptide-displaying M13 bacteriophage/PLGA nanofibers as cell-adhesive matrices for smooth muscle cells. *J. Korean Phys. Soc.* **2015**, *66*, 12–16. [CrossRef]

79. Shin, Y.C.; Lee, J.H.; Jin, L.; Kim, M.J.; Kim, C.; Hong, S.W.; Oh, J.W.; Han, D.W. Cell-adhesive matrices composed of RGD peptide-displaying M13 bacteriophage/poly(lactic-co-glycolic acid) nanofibers beneficial to myoblast differentiation. *J. Nanosci. Nanotechnol.* **2015**, *15*, 7907–7912. [CrossRef]

80. Shin, Y.C.; Kim, C.; Song, S.J.; Jun, S.; Kim, C.S.; Hong, S.W.; Hyon, S.H.; Han, D.W.; Oh, J.W. Ternary aligned nanofibers of RGD peptide-displaying M13 bacteriophage/PLGA/graphene oxide for facilitated myogenesis. *Nanotheranostics* **2018**, *2*, 144–156. [CrossRef]

81. Lauria, I.; Dickmeis, C.; Roder, J.; Beckers, M.; Rutten, S.; Lin, Y.Y.; Commandeur, U.; Fischer, H. Engineered Potato virus X nanoparticles support hydroxyapatite nucleation for improved bone tissue replacement. *Acta Biomater.* **2017**, *62*, 317–327. [CrossRef] [PubMed]

82. Maturavongsadit, P.; Luckanagul, J.A.; Metavarayuth, K.; Zhao, X.; Chen, L.; Lin, Y.; Wang, Q. Promotion of in vitro chondrogenesis of mesenchymal stem cells using in situ hyaluronic hydrogel functionalized with rod-like viral nanoparticles. *Biomacromolecules* **2016**, *17*, 1930–1938. [CrossRef] [PubMed]

83. Shah, R.N.; Shah, N.A.; Del Rosario Lim, M.M.; Hsieh, C.; Nuber, G.; Stupp, S.I. Supramolecular design of self-assembling nanofibers for cartilage regeneration. *Proc. Natl. Acad. Sci. USA* **2010**, *107*, 3293–3298. [CrossRef] [PubMed]

84. Caprini, A.; Silva, D.; Zanoni, I.; Cunha, C.; Volonte, C.; Vescovi, A.; Gelain, F. A novel bioactive peptide: Assessing its activity over murine neural stem cells and its potential for neural tissue engineering. *New Biotechnol.* **2013**, *30*, 552–562. [CrossRef] [PubMed]

85. Gungormus, M.; Oren, E.E.; Horst, J.A.; Fong, H.; Hnilova, M.; Somerman, M.J.; Snead, M.L.; Samudrala, R.; Tamerler, C.; Sarikaya, M. Cementomimetics-constructing a cementum-like biomineralized microlayer via amelogenin-derived peptides. *Int. J. Oral Sci.* **2012**, *4*, 69–77. [CrossRef] [PubMed]

86. Ramaraju, H.; Miller, S.J.; Kohn, D.H. Dual-functioning peptides discovered by phage display increase the magnitude and specificity of BMSC attachment to mineralized biomaterials. *Biomaterials* **2017**, *134*, 1–12. [CrossRef]

87. Wang, T.; Petrenko, V.A.; Torchilin, V.P. Optimization of landscape phage fusion protein-modified polymeric PEG-PE micelles for improved breast cancer cell targeting. *J. Nanomed. Nanotechnol.* **2012**, *4*, 008.

88. Frenkel, D.; Solomon, B. Filamentous phage as vector-mediated antibody delivery to the brain. *Proc. Natl. Acad. Sci. USA* **2002**, *99*, 5675–5679. [CrossRef]

89. Enshell-Seijffers, D.; Smelyanski, L.; Gershoni, J.M. The rational design of a 'type 88' genetically stable peptide display vector in the filamentous bacteriophage fd. *Nucleic Acids Res.* **2001**, *29*, e50. [CrossRef]

90. Wang, W.; Chen, X.; Li, T.; Li, Y.; Wang, R.; He, D.; Luo, W.; Li, X.; Wu, X. Screening a phage display library for a novel FGF8b-binding peptide with anti-tumor effect on prostate cancer. *Exp. Cell Res.* **2013**, *319*, 1156–1164. [CrossRef]

91. Fan, X.; Venegas, R.; Fey, R.; van der Heyde, H.; Bernard, M.A.; Lazarides, E.; Woods, C.M. An in vivo approach to structure activity relationship analysis of peptide ligands. *Pharm. Res.* **2007**, *24*, 868–879. [CrossRef] [PubMed]

92. Bean, J.E.; Alves, D.R.; Laabei, M.; Esteban, P.P.; Thet, N.T.; Enright, M.C.; Jenkins, A.T.A. Triggered release of bacteriophage K from agarose/hyaluronan hydrogel matrixes by Staphylococcus aureus virulence factors. *Chem. Mater.* **2014**, *26*, 7201–7208. [CrossRef]

93. Schmidt, A.; Haas, S.J.; Hildebrandt, S.; Scheibe, J.; Eckhoff, B.; Racek, T.; Kempermann, G.; Wree, A.; Putzer, B.M. Selective targeting of adenoviral vectors to neural precursor cells in the hippocampus of adult mice: New prospects for in situ gene therapy. *Stem Cells* **2007**, *25*, 2910–2918. [CrossRef] [PubMed]

94. Kelly, K.A.; Waterman, P.; Weissleder, R. In vivo imaging of molecularly targeted phage. *Neoplasia* **2006**, *8*, 1011–1018. [CrossRef] [PubMed]

 *nanomaterials*

Article

Evaluation of Three Morphologically Distinct Virus-Like Particles as Nanocarriers for Convection-Enhanced Drug Delivery to Glioblastoma

Joel A. Finbloom [1,†], Ioana L. Aanei [1,†], Jenna M. Bernard [1,†], Sarah H. Klass [1], Susanna K. Elledge [1], Kenneth Han [1], Tomoko Ozawa [2], Theodore P. Nicolaides [3], Mitchel S. Berger [2] and Matthew B. Francis [1,4,*]

[1] Department of Chemistry, University of California, Berkeley, CA 94720, USA; jaf@berkeley.edu (J.A.F.); aanei.ioana@gmail.com (I.L.A.); bernarje@gmail.com (J.M.B.); sklass@berkeley.edu (S.H.K.); susannaelle@berkeley.edu (S.K.E.); kennethhan@berkeley.edu (K.H.)

[2] Department of Neurological Surgery, University of California, San Francisco, CA 94158, USA; tomoko.ozawa@ucsf.edu (T.O.); Mitchel.Berger@ucsf.edu (M.S.B.)

[3] Department of Pediatrics, NYU Langone Medical Center, New York, NY 10016, USA; Theodore.nicolaides@nyumc.org

[4] Materials Sciences Division, Lawrence Berkeley National Laboratories, Berkeley, CA 94720, USA

* Correspondence: mbfrancis@berkeley.edu

† These authors contributed equally to this work.

Received: 13 November 2018; Accepted: 1 December 2018; Published: 5 December 2018

Abstract: Glioblastoma is a particularly challenging cancer, as there are currently limited options for treatment. New delivery routes are being explored, including direct intratumoral injection via convection-enhanced delivery (CED). While promising, convection-enhanced delivery of traditional chemotherapeutics such as doxorubicin (DOX) has seen limited success. Several studies have demonstrated that attaching a drug to polymeric nanoscale materials can improve drug delivery efficacy via CED. We therefore set out to evaluate a panel of morphologically distinct protein nanoparticles for their potential as CED drug delivery vehicles for glioblastoma treatment. The panel consisted of three different virus-like particles (VLPs), MS2 spheres, tobacco mosaic virus (TMV) disks and nanophage filamentous rods modified with DOX. While all three VLPs displayed adequate drug delivery and cell uptake in vitro, increased survival rates were only observed for glioma-bearing mice that were treated via CED with TMV disks and MS2 spheres conjugated to doxorubicin, with TMV-treated mice showing the best response. Importantly, these improved survival rates were observed after only a single VLP–DOX CED injection several orders of magnitude smaller than traditional IV doses. Overall, this study underscores the potential of nanoscale chemotherapeutic CED using virus-like particles and illustrates the need for further studies into how the overall morphology of VLPs influences their drug delivery properties.

Keywords: virus-like particles; glioblastoma; convection-enhanced delivery; tobacco mosaic virus; bioconjugation; doxorubicin; drug delivery; protein-based nanomaterials; viral capsid

---

1. Introduction

Protein-based nanomaterials are a promising class of nanocarriers for drug delivery and diagnostic applications. These materials are formed through the self-assembly of protein monomers into larger nanoscale scaffolds of varying morphologies and properties [1–4]. Protein-based nanomaterials and specifically virus-like particles (VLPs) based on naturally occurring viruses, have demonstrated effectiveness in drug delivery and imaging applications [1–12]. Unlike other synthetic delivery vehicles, such as polymeric micelles and liposomes, VLPs are homogenous in their size distribution and are produced by inexpensive recombinant expression [1,2]. Further, VLPs are degradable in the body and have demonstrated few toxicity issues [1,5]. These protein-based nanomaterials also allow for site-selective modification through amino acid mutagenesis of natural or noncanonical amino acids into the protein backbone [6,13–15]. This site-selective conjugation allows for greater control over the location and amount of cargo loaded onto the VLPs, which can have significant effects on cancer targeting and delivery efficiencies.

Glioblastoma multiforme (GBM) is one of the deadliest and hardest to treat cancers, with over 17,000 new diagnoses per year [16]. While traditional small molecule chemotherapeutic approaches have largely failed to treat GBM, many nanomaterial-based approaches are being developed to enhance GBM treatment [16–20], although the reports of such approaches with virus-like particles is limited [21]. Alternative delivery routes are also being investigated for GBM treatment. Particularly, intratumoral injection of chemotherapeutics via convection-enhanced delivery has seen success in enhancing GBM treatment [22–24]. A constant pressure is maintained during the injection through a microfluidic pump that creates a fluid convection to facilitate a homogenous diffusion of the drug molecule throughout the targeted area [22]. After CED infusion, it is important for the chemotherapeutic agent to remain in the tumor tissue. We therefore hypothesized that VLP nanocarriers might be retained better in the tumor tissue after CED, as has been observed for polymeric constructs [22,25]. Additionally, the differing morphologies of VLPs could influence their tumor treatment efficacies in complex ways. This may be particularly important when developing drug delivery systems for glioblastoma treatment, as gliomas are known to form nanodimensional (50–200 nm) pores in their associated vasculature [26,27]. These pores may critically influence nanoparticle extravasation in a morphologically-dependent manner. Intriguingly, one report has noted differences in the pore sizes of malignant and benign glioma [27].

To determine the efficacies of VLPs as nanocarriers for GBM treatment, a panel of distinct VLPs was evaluated for their drug delivery potential. This panel consisted of a 27 nm MS2 sphere, an 18 nm tobacco mosaic virus (TMV) disk and a 50 nm nanophage filamentous rod (Figure 1). MS2 VLPs have been widely used by our lab and others as drug delivery and diagnostic nanocarriers, as MS2 has a series of 2 nm pores along the surface that allow for interior capsid modification and subsequent drug release [2,6,13,28,29]. While MS2 has been extensively studied, it has not been tested for glioblastoma drug delivery. We recently reported a stable nanodisk composed of a double-arginine mutant of the tobacco mosaic virus [30]. These TMV disks maintained their structural assembly within all biologically-relevant conditions tested, which is uncommon for recombinantly expressed TMV mutants that lack their genomic material [31,32]. The TMV disks were further functionalized with chemotherapeutic cargo to showcase their potential as drug delivery vehicles, as disk-shaped nanomaterials have shown promise in other studies for their enhanced tumor accumulation and cell penetration [33,34]. The last member of the VLP panel was a nanoscale variation of the filamentous fd phage. The nanophage (NP) VLP was recently reported as a new potential nanocarrier [35] but has not previously been functionalized and evaluated for its drug delivery potential. This panel therefore encompasses disparate VLP morphologies, while maintaining relatively consistent dimensions of 15–50 nm. All three of the VLPs presented herein were functionalized with chemotherapeutic cargo to evaluate the efficacies of our VLP panel for drug delivery in glioblastoma models.

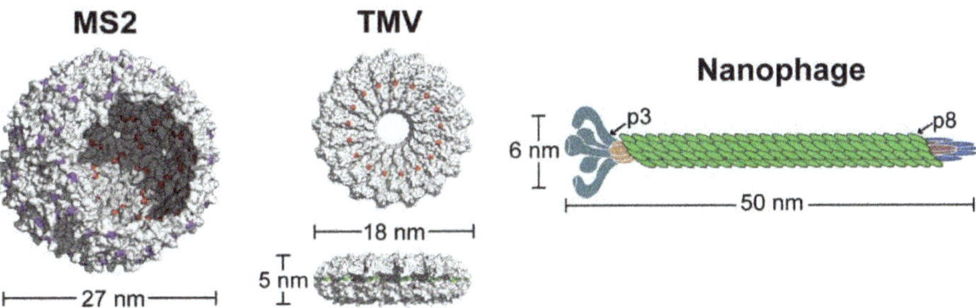

Figure 1. Panel of virus-like particles under evaluation. Three different nanocarriers composed of MS2 spheres, TMV disks and nanophage filamentous rods were tested to determine drug delivery efficacies. Each VLP contains reactive handles for bioconjugation, such as cysteines (**red**), reactive amines (**green**), or noncanonical p-aminophenylalanine moieties (**purple**).

2. Materials and Methods

2.1. Reagents and Instruments

Unless otherwise noted, reagents were purchased from Sigma (St. Louis, MO, USA) and used without further purification. Sulfo-LC-SPDP was purchased from Thermo (Waltham, MA, USA). Water was deionized using the NANOpure purification system (Thermo). *N*-methylpyridinium-4-carboxaldehyde benzenesulfonate hydrate (Rapoport's salt, RS) was obtained from Alfa Aesar (Ward Hill, MA, USA). NAP desalting columns were purchased from GE Healthcare (Marlborough, MA, USA). Spin concentrators with different molecular weight cutoffs (MWCO) were from Millipore (Billerica, MA, USA). Aminophenol-PEG$_{5k}$-OMe [14], alkoxyamine-PEG$_{5k}$-OMe [36] and DOX-EMCH [37] were synthesized as reported previously. EMEM media for cell culture was purchased from ATCC (Manassas, VA, USA). MTS Assay ((3-(4,5-dimethylthiazol-2-yl)-5-(3-carboxymethoxyphenyl)-2-(4-sulfophenyl)-2*H*-tetrazolium)) was purchased from Promega (Madison, WI, USA). Doxorubicin was purchased from Pfizer as a 2 mg/mL stock solution in saline. Liposomal Dox was purchased from SunPharma (Mumbai, India) as a 2 mg/mL solution.

Liquid chromatography mass spectrometry (LC/MS) analysis was performed using acetonitrile (Optima grade, 99.9%, Thermo Fisher), formic acid (99+%, Pierce, Rockford, IL, USA) and dd-H$_2$O as mobile phase solvents. Electrospray ionization mass spectrometry (ESI-MS) of proteins was performed using an Agilent 1260 series liquid chromatograph outfitted with an Agilent 6224 time-of-flight (TOF) LC/MS system (Santa Clara, CA, USA). The LC was equipped with a Poroshell 300SB-C18 (5 μm particles, 1.0 mm × 75 mm, Agilent) analytical column. Data were collected and analyzed using Agilent MassHunter Qualitative Analysis B.05.00.

Sodium dodecyl sulfate-polyacrylamide gel electrophoresis (SDS-PAGE) was carried out on TRIS gels in a Mini-Protean apparatus from Bio-Rad (Hercules, CA, USA) or on Bis-TRIS gels in a Mini Gel Tank apparatus (Thermo), following the protocol from the manufacturer. The protein electrophoresis samples were heated for 10 min at 95 °C in the presence of β-mercaptoethanol to ensure reduction of any disulfide bonds. Gels were run for 35–60 min at 150–200 V in 2-(*N*-morpholino)ethanesulfonic acid (MES)—SDS buffer to allow good separation of the bands. Commercially available markers (Bio-Rad) were applied to at least one lane of each gel for assignment of apparent molecular masses. Visualization of protein bands was accomplished by staining with Coomassie Brilliant Blue R-250 (Bio-Rad). Quantification of the degree of modification was obtained by imaging on a Gel Doc™ EZ imager (Bio-Rad) and subsequent optical densitometry using the ImageJ (National Institutes of Health, Bethesda, MD, USA) or Image Lab (Bio-Rad) software.

UV-Vis spectrophotometer readings were carried out using a Cary 50 Bio Spectrophotometer (Agilent, Santa Clara, CA, USA) or a NanoDrop 1000 (Thermo Scientific). Analytical size exclusion was performed on an Agilent 1100 series HPLC equipped with a PolySep-GFC-P 5000 column (Phenomenex, Torrance, CA, USA), at a flow rate of 1 mL/min. Incucyte live cell imaging (Essen Bioscience, Ann Arbor, MI, USA) was used to monitor the cellular uptake of DOX-protein conjugates. An IVIS 50 Lumina imaging system (Perkin Elmer, Waltham, MA, USA) was used to measure in vivo and ex vivo bioluminescence.

2.2. Protein Purification and Expression

MS2 and TMV proteins were expressed and purified using previously published methods [15,28,30]. Nanophage was expressed and purified using an adapted method from a previously published report [35]. All proteins were purified using anion exchange chromatography with a diethylaminoethanol (DEAE) Sepharose column followed by size exclusion chromatography. Purity was confirmed by SDS-PAGE, LC/MS and HPLC-SEC. Detailed protocols of nanophage expression and purification are available in the Supporting Information.

2.3. Protein Modification

Protein modification reactions varied depending on the protein and the synthetic cargo. Full modification protocols of each protein and each modification are available in the Supporting Information. Typical modifications involved the use of 10 equiv maleimide, 40 equiv of isothiocyanate, 10–20 equiv of PEG and 1–2 equiv of DOX-EMCH. In the case of nanophage, the p8 monomer of the coat protein was the primary target for bioconjugation. All protein conjugates were purified via elution through a Nap desalting column followed by multiple rounds of spin concentration with a designated molecular weight cut-off (MWCO) that would allow for small-molecule flow-through but prevent protein flow-through. Protein conjugates were characterized by a combination of LC/MS, gel electrophoresis and HPLC-SEC (Supporting Figures S1–S4).

2.4. Cell Culture

U87-MG human glioblastoma cells were obtained from the UC Berkeley Cell Culture Facility. U87-MG human glioblastoma cells bearing the luciferase reporter (U87-Luc) were acquired from UCSF. Cells were cultured in DMEM containing phenol red (ATCC, Manassas, VA, USA) or DMEM without phenol red (Thermo, Waltham, MA, USA) with 10% fetal bovine serum (Omega Scientific, Tarzana, CA, USA) and 1% penicillin/streptomycin (Thermo) at 37 °C and 5% CO_2.

2.5. Cell Viability Assays

U87-MG cells were trypsinized and diluted to a density of 50,000 cells/mL. An aliquot of 100 µL of cell stock was placed in each well of a 96 well plate (Corning, Corning, NY, USA) for a density of 5000 cells/well. The plate was incubated at 37 °C, 5% CO_2 overnight. Following this, media was removed from the plate and 100 µL of appropriate sample stocks media was added. The cells were incubated at 37 °C, 5% CO_2 for 3 d. The media containing the sample was removed from the well and 100 µL of MTS media (20% MTS in media) was added to each well and incubated for 1–3 h. An Infinite 200 Pro plate reader (Tecan, Switzerland) was used to measure the MTS absorbance at 490 nm. Cell viability was calculated as an absorbance percent relative to the untreated cell control. The experiment was performed in triplicate.

2.6. Cell Uptake Studies

Cell uptake studies were performed as previously described [30]. U87-MG cells were trypsinized and diluted to a density of 50,000 cells/mL. An aliquot of 200 µL of cell stock was placed in a well of a 96 well plate (Corning) for a density of 10,000 cells/well. The plate was incubated at 37 °C, 5% CO_2 for

48 h. Following this, media was removed from the plate and 200 µL of appropriate sample stocks (standardized to 1 µM DOX) in phenol free media was added. The cells were incubated at 37 °C, 5% CO_2 for 48 h. Incucyte Zoom Live-Cell Analysis System (EssenBio, Ann Arbor, MI, USA) was used to collect images every hour post incubation. Phase and green fluorescence images at 20× magnification were collected, capitalizing on the intrinsic fluorescence of doxorubicin. The images were processed using Incucyte Zoom proprietary software v.2016A and the Top-Hat background subtraction algorithm (radius 10 µm, threshold 0.5% of green calibration dye signal, GCU) was used to define the boundaries of the cells (green objects). The mean fluorescence intensities of 4 images taken in each well were plotted against the incubation time.

2.7. Tumor Growth and Survival Studies in Glioblastoma Models

All animal procedures were performed according to a protocol approved by the University of California San Francisco Institutional Animal Care and Use Committee (IACUC). Five-week old athymic (nude) female mice weighing 18–23 g were purchased from Simonsen Labs (Gilroy, CA, USA). For tumor inoculation, 3×10^5 U87-Luc glioblastoma cells with luciferase reporter gene were implanted intracranially [38].

Mice bearing U87-Luc intracranial tumors in sets of 9 animals per study group were injected via convection enhanced delivery (CED) with 10 µL of PBS, DOX, Lipo-Dox, or VLP-DOX conjugates in sterile saline, at day 11 post-implantation. CED infusion cannula were made with silica tubing (Polymicro Technologies, Phoenix, AZ, USA) fused to a 0.1 mL syringe (Plastic One, Roanoke, VA, USA) with a 0.5 mm stepped-tip needle that protruded from the silica guide base. Syringes were loaded with the agents and attached to a microinfusion pump (Bioanalytical Systems, Lafayette, IN, USA). The silica cannula attached with a infusion pump was lowered to a 4 mm depth through a skull hole, which was made by skull puncture with a coordination of 3 mm to the right from bregma and just on top of the coronal suture (the same region in the caudate putamen at which tumor cells were injected). The agents were infused at a rate of 1 µL/min until a volume of 10 µL had been delivered. Cannulae were removed 5 min post completion of infusion.

Tumor size was monitored using the luciferase reporter system and measuring bioluminescence on an IVIS 50 Lumina system (Supporting Figure S5). Mice were euthanized when tumor burden reached levels determined by IACUC guidelines. Kaplan Meier survival curves were plotted based on these survival points. Log rank tests were performed on the Kaplan Meier curves to gauge statistical significance. When dividing mice into small and large tumor cohorts, a Bioluminescence Intensity (BLI) of 10^7 was taken as the cutoff, as this was the standard midpoint across groups when assessing tumor size prior to CED injection. Statistical comparison between small and large tumor cohorts was performed using log rank tests of the Kaplan Meier survival curves for each cohort. A full listing of tumor size in all mice that were sacrificed due to tumor burden is available in Supporting Figure S6. Detailed side-by-side survival curves of key cohorts are available in Supporting Figure S7.

3. Results and Discussion

3.1. Chemical Modification of VLPs for Drug Delivery

In order to investigate the potential of our VLP panel as nanocarriers for glioblastoma treatment, we synthesized a series of conjugates with the chemotherapeutic molecule doxorubicin (DOX). Doxorubicin is an anthracycline antibiotic commonly used for the treatment of many types of cancer. However, its systemic toxicity (especially cardiac toxicity) and poor penetration through the blood-brain barrier limit its use in the treatment of brain tumors [16,39]. Several different formulations of doxorubicin have been developed, including PEGylated liposomal doxorubicin (Doxil) [40]. Unfortunately, none of these agents showed activity in clinical trials investigating its use to treat brain cancer [16]. We hypothesized that VLP carriers could enhance the delivery of DOX for glioblastoma treatment. By attaching doxorubicin to the protein scaffolds through acid-labile linkers, the release and distribution of DOX can be controlled to achieve more effective drug release and tumor treatment.

The ketone moiety on DOX was modified via a condensation reaction with N-ε-maleimidocaproic acid hydrazide (EMCH), as previously described (Figure S1) [30,37]. The maleimide moiety of DOX–EMCH was reacted with thiol residues on MS2, TMV and Nanophage VLPs to produce VLP–DOX conjugates (Figure 2). This modification was shown to retain the cytotoxic activity of doxorubicin upon hydrolysis of the hydrazone linkage between EMCH and DOX, which releases DOX in its native state. It is anticipated that the DOX will release upon endocytosis of the VLP–DOX conjugates, as the late endosome and subsequent lysosome have internal pH in the range of pH 4.5–5.5, which is sufficient to cleave the hydrazone linkage. For the nanophage, native Lys residues were first converted to thiols using Traut's reagent [41]. The VLPs were further modified with PEG_{5k}, as this is a standard modification for improving biodistribution and minimizing the immune response against VLPs [3,4,28]. Additionally, VLP PEGylation helped improve the solubility of the protein constructs, as DOX is a relatively hydrophobic molecule.

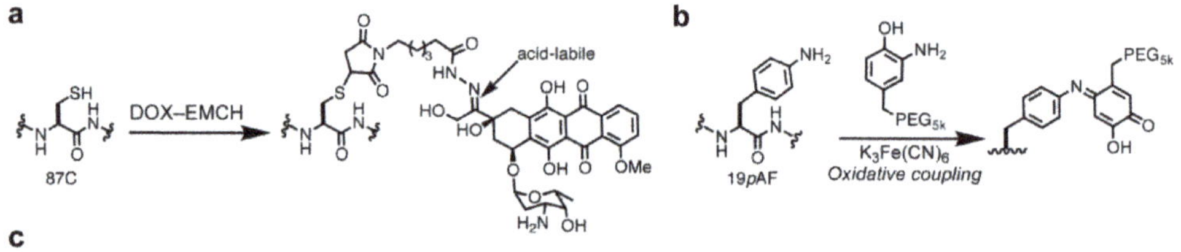

Virus-like Particle	DOX attachment site	DOX modification %	PEG attachment site	PEG modification %
MS2 spheres	87C (interior)	35% (63 per capsid)	19pAF (exterior)	70% (126 per capsid)
TMV disks	123C (exterior face)	30% (10 per capsid)	N terminus (exterior edge)	50% (17 per capsid)
Nanophage rods	Lys (exterior)	20% (20 per capsid)	Lys (exterior)	15% (15 per capsid)

Figure 2. Doxorubicin (DOX) and PEG conjugation to virus-like particles for drug delivery studies. (**a**) Bioconjugation scheme for the modification of MS2 with (**a**) DOX–EMCH and (**b**) PEG. DOX–VLP conjugates contain an acid-labile hydrazone linkage, which is anticipated to cleave upon VLP endocytosis and degradation. Modification schemes for TMV and nanophage are available in the Supporting Information. (**c**) Each VLP was modified with DOX–EMCH using maleimide addition to cysteine residues (MS2, TMV) or to thiols synthetically installed onto lysine residues (nanophage) as shown in Figure S1. All VLPs were modified with PEG using either oxidative couplings (MS2, TMV) or NHS ester reactions (nanophage) to improve biodistribution.

3.2. Evaluation of VLP–DOX Conjugates In Vitro

After chemical modification, the VLP–DOX conjugates were evaluated for their effects on cell viability. U87-MG glioblastoma cells were incubated with the agents at different concentrations for 72 h to allow sufficient time for DOX release from the VLP conjugates upon endocytosis and a cell viability assay was performed (Figure 3). All nanocarriers showed a similar dose-response curve and had comparable efficiency compared to the free drug, suggesting that the drug was released and was able to reach the nucleus and induce cell death.

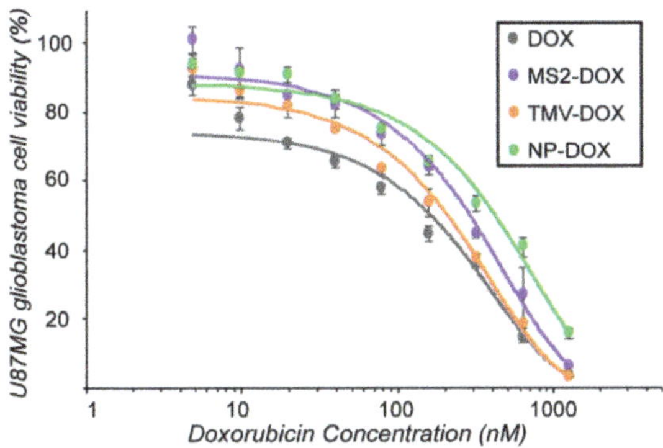

Figure 3. VLP delivery of doxorubicin to U87-MG glioblastoma cells. U87-MG glioblastoma cells were treated with varying amounts of VLP–DOX conjugates for 72 h. Significant cell death was observed for all VLPs tested. Treatment with the VLPs alone displayed no toxicity (data not shown).

In order to understand the uptake kinetics of DOX conjugates in glioblastoma cells, cell uptake was monitored using live cell imaging. Cells were incubated with VLPs bearing a standard dose of 1 µM DOX and images were collected every hour for 48 h to assess the rate of VLP–DOX accumulation. Both MS2–DOX and TMV–DOX displayed fast uptake into cells, while the nanophage–DOX conjugates displayed slower uptake in comparison (Figure 4). This may be due to different cellular uptake efficiencies between nanoparticle morphologies, as has been observed previously [33,42–45]. Overall, despite subtle differences in uptake kinetics and drug delivery efficacies, all three VLPs were amenable to DOX modification and were able to deliver their chemotherapeutic payload in cell culture effectively.

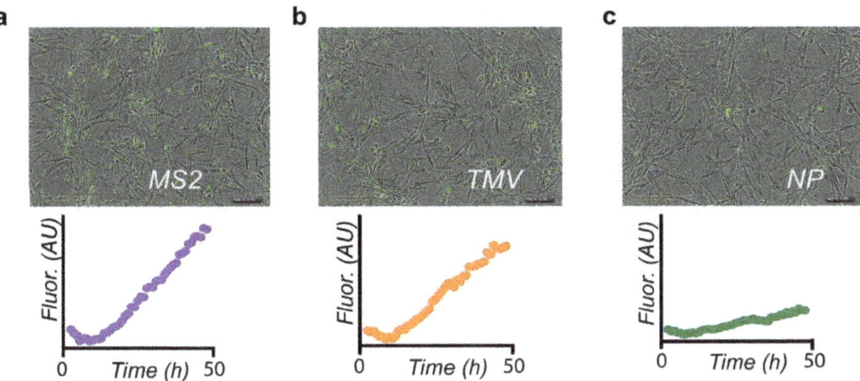

Figure 4. VLP-DOX uptake into U87-MG cells at 48 h. Uptake kinetics of VLP-DOX conjugates into U87-MG glioblastoma cells were monitored in the green channel to detect DOX fluorescence. Data are presented with consistent scaling. Initial intensities are artificially high due to the presence of autofluorescent compounds in the cell media. Upon photobleaching, the FLI drops to more accurate baselines. Both (**a**) MS2 and (**b**) TMV demonstrated significant cellular uptake, while the uptake of (**c**) nanophage (NP) was markedly slower. The TMV data appeared previously in reference [30].

3.3. Convection-Enhanced Delivery of VLP–DOX Conjugates

Each of the three VLP–DOX conjugates was evaluated for its ability to treat glioblastoma after CED injection. VLP-DOX conjugates were compared to PBS, DOX and liposomal DOX (Lipo–DOX) treatment via CED injection 11 days after intracranial tumor implantation (Figure 5). While the payload of DOX per nanocarrier was variable, each glioma-bearing mouse was injected with a standardized DOX payload of 20 µg/kg and tumor size was monitored via bioluminescence intensity (BLI, Figure 5a). The tumor growth of MS2 and TMV-treated mice demonstrated some growth inhibition, although the large variability in tumor reduction within groups led to no statistical difference between treatment groups. Kaplan Meier survival analysis of treated mice suggested improved outcomes for mice treated with TMV–DOX conjugates, as three out of eight mice survived past 40 days, whereas for the PBS-treated group, no mouse survived past 34 days (Figure 5b). This is particularly encouraging, as the CED injected dose of DOX was orders of magnitude lower than the milligrams per kilogram of doxorubicin typically injected intravenously [18,46]. However, log rank analysis of survival curves of TMV and PBS-treated mice did not demonstrate statistical significance between the groups for the full set of animals ($p = 0.052$).

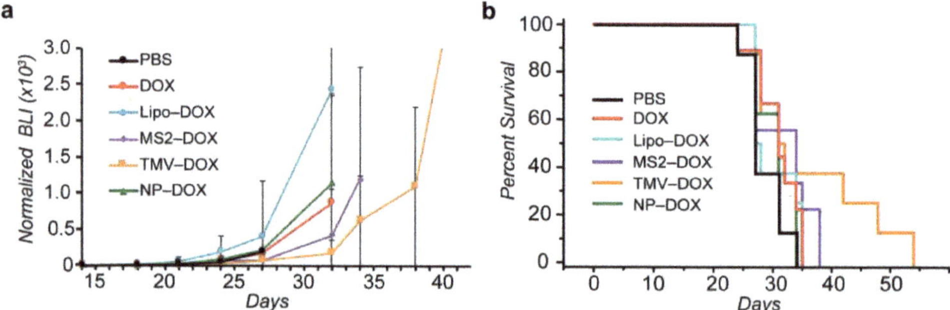

Figure 5. Evaluation of VLP panel for glioblastoma treatment. MS2 spheres, TMV disks and nanophage rods bearing DOX payloads of 20 µg/kg were injected via convection-enhanced delivery infusion (CED) into U87-MG glioma-bearing mice. (**a**) Tumor growth analysis suggested modest tumor growth inhibition from MS2 and TMV-treated groups. (**b**) Kaplan-Meier survival curves of mice with U87-MG glioma. The survival curve results suggest improved efficacy with TMV–DOX treatment.

The glioma-bearing mice under study could be divided into two cohorts: one with small tumors prior to CED injection and one with larger tumors. The cutoff between large and small tumor size was determined by analyzing the bioluminescence intensities (BLI) of treatment groups and arriving at a general midpoint of 10^7 BLI (Supporting Figure S6). This cutoff allowed for relatively even distributions between large and small tumors within each treatment group, with the exception of Lipo–DOX, which only had two large tumors within that group. Analysis of survival times between large and small tumors within treatment groups revealed significant differences for MS2–DOX and TMV–DOX treated groups (Supporting Figure S7), with smaller tumors responding better to the VLP treatment (Figure 6). This analysis reveals that the size of the glioblastoma could significantly influence the drug delivery efficacy of VLPs. This may be due to differences in VLP diffusion within large versus small tumors, or due to vasculature differences between large and small tumors that could influence drug delivery efficacies [47–49]. It is also possible that larger tumors would require higher injected doses to reduce the tumor burden and increase survival times. Further analysis of these tumor size cohorts revealed that both MS2–DOX and TMV–DOX treatments significantly increased the survival times of small tumor-bearing mice, when compared to PBS controls, with TMV–DOX treatment displaying the highest improved survival (Figure 6). Lastly, mice treated with MS2–DOX displayed significantly improved survival rates when compared to DOX treatment alone. Taken together, these findings underscore the potential of VLPs as nanocarriers for GBM treatment via CED, especially for the treatment of smaller tumors and illustrate some of the morphological effects that are at play in VLP drug delivery.

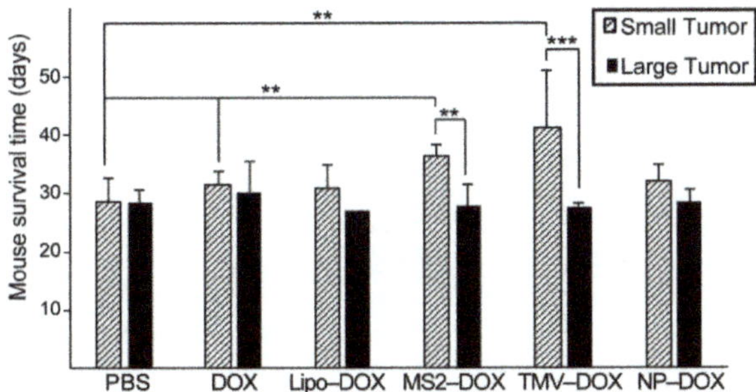

Figure 6. Tumor size dependence of mouse response to VLP treatment. Treatment groups could be divided into two categories based on tumor size as measured by bioluminescence intensity (BLI). Small tumors (as defined by a BLI < 107 AU) showed significantly increased response to CED treatment in the case of TMV and MS2 treated mice when compared to larger tumors (BLI > 107 AU). This BLI cutoff led to a relatively even distribution of mice within each treatment group with the exception of Lipo–DOX, where only two mice had large tumors. Both MS2–DOX and TMV–DOX treatment of small tumors led to significantly enhanced survival when compared to PBS treatment. ** $p < 0.05$. *** $p < 0.01$ as measured by log rank test of survival points. See Supporting Information Figures S5–S7 for the tumor size and survival data for individual animals.

4. Conclusions

We present here an analysis of three morphologically distinct virus-like particles on their ability to deliver drugs to glioblastoma. A panel of virus-like particles representing spheres, disks and filamentous rods was modified with chemotherapeutic cargo and PEGylated via orthogonal bioconjugation strategies. These VLPs were taken up into cancer cells and effectively released their chemotherapeutic payload in cell culture. The VLPs were further tested in glioma-bearing mouse models and showed significant differences in their efficacies for glioblastoma treatment via CED injection. Importantly, glioblastoma-bearing mice treated with TMV-DOX demonstrated improved survival. This is especially encouraging given the low dose of DOX delivered to the mice. Lastly, tumor size dependence of mouse survival was observed for VLP-treated mice, indicating that the tumor environment could have significant implications for GBM treatment via protein-based nanomaterials. Future studies will focus on the evaluation of these VLPs in other tumor models to determine if different tumor types require different VLPs for effective delivery of a chemotherapeutic payload. These studies will help expand our understanding of the effects of virus-like particle morphology on drug delivery, which may have significant implications in the growing field of nanomedicine and help lead to the development of functional protein-based nanomaterials for cancer treatment.

Supplementary Materials: The following are available online at http://www.mdpi.com/2079-4991/8/12/1007/s1, Figure S1: Modification Scheme for TMV and Nanophage Conjugation, Figure S2: LC/MS characterization of DOX-modified VLPs, Figure S3: HPLC-SEC of Nanophage-PEG$_{5k}$-DOX, Figure S4: Protein gel characterization of VLP-PEG$_{5k}$-DOX conjugates, Figure S5: Individual tumor growth plots for each treatment group, Figure S6: Tumor size distribution of glioblastoma-bearing mice and survival points, Figure S7: Kaplan Meier Survival Curves of small and large tumor cohorts.

Author Contributions: J.A.F., I.L.A. and J.M.B. contributed equally toward this work. Conceptualization, J.A.F., I.L.A., J.M.B., S.H.K. and M.B.F.; Formal analysis, J.A.F., I.L.A., J.M.B., S.H.K., T.O., T.P.N. and M.B.F.; Funding acquisition, T.P.N., M.S.B. and M.B.F.; Methodology, J.A.F., I.L.A., J.M.B., S.H.K., S.K.E., K.H. and T.O.; Supervision, T.P.N., M.S.B. and M.B.F.; Writing—original draft, J.A.F., I.L.A., J.M.B. and M.B.F.

Funding: This work was supported by the UCSF Hana Jabsheh Fund. J.A.F. was supported under contract FA9550-11-C-0028 and awarded by the Department of Defense, Air Force Office of Scientific Research, National Defense Science and Engineering Graduate (NDSEG) Fellowship, 32 CFR 168a.

Acknowledgments: We thank Youngho Seo for scientific guidance. We thank Raquel Santos and Edgar Lopez Lepe for technical assistance.

Conflicts of Interest: The authors declare no conflict of interest.

References

1. Wen, A.M.; Steinmetz, N.F. Design of virus-based nanomaterials for medicine, biotechnology, and energy. *Chem. Soc. Rev.* **2016**, *45*, 4074–4126. [CrossRef] [PubMed]
2. Dedeo, M.T.; Finley, D.T.; Francis, M.B. Viral capsids as self-assembling templates for new materials. *Prog. Mol. Biol. Transl. Sci.* **2011**, *103*, 353–392. [PubMed]
3. Shukla, S.; Ablack, A.L.; Wen, A.M.; Lee, K.L.; Lewis, J.D.; Steinmetz, N.F. Increased tumor homing and tissue penetration of the filamentous plant viral nanoparticle potato virus X. *Mol. Pharm.* **2013**, *10*, 33–42. [CrossRef] [PubMed]
4. Shukla, S.; Eber, F.J.; Nagarajan, A.S.; DiFranco, N.A.; Schmidt, N.; Wen, A.M.; Eiben, S.; Twyman, R.M.; Wege, C.; Steinmetz, N.F. The Impact of Aspect Ratio on the Biodistribution and Tumor Homing of Rigid Soft-Matter Nanorods. *Adv. Healthc. Mater.* **2015**, *4*, 874–882. [CrossRef] [PubMed]
5. Czapar, A.E.; Steinmetz, N.F. Plant viruses and bacteriophages for drug delivery in medicine and biotechnology. *Curr. Opin. Chem. Biol.* **2017**, *38*, 108–116. [CrossRef] [PubMed]
6. Wu, W.; Hsiao, S.C.; Carrico, Z.M.; Francis, M.B. Genome-free viral capsids as multivalent carriers for taxol delivery. *Angew. Chem. Int. Ed.* **2009**, *48*, 9493–9497. [CrossRef]
7. Qazi, S.; Liepold, L.O.; Abedin, M.J.; Johnson, B.; Prevelige, P.; Frank, J.A.; Douglas, T. P22 Viral Capsids as Nanocomposite High-Relaxivity MRI Contrast Agents. *Mol. Pharm.* **2013**, *10*, 11–17. [CrossRef] [PubMed]
8. Rhee, J.-K.; Baksh, M.; Nycholat, C.; Paulson, J.C.; Kitagishi, H.; Finn, M.G. Glycan-targeted virus-like nanoparticles for photodynamic therapy. *Biomacromolecules* **2012**, *13*, 2333–2338. [CrossRef]
9. Zhao, X.; Chen, L.; Luckanagul, J.A.; Zhang, X.; Lin, Y.; Wang, Q. Enhancing Antibody Response against Small Molecular Hapten with Tobacco Mosaic Virus as a Polyvalent Carrier. *ChemBioChem* **2015**, *16*, 1279–1283. [CrossRef]
10. Czapar, A.E.; Zheng, Y.-R.; Riddell, I.A.; Shukla, S.; Awuah, S.G.; Lippard, S.J.; Steinmetz, N.F. Tobacco mosaic virus delivery of phenanthriplatin for cancer therapy. *ACS Nano* **2016**, *10*, 4119–4126. [CrossRef]
11. Chariou, P.L.; Lee, K.L.; Pokorski, J.K.; Saidel, G.M.; Steinmetz, N.F. Diffusion and Uptake of Tobacco Mosaic Virus as Therapeutic Carrier in Tumor Tissue: Effect of Nanoparticle Aspect Ratio. *J. Phys. Chem. B* **2016**, *120*, 6120–6129. [CrossRef] [PubMed]
12. Douglas, T.; Young, M. Host–guest encapsulation of materials by assembled virus protein cages. *Nature* **1998**, *393*, 152–155. [CrossRef]
13. Witus, L.S.; Francis, M.B. Using Synthetically Modified Proteins to Make New Materials. *Acc. Chem. Res.* **2011**, *44*, 774–783. [CrossRef] [PubMed]
14. Obermeyer, A.C.; Jarman, J.B.; Francis, M.B. N-terminal modification of proteins with o-aminophenols. *J. Am. Chem. Soc.* **2014**, *136*, 9572–9579. [CrossRef] [PubMed]
15. Obermeyer, A.C.; Capehart, S.L.; Jarman, J.B.; Francis, M.B. Multivalent viral capsids with internal cargo for fibrin imaging. *PLoS ONE* **2014**, *9*, e100678. [CrossRef] [PubMed]
16. Omuro, A.; DeAngelis, L.M. Glioblastoma and Other Malignant Gliomas. *JAMA* **2013**, *310*, 1842–1850. [CrossRef]
17. Jensen, S.A.; Day, E.S.; Ko, C.H.; Hurley, L.A.; Luciano, J.P.; Kouri, F.M.; Merkel, T.J.; Luthi, A.J.; Patel, P.C.; Cutler, J.I.; et al. Spherical nucleic acid nanoparticle conjugates as an RNAi-based therapy for glioblastoma. *Sci. Transl. Med.* **2013**, *5*, 209ra152. [CrossRef]
18. Wohlfart, S.; Gelperina, S.; Kreuter, J. Transport of drugs across the blood–brain barrier by nanoparticles. *J. Control. Release* **2012**, *161*, 264–273. [CrossRef]
19. Shatsberg, Z.; Zhang, X.; Ofek, P.; Malhotra, S.; Krivitsky, A.; Scomparin, A.; Tiram, G.; Calderón, M.; Haag, R.; Satchi-Fainaro, R. Functionalized nanogels carrying an anticancer microRNA for glioblastoma therapy. *J. Control. Release* **2016**, *239*, 159–168. [CrossRef]
20. Chen, W.; Zou, Y.; Zhong, Z.; Haag, R. Cyclo(RGD)-Decorated Reduction-Responsive Nanogels Mediate Targeted Chemotherapy of Integrin Overexpressing Human Glioblastoma In Vivo. *Small* **2017**, *13*, 1601997. [CrossRef]

21. Chao, C.-N.; Yang, Y.-H.; Wu, M.-S.; Chou, M.-C.; Fang, C.-Y.; Lin, M.-C.; Tai, C.-K.; Shen, C.-H.; Chen, P.-L.; Chang, D.; et al. Gene therapy for human glioblastoma using neurotropic JC virus-like particles as a gene delivery vector. *Sci. Rep.* **2018**, *8*, 2213. [CrossRef]

22. Allard, E.; Passirani, C.; Benoit, J.-P. Convection-enhanced delivery of nanocarriers for the treatment of brain tumors. *Biomaterials* **2009**, *30*, 2302–2318. [CrossRef] [PubMed]

23. Xi, G.; Robinson, E.; Mania-Farnell, B.; Vanin, E.F.; Shim, K.-W.; Takao, T.; Allender, E.V.; Mayanil, C.S.; Soares, M.B.; Ho, D.; et al. Convection-enhanced delivery of nanodiamond drug delivery platforms for intracranial tumor treatment. *Nanomed. Nanotechnol. Biol. Med.* **2014**, *10*, 381–391. [CrossRef] [PubMed]

24. Yamashita, Y.; Krauze, M.T.; Kawaguchi, T.; Noble, C.O.; Drummond, D.C.; Park, J.W.; Bankiewicz, K.S. Convection-enhanced delivery of a topoisomerase I inhibitor (nanoliposomal topotecan) and a topoisomerase II inhibitor (pegylated liposomal doxorubicin) in intracranial brain tumor xenografts. *Neuro. Oncol.* **2007**, *9*, 20–28. [CrossRef] [PubMed]

25. Zhan, W.; Wang, C.-H. Convection enhanced delivery of liposome encapsulated doxorubicin for brain tumour therapy. *J. Control. Release* **2018**, *285*, 212–229. [CrossRef] [PubMed]

26. Hashizume, H.; Baluk, P.; Morikawa, S.; McLean, J.W.; Thurston, G.; Roberge, S.; Jain, R.K.; McDonald, D.M. Openings between Defective Endothelial Cells Explain Tumor Vessel Leakiness. *Am. J. Pathol.* **2000**, *156*, 1363–1380. [CrossRef]

27. Mittapalli, R.K.; Adkins, C.E.; Bohn, K.A.; Mohammad, A.S.; Lockman, J.A.; Lockman, P.R. Quantitative Fluorescence Microscopy Measures Vascular Pore Size in Primary and Metastatic Brain Tumors. *Cancer Res.* **2017**, *77*, 238–246. [CrossRef]

28. Farkas, M.E.; Aanei, I.L.; Behrens, C.R.; Tong, G.J.; Murphy, S.T.; O'Neil, J.P.; Francis, M.B. PET Imaging and biodistribution of chemically modified bacteriophage MS2. *Mol. Pharm.* **2013**, *10*, 69–76. [CrossRef]

29. Aanei, I.L.; ElSohly, A.M.; Farkas, M.E.; Netirojjanakul, C.; Regan, M.; Taylor Murphy, S.; O'Neil, J.P.; Seo, Y.; Francis, M.B. Biodistribution of Antibody-MS2 Viral Capsid Conjugates in Breast Cancer Models. *Mol. Pharm.* **2016**, *13*, 3764–3772. [CrossRef]

30. Finbloom, J.A.; Han, K.; Aanei, I.L.; Hartman, E.C.; Finley, D.T.; Dedeo, M.T.; Fishman, M.; Downing, K.H.; Francis, M.B. Stable Disk Assemblies of a Tobacco Mosaic Virus Mutant as Nanoscale Scaffolds for Applications in Drug Delivery. *Bioconjug. Chem.* **2016**, *27*, 2480–2485. [CrossRef]

31. Miller, R.A.; Presley, A.D.; Francis, M.B. Self-assembling light-harvesting systems from synthetically modified tobacco mosaic virus coat proteins. *J. Am. Chem. Soc.* **2007**, *129*, 3104–3109. [CrossRef] [PubMed]

32. Dedeo, M.T.; Duderstadt, K.E.; Berger, J.M.; Francis, M.B. Nanoscale protein assemblies from a circular permutant of the tobacco mosaic virus. *Nano Lett.* **2010**, *10*, 181–186. [CrossRef] [PubMed]

33. Gratton, S.E.; Ropp, P.; Pohlhaus, P.D.; Luft, J.C.; Madden, V.J.; Napier, M.E.; DeSimone, J.M. The effect of particle design on cellular internalization pathways. *Proc. Natl. Acad. Sci. USA* **2008**, *105*, 11613–11618. [CrossRef] [PubMed]

34. Adriani, G.; de Tullio, M.D.; Ferrari, M.; Hussain, F.; Pascazio, G.; Liu, X.; Decuzzi, P. The preferential targeting of the diseased microvasculature by disk-like particles. *Biomaterials* **2012**, *33*, 5504–5513. [CrossRef] [PubMed]

35. Sattar, S.; Bennett, N.J.; Wen, W.X.; Guthrie, J.M.; Blackwell, L.F.; Conway, J.F.; Rakonjac, J. Ff-nano, short functionalized nanorods derived from Ff (f1, fd, or M13) filamentous bacteriophage. *Front. Microbiol.* **2015**, *6*, 316. [CrossRef] [PubMed]

36. Schlick, T.L.; Ding, Z.; Kovacs, E.W.; Francis, M.B. Dual-Surface Modification of the Tobacco Mosaic Virus. *J. Am. Chem. Soc.* **2005**, *127*, 3718–3723. [CrossRef] [PubMed]

37. Toita, R.; Murata, M.; Abe, K.; Narahara, S.; Piao, J.S.; Kang, J.-H.; Hashizume, M. A nanocarrier based on a genetically engineered protein cage to deliver doxorubicin to human hepatocellular carcinoma cells. *Chem. Commun.* **2013**, *49*, 7442–7444. [CrossRef]

38. Ozawa, T.; James, C.D. Establishing Intracranial Brain Tumor Xenografts With Subsequent Analysis of Tumor Growth and Response to Therapy using Bioluminescence Imaging. *J. Vis. Exp.* **2010**, e1986. [CrossRef]

39. Pardridge, W.M. The blood-brain barrier: Bottleneck in brain drug development. *NeuroRx* **2005**, *2*, 3–14. [CrossRef]

40. Gabizon, A.; Shmeeda, H.; Barenholz, Y. Pharmacokinetics of Pegylated Liposomal Doxorubicin. *Clin. Pharmacokinet.* **2003**, *42*, 419–436. [CrossRef]

41. Jue, R.; Lambert, J.M.; Pierce, L.R.; Traut, R.R. Addition of sulfhydryl groups of Escherichia coli ribosomes by protein modification with 2-iminothiolane (methyl 4-mercaptobutyrimidate). *Biochemistry* **1978**, *17*, 5399–5406. [CrossRef] [PubMed]

42. Agarwal, R.; Singh, V.; Jurney, P.; Shi, L.; Sreenivasan, S.V.; Roy, K. Mammalian cells preferentially internalize hydrogel nanodiscs over nanorods and use shape-specific uptake mechanisms. *Proc. Natl. Acad. Sci. USA* **2013**, *110*, 17247–17252. [CrossRef] [PubMed]

43. Chithrani, B.D.; Ghazani, A.A.; Chan, W.C. Determining the Size and Shape Dependence of Gold Nanoparticle Uptake into Mammalian Cells. *Nano Lett.* **2006**, *6*, 662–668. [CrossRef] [PubMed]

44. He, C.; Hu, Y.; Yin, L.; Tang, C.; Yin, C. Effects of particle size and surface charge on cellular uptake and biodistribution of polymeric nanoparticles. *Biomaterials* **2010**, *31*, 3657–3666. [CrossRef] [PubMed]

45. Jiang, W.; Kim, B.Y.S.; Rutka, J.T.; Chan, W.C.W. Nanoparticle-mediated cellular response is size-dependent. *Nat. Nanotechnol.* **2008**, *3*, 145–150. [CrossRef] [PubMed]

46. Ulbrich, K.; Etrych, T.; Chytil, P.; Pechar, M.; Jelinkova, M.; Rihova, B. Polymeric anticancer drugs with pH-controlled activation. *Int. J. Pharm.* **2004**, *277*, 63–72. [CrossRef] [PubMed]

47. Holash, J.; Wiegand, S.J.; Yancopoulos, G.D. New model of tumor angiogenesis: Dynamic balance between vessel regression and growth mediated by angiopoietins and VEGF. *Oncogene* **1999**, *18*, 5356–5362. [CrossRef] [PubMed]

48. Alfonso, J.C.L.; Köhn-Luque, A.; Stylianopoulos, T.; Feuerhake, F.; Deutsch, A.; Hatzikirou, H. Why one-size-fits-all vaso-modulatory interventions fail to control glioma invasion: In silico insights. *Sci. Rep.* **2016**, *6*, 37283. [CrossRef] [PubMed]

49. Gevertz, J.L.; Torquato, S. Modeling the effects of vasculature evolution on early brain tumor growth. *J. Theor. Biol.* **2006**, *243*, 517–531. [CrossRef]

 *nanomaterials*

Article

Calcium Phosphate Nanoparticle-Based Vaccines as a Platform for Improvement of HIV-1 Env Antibody Responses by Intrastructural Help

Dominik Damm [1], Leonardo Rojas-Sánchez [2], Hannah Theobald [1], Viktoriya Sokolova [2], Richard T. Wyatt [3], Klaus Überla [1], Matthias Epple [2] and Vladimir Temchura [1,*]

[1] Institute of Clinical and Molecular Virology, Friedrich-Alexander University Erlangen-Nürnberg, 91054 Erlangen, Germany; dominik.damm@uk-erlangen.de (D.D.); hannah.theobald@fau.de (H.T.); klaus.ueberla@fau.de (K.Ü.)

[2] Inorganic Chemistry and Center for Nanointegration Duisburg-Essen (CeNIDE), University of Duisburg-Essen, 45141 Essen, Germany; leonardo.rojas-sanchez@uni-due.de (L.R.-S.); viktoriya.sokolova@uni-due.de (V.S.); matthias.epple@uni-due.de (M.E.)

[3] Department of Immunology and Microbial Science, The Scripps Research Institute, La Jolla, CA 92037, USA; wyatt@scripps.edu

* Correspondence: vladimir.temchura@fau.de; Tel.: +49-9131-85-43652

Received: 5 September 2019; Accepted: 25 September 2019; Published: 27 September 2019

Abstract: Incorporation of immunodominant T-helper epitopes of licensed vaccines into virus-like particles (VLP) allows to harness T-helper cells induced by the licensed vaccines to provide intrastructural help (ISH) for B-cell responses against the surface proteins of the VLPs. To explore whether ISH could also improve antibody responses to calcium phosphate (CaP) nanoparticle vaccines we loaded the nanoparticle core with a universal T-helper epitope of Tetanus toxoid (p30) and functionalized the surface of CaP nanoparticles with stabilized trimers of the HIV-1 envelope (Env) resulting in Env-CaP-p30 nanoparticles. In contrast to soluble Env trimers, Env containing CaP nanoparticles induced activation of naïve Env-specific B-cells in vitro. Mice previously vaccinated against Tetanus raised stronger humoral immune responses against Env after immunization with Env-CaP-p30 than mice not vaccinated against Tetanus. The enhancing effect of ISH on anti-Env antibody levels was not attended with increased Env-specific IFN-γ CD4 T-cell responses that otherwise may potentially influence the susceptibility to HIV-1 infection. Thus, CaP nanoparticles functionalized with stabilized HIV-1 Env trimers and heterologous T-helper epitopes are able to recruit heterologous T-helper cells induced by a licensed vaccine and improve anti-Env antibody responses by intrastructural help.

Keywords: nano-vaccines; HIV-1 Env trimers; B-cell targeting; intrastructural help

---

1. Introduction

The unique characteristic of HIV-1 to infect activated CD4 T-cells requires alternative vaccination strategies that differ from the classical approaches used [1]. Polyfunctional serum antibodies against envelope glycoprotein (Env) of HIV-1 correlate with spontaneous HIV-1 control by elite controllers [2] and appear to be crucial for vaccine-mediated protection against HIV-1 [3–5]. However, limited breadth, poor persistence of antibody responses to Env, and potential enhancement of susceptibility to HIV infection by vaccine-induced HIV-specific T-helper cell responses compromise HIV-1 vaccines previously designed for clinical trials [6,7]. Recently we demonstrated how T-helper cells induced by a licensed vaccine against Tetanus toxoid (TT) can be harnessed to provide help for Env-specific B-cells in mice immunized with HIV-1 virus-like particles (VLP) containing T-helper epitopes of TT [8]. This "intrastructural help" [9] (ISH) can be explained by uptake of entire VLPs by Env-specific B-cells and subsequent presentation of epitopes from all proteins of the VLPs on their major histocompatibility complex class II (MHC-II) molecules to harness corresponding T-cell help [8,10].

Biodegradable calcium phosphate (CaP) nanoparticles have several advantages compared to biological and polymer-based nanoparticles and have been used for experimental vaccination during the past decade [11]. Previously we designed CaP nanoparticles covered with a model antigen in order to achieve efficient targeting and activation of cognate B-cells in vitro [12] as well as induction of humoral immune responses in vivo [13]. In animal models, where CD4 T-cells were genetically non-reactive to the surface antigen of the CaP nanoparticles, incorporation of the p30 peptide, a promiscuous (universal) T-helper epitope from TT [14], overcame the lack of functionally active CD4 T-cell epitopes [13]. These data indicate that CaP nanoparticle-based vaccines might be potentially used for improvement of HIV-Env antibody responses by the ISH approach.

A gp160 precursor of HIV-1 Env is proteolytically cleaved into non-covalently linked gp120 and gp41 subunits, which assemble into a trimer of heterodimers [15]. Historically, the initial candidate vaccines for immunization with Env-based antigens used monomeric gp120 [16]. Recent advances in design and manufacturing of soluble HIV-1 Env trimers stabilized in the closed pre-fusion state provide an antigenic form of Env that induces Env antibodies recognizing the quaternary conformation associated with neutralizing activity against autologous HIV-1 [15,16].

In this study, we therefore designed CaP nanoparticles with stabilized HIV-1 Env trimers coupled to the surface and universal T-helper peptides of TT in the core. These nanoparticles were used to demonstrate effects of ISH in mice previously immunized with a licensed vaccine against Tetanus toxoid.

2. Materials and Methods

2.1. Mice, Ethical Statement

Six- to eight-week-old female wild-type (wt) C57bl/6NRj (Bl6) mice (Janvier, Le Genest-Saint-Isle, France), as well as mice with transgenic B-cell receptors (BCR) specific for HIV-1 Env (PGT-121 mice [17], in-house breeding, kindly provided by Dr. M. Nussenzweig, The Rockefeller University, New York, NY, USA) were used in this study. Mice were accommodated in the animal facility of the Faculty of Medicine, FAU (Erlangen, Germany), in accordance with the national law and were handled according to instructions of the Federation of European Laboratory Animal Science Associations. All animal experiments were approved by an external ethics committee of the North Rhine-Westphalian Ministry for Nature, Environment and Consumer Protection (license AZ 84-02.04.2014. A191) and confirmed by the Government of Lower Franconia (license 55.2-2532-2-96).

2.2. Plasmids and VLP Production

The plasmid encoding for BG505 NFL2P gp140 [18] was used for expression and purification of stabilized Env trimers (see 2.3). We introduced recombinant transmembrane (TM) and cytoplasmic (CD) domains of the surface protein G from the vesicular stomatitis virus (VSV-G; GenBank Accession Number: GU177825.1) downstream of the sequence encoding for the soluble Env protein to elicit pseudo-typed, membrane-bound Env trimers (BG505 NFL2P gp140-GTMCD). We further introduced the nucleotide sequence ttcaacaacttcaccgttagcttctggctgcgcgttccgaaagtttctgcttcccacctggaa, that encodes for the TT-derived peptide p30 (FNNFTVSFWLRVPKVSASHLE) into the Gag open reading frame of the plasmid Hgpsyn [19] (Hgpsyn-TTp30) that contains the codon-optimized genetic information for the HIV-1 structural precursor protein (Gag) and viral enzymes (Pol). The p30 sequence was inserted between the p17 matrix protein and the p24 capsid protein. VLPs lacking Env (Δ-VLP), VLPs bearing BG505 NFL2P gp140-GTMCD Env trimers on the surface (Env-VLP) and Env-VLP with the Gag-p30 fusion protein inside (Env-VLP-p30) were produced by co-transfection of 293T cells with equal amounts of the corresponding plasmids and purified as described previously [20].

2.3. Production and Analyses of HIV-1 Env Trimers

2.3.1. Production of HIV-1 Env Trimers

The HIV-1 subtype A envelope protein BG505 NFL2P gp140 (Env) was used as the primary antigen. In brief, this antigen consists of three heterodimeric gp120-gp41$_{ecto}$ subunits and was stabilized for the soluble, trimeric conformation by truncation of the transmembrane and cytoplasmic domains, introduction of a flexible linker (2 x G4S) between the globular gp120 subunit and the gp41 ectodomain as well as a point mutation at amino acid position 559 (I559P) [18]. Env trimers were produced by transfection of 293F cells at a density of 1.0×10^6 cells per mL with the plasmid encoding for BG505 NFL2P gp140 (1 μg/mL DNA). Linear polyethylenimine (PEI, Polysciences Inc., Warrington, PA, USA) was used as transfection reagent in a threefold excess compared to the mass of DNA. Cell culture medium was changed 6 h after the transfection. The supernatant was harvested after three days, sterile-filtered and run over a lectin affinity column (Agarose-bound *Galanthus nivalis* lectin, Vector Laboratories Inc., Burlingame, CA, USA). Env trimers were eluted using 1M Methyl-α-D-mannopyranoside and concentrated with a 10 kDa Amicon cutoff filter (Sigma-Aldrich, St. Louis, MO, USA). Between the centrifugation steps, buffer changes were performed by refilling with DPBS without bivalent cations (Thermo Fisher Scientific, Waltham, MA, USA). The final Env concentration was determined by photometric measurement with a nanodrop device (Thermo Fisher Scientific, Waltham, MA, USA). As a control, we introduced a stop codon downstream of the flexible linker and deleted the nucleotide sequence for the gp41 ectodomain in order to produce gp120 monomers. These proteins were purified in the same way as described above for the trimers. All purified proteins were analyzed by NativePAGE and western blot (WB). For the UV-Vis spectroscopy (see 2.4.4) Env trimers were labelled with AlexaFluor®-488 fluorescent dye using a protein labeling kit (Thermo Fisher Scientific, Waltham, MA, USA) following the manufacturer's instructions.

2.3.2. NativePAGE Analysis of HIV-1 Env Trimers

The native conformation of Env trimers and monomers was addressed using the NativePAGE system (Thermo Fisher Scientific, Waltham, MA, USA) following the manufacturer's guidelines. In brief, 1 μg of each protein was mixed with G-250 additive and loaded onto a 4–16% Native Page Bis-Tris gel. The proteins were separated according to their native size by gel electrophoresis. Afterwards, excessive Coomassie stain was removed from the native gel by overnight fixation in 10% acetic acid and 30% ethanol in ultra-pure H_2O. Faint protein bands on the gel were then developed by staining with silver nitrate using a Silver Stain Kit (Pierce Biotechnology Inc., Rockford, IL, USA).

2.3.3. Western Blot Analysis of HIV-1 Env Trimers

For the antigen-specific immunoblotting of purified Env, 1 μg of each soluble protein sample or 300 ng Env on VLPs were mixed with house-made, reducing SDS sample buffer, boiled and then loaded onto a 12% SDS gel. After gel electrophoresis, proteins were transferred onto a nitrocellulose membrane which was subsequently blocked with 5% skimmed milk in DPBS supplemented with 0.1% Tween20 (PBS-T). The blocked membranes were incubated with polyclonal goat anti-gp120 (Acris Antibodies GmbH, Herford, Germany) and horseradish peroxidase-coupled secondary anti-goat IgG antibody (Dianova, Hamburg, Germany) with multiple washing steps in between. The membranes were finally developed with house-made ECL solution and protein bands were imaged using an Advanced Fluorescence Imager (Intas, Göttingen, Germany).

2.4. Production of Calcium Phosphate Nanoparticles

2.4.1. Instruments

Dynamic light scattering (DLS) and zeta potential were measured with a Zetasizer Nano ZS instrument (laser wavelength λ = 633 nm, Malvern Instruments, Malvern, UK) with the Smoluchowski approximation. Data obtained from Malvern software were used without further treatment; the particle size results refer to the z-average. Scanning electron microscopy (SEM) was performed with an ESEM Quanta 400 instrument (FEI Co., Hillsboro, OR, USA) and gold/palladium-sputtered samples. Calcium concentrations were measured by atomic absorption spectroscopy (AAS) with an M-Series AA spectrometer (Thermo Electron Corporation, Schwerte, Germany). UV-Vis absorption spectra were measured with a DS-11 FX+ spectrophotometer ("Nanodrop", DeNovix, Wilmington, DE, USA) and a Cary 300 Bio spectrophotometer (Agilent Technologies, Santa Clara, CA, USA). Ultracentrifugation was done at 20 °C with a Sorvall WX Ultra Series centrifuge (Thermo Electron Corporation, Schwerte, Germany). Freeze-drying (lyophilization) was carried out with a Christ Alpha 2-4 LSC instrument (Martin Christ GmbH, Osterode am Harz, Germany). The endotoxin concentration was measured with an Endosafe Nexgen-PTS handheld spectrophotometer (Charles River, Boston, MA USA). Ultrapure water (Purelab, ELGA LabWater, Celle, Germany) was used for all preparations. All nanoparticles were prepared and analyzed at room temperature.

2.4.2. Synthesis of Calcium Phosphate Nanoparticles

The CaP nanoparticle synthesis was performed according to our previously described method [21]. In brief, aqueous solutions of calcium lactate (18 mM, pH = 10, p.a., Sigma-Aldrich Corp., St. Louis, MO, USA), diammonium hydrogen phosphate (10.8 mM, pH = 10, p.a., VWR Life-Sciences) and branched polyethyleneimine (PEI, Mw = 25 kDa, Sigma-Aldrich Corp., St. Louis, MO, USA) were simultaneously pumped at a volume ratio of 5:5:7 mL during one minute into a stirred vessel with 20 mL of ultrapure water. The dispersion was stirred for 20 min. and the CaP/PEI nanoparticle dispersion was used immediately for the following steps. For the synthesis of adjuvant-containing nanoparticles, 1 mL CaP/PEI nanoparticle dispersion was mixed with aqueous solutions of either 60 µL p30 peptide (1 mg/mL) or 40 µL CpG (1 mg/mL) under stirring, followed by 30 min. stirring at room temperature (RT). The adjuvant loading was determined by measuring the residual concentration in the supernatant by UV microvolume spectroscopy ("nanodrop"). For further surface modifications a silica shell was added to the nanoparticles. To this end, 1 mL of either CaP/PEI or of adjuvant-containing nanoparticle dispersion was added to a mixture of 4 mL ethanol (Fisher Chemicals, Hampton, NH, USA), 5 µL tetraethylorthosilicate (TEOS, Sigma-Aldrich Corp., St. Louis, MO, USA) and 10 µL aqueous ammonia solution (7.8 wt.%) and stirred for 16 h. After this time, the nanoparticles were isolated by ultracentrifugation (66,000× g, 30 min, 20 °C) and redispersed with 1 mL ultrapure water followed by ultrasonication (UP50H, Hielscher, Teltow, Germany, sonotrode MS7m cycle 0.8, amplitude 70%, 4 s). Silica-terminated calcium phosphate nanoparticles (CaP/PEI/SiO$_2$) were obtained. To prepare the thiol-terminated calcium phosphate nanoparticles a surface modification with (3-mercaptopropyl)trimethoxysilane (MPS, Sigma-Aldrich Corp., St. Louis, MO, USA) was performed. For this, 1 mL of CaP/PEI/SiO$_2$ nanoparticles was added to a mixture of 4 mL ethanol and 50 µL (3-mercaptopropyl)trimethoxysilane and stirred for 8 h. The nanoparticles were isolated by ultracentrifugation (66,000× g, 30 min, 20 °C), redispersed with 1 mL H$_2$O and ultrasonicated. After this step, thiol-terminated calcium phosphate nanoparticles (CaP/PEI/SiO$_2$-SH) were obtained.

2.4.3. Functionalization of CaP Nanoparticles with Env Trimers

The Env trimers (Mw = 140 kDa) were coupled to the nanoparticle surface via a sulfo-SMCC cross-linker (sulfosuccinimidyl-*trans*-4-(N-maleimidomethyl)cyclohexane-1-carboxylate, Merck, Darmstadt, Germany). One end of the linker reacts with the primary amines in Env and the other end with the thiol groups on the nanoparticle surface. For this, 300 µL aqueous sulfo-SMCC solution (1.78 mg/mL) were given to 600 µL Env in DPBS (1 mg/mL) and incubated for 2 h at 4 °C. After incubation, the protein was purified from the unreacted cross-linker with a 3 kDa Amicon ultracentrifuge filter (Merck, Darmstadt, Germany), following the manufacturer recommendations. To attach the Env trimers to the nanoparticle surface, 330 µL activated protein (1 mg/mL) were given to 4 mL CaP/PEI/SiO$_2$-SH nanoparticle dispersion and incubated for 24 h at 4 °C. After this time, the particles were isolated by centrifugation at 21,000× g and 8 °C and washed once with 1 mL H_2O. Finally, the nanoparticles were redispersed in 4 mL H_2O and ultrasonicated (UP50H, Hielscher, sonotrode MS7m cycle 0.8, amplitude 70%, 4 s). The different syntheses were carried out with sterile-filtered solutions.

2.4.4. CaP nanoparticle Characterization and Storage

For the final nanoparticle dispersion, an endotoxin quantification assay was performed with an Endosafe Nexgen-PTS device. No endotoxin was detected in all synthesized nanoparticles (<0.1 EU/mL). To calculate the number of nanoparticles per volume unit, the Ca^{2+} concentration was measured by AAS and then tentatively expressed as the most common calcium phosphate, i.e., hydroxyapatite, $Ca_{10}(PO4)_6(OH)_2$. The nanoparticle concentration in 1 mL dispersion is calculated from the measured calcium phosphate concentration and the density of hydroxyapatite (3140 kg/m^3) assuming a complete spherical morphology. Additionally, the number of Env units on the nanoparticle surface was determined by UV-Vis spectroscopy with an AlexaFluor®-488 labelled Env protein and UV microvolume spectroscopy ("Nanodrop") (see [22] for typical calculation steps to obtain these data). For storage and transportation, the nanoparticle dispersion was lyophilized according to our previously reported protocol [23]. 20 mg D-(+)-trehalose dihydrate (Sigma-Aldrich Corp., St. Louis, MO, USA) were added to 1 mL of the nanoparticle dispersions as cryoprotectant followed by shock-freezing with liquid nitrogen and lyophilization for 72 h at 0.31 mBar and −10 °C. Immediately before the application, the nanoparticles were redispersed in 1 mL ultrapure water and gently sonicated with an ultrasonication bath.

2.5. B-cell Activation In Vitro

B-cells from wt Bl6 mice or PGT-121 BCR-transgenic mice were isolated from the spleen by magnetic cell separation (Miltenyi Biotec, Bergisch Gladbach, Germany, #130-090-862,). 2.0×10^5 cells were incubated with different concentrations of Env-coupled nanoparticles in U-bottom 96-well plates. As controls, we additionally incubated B cells with soluble Env trimers, with Env-VLP (0.2 µg of Env/mL) and with 2 µg/mL of LPS (Sigma-Aldrich, Corp., St. Louis, MO, USA). After 18 h incubation at 37 °C and 5% CO_2, cells were stained with Fixable Viability Dye (Thermo Fisher Scientific, Waltham, MA, USA) and with antibodies against the B-cell surface antigen CD19 (Thermo Fisher Scientific, Waltham, MA, USA) and the early activation marker CD69 (Thermo Fisher Scientific, Waltham, MA, USA). B-cell activation in living B-cells was subsequently measured on a benchtop flow cytometer BD™ LSR II (BD Biosciences, Franklin Lakes, NJ, USA) and analyzed with the FlowJo software (BD Biosciences, Franklin Lakes, NJ, USA).

2.6. Analyses of In Vivo Induced Immune Responses

2.6.1. Immunization, Collection of Blood and Organ Samples

All immunizations were performed intramuscularly in both hind legs in Bl6 mice that were at the age of 6–8 weeks by the time of the first injection. For the induction of intrastructural help, mice were immunized on day 0 and day 28 with Tetanus toxoid vaccine (Tetanol®pur, GSK Vaccines GmbH, Marburg, Germany) diluted 1:10 in sterile DPBS to induce CD4 T-cell responses against the TT peptide p30 (ISH group) or with DPBS alone (control group). All mice were then boosted on day 56, day 84 and day 112 with one of the following particle types: Env trimer-coupled CaP nanoparticles (Env-CaP); Env-CaP with encapsulated p30 peptide (Env-CaP-p30); or Env-VLP-p30 with the membrane-bound form of the Env trimer on the surface and the Gag-p30 fusion protein inside. To evaluate the influence of encapsulated CpG as an adjuvant, mice were boosted thrice with Env-CaP-CpG particles in the same immunization protocol as described above. The injection doses were normalized to 10 µg of Env delivered with CaP nanoparticles and 300 ng of Env delivered with VLPs per mouse. To investigate humoral immune responses against Env, blood samples were collected 2 weeks after each particle immunization. To investigate humoral immune responses against TT, blood samples were collected on day 49. All blood samples were taken under isoflurane anesthesia from the retrobulbar venous plexus with non-heparinized, single-use capillaries (minicaps®, Hirschmann, Eberstadt, Germany). Collected blood samples were centrifuged for 5 min at 5000 rpm. The upper serum fractions were isolated and stored at −20 °C. Mice were sacrificed on day 126 and spleens were isolated for further assessment of the cellular immune response against both Env and p30.

2.6.2. Analyses of Humoral Immune Responses

Serum ELISAs to address the humoral immune response against Env were performed as described previously [8]. Briefly, white opaque MaxiSorp 96-well plates (Greiner Bio One, Frickenhausen, Germany) were coated with 100 ng Env trimer in coating buffer (0.1 M Na_2CO_3, 0.1 M $NaHCO_3$ in H_2O, pH 9.6) per well at 4 °C overnight. After 1 h blocking with 5% skimmed milk (diluted in DPBS containing 0.05% Tween-20), the wells were incubated with serum from immunized mice diluted 1:1000 in 2% skimmed milk for 1 h at RT. Plates were washed and then incubated for 1 h at RT with 1:4000 dilutions of HRP-coupled secondary antibodies specific for different murine IgG subtypes as well as total IgG (Southern Biotech, Birmingham, AL, USA). After thorough washing, serum binding was detected by addition of ECL solution and measurement of the relative light units per second (RLU/s) with an Orion microplate luminometer (Berthold Detection Systems GmbH, Pforzheim, Germany).

2.6.3. Analyses of Cellular Immune Responses

For the analysis of the cytokine profiles of both Env- and p30-specific CD4 T-cells after immunizations, we performed in vitro intracellular cytokine staining (ICS) of CD4 T-cells for interferon gamma (IFN-γ), tumor necrosis factor alpha (TNF-α) and interleukin-2 (IL-2), as well as cytokine ELISA for IL-5. Isolated spleens were dissociated by using a gentleMACS™ Dissociator (Miltenyi Biotech, Bergisch Gladbach, Germany) following the manufacturer's guidelines. The cell suspensions were run through 70 µm cell strainers. We removed erythrocytes by incubation with ACK lysis buffer (150 mM NH_4Cl, 10 mM $KHCO_3$, 0.1 mM EDTA in H_2O, pH 7.2) for 8 min at RT; the lysis reaction was stopped by the addition of R10 medium (RPMI 1640, 10% FCS, 1% Penicillin-Streptomycin, 10 mM HEPES, 2 mM L-glutamine, 50 µM 2-mercaptoethanol). The isolated splenocytes were washed twice by centrifugation and resuspension in R10 and subsequently counted using a Countess Automated Cell Counter (Thermo Fisher Scientific, Waltham, MA, USA). Isolated splenocytes were seeded in 96-well U-bottom plates (1.0×10^6 cells per well). For antigen-driven cytokine production by p30-specific CD4 T-cells, splenocytes were re-stimulated with 5 µg/mL of p30 peptide. Since there is no identified immunodominant MHC-II restricted Env peptides for Bl6 mice, we established a re-stimulation protocol to induce Env-specific cellular responses. To this end, we purified PGT121 B-cells (see above) and incubated them for 3 h at 37 °C with Env-VLPs containing 0.2 µg/mL Env or with the same amount of Δ-VLP. Thereafter, the B-cells, that were supposed to take up the Env-VLP in a BCR-dependent manner and subsequently present an array of different Env peptides to the CD4 T-cells, were washed with R10 and added to splenocyte suspensions (1.5×10^5 PGT121 B-cells per well).

Intracellular Cytokine Staining

Antigen re-stimulated and unstimulated (control) splenocytes were incubated in the presence of 2 µg/mL anti-mouse CD28 antibody and 3 µg/mL Brefeldin A (eBioscience, San Diego, CA, USA) for 6 h at 37 °C and 5% CO_2. After incubation, cells were washed with FACS buffer (1% FCS, 1 mM EDTA in DPBS without bivalent cations) and stained with anti-mouse CD4 antibody (eBioscience, San Diego, CA, USA) and Fixable Viability Dye (Thermo Fisher Scientific, Waltham, MA, USA). Then, the cells were fixed with 2% paraformaldehyde in DPBS and subsequently permeabilized using 0.5% Saponin in FACS buffer and intracellularly stained with anti-mouse IL-2, IFN-γ and TNF-α (eBioscience, San Diego, CA, USA). The cells were washed twice with permeabilization buffer and twice with FACS buffer. The cytokine accumulation in the CD4 T-cells was then analyzed by flow cytometry. The background values of unstimulated cultures (Δ-VLP for Env) were subtracted for each individual mouse.

Cytokine ELISA

Antigen re-stimulated and control splenocytes were incubated in the presence of 2 µg/mL anti-mouse CD28 antibody for 60 h at 37 °C and 5% CO_2. The culture media were harvested and the IL-5 cytokine secretion was analyzed using a Mouse IL-5 ELISA kit (Invitrogen, Carlsbad, CA, USA) following the manufacturer's instructions. The background values of unstimulated cultures were subtracted for each individual mouse.

2.7. Statistical Analysis

Statistical analyses were performed as indicated in the figure legends with the GraphPad Prism 7 software (Graphpad Software Inc., San Diego, CA, USA).

3. Results and Discussion

3.1. Production and Characterization of Soluble HIV-1 Env Trimers and Env-VLP-p30

For the coupling onto the surface of the CaP nanoparticles we needed faithful mimetics of the HIV-1 Env spike. During the last decade, there were great efforts to create soluble Env trimers that are stabilized in a closed pre-fusion conformation. This was achieved by truncation of the protein at amino acid position 664 within the gp41 subunit as well as by the introduction of various stabilizing mutations i.e., I559P [24].

In this study, we used an Env trimer that is derived from the membrane-embedded native form (Figure 1A; left) of a subtype A HIV-1 isolate from an infected baby (BG505). In addition to the modifications mentioned above, a flexible linker (2xG4S) was more recently introduced between the gp120 subunit and the gp41 ectodomain to overcome the need for proper cleavage of the precursor protein in order to receive stabilized well-folded proteins [18].

We expressed these Env trimers (BG505 NFL2P gp140) (Figure 1A, middle left) as well as monomeric gp120 subunits (BG505 NFL2P gp120) (Figure 1A, middle right) in 293F cells and purified them by Lectin affinity chromatography. The purified and concentrated proteins were analyzed on a native gel followed by a silver staining. A prominent band at approximately 700 kDa represented the globular native form of an Env trimer that is formed by three non-covalently assembled gp120-gp41$_{ecto}$ (gp140) heterodimers (Figure 1B). Two minor bands showed a low percentage of gp140 dimers and monomers. The purified gp120 resulted in two major bands on the native gel: a monomeric protein at 200 kDa and a gp120 dimer at 480 kDa that is formed by aberrant intermolecular di-sulfide bridges [25] as well as a large aggregate fraction (Figure 1B). As expected, the expression of gp140 (but not of gp120) results in the production of Env trimers, since the trimerization domain is located in the gp41 ectodomain and stabilized by the I559P mutation.

To compare intrastructural help for CaP nanoparticles side by side with that for HIV-1 VLPs, we generated VLPs with a membrane-bound form of the stabilized Env trimer on the surface and the TT-derived peptide p30 inside (Env-VLP-p30). In order to array stabilized BG505 NFL2P gp140 trimers on HIV-1 VLPs, the transmembrane and cytoplasmic domains of the vesicular stomatitis virus G-protein (VSV-G) were fused to the open reading frame encoding Env. Figure 1C represents the BG505 NFL2P gp140-GTMCD expression construct and the corresponding protein (Figure 1A; right) has an estimated size between 150 and 160 kDa. To produce HIV-1 VLPs that contain the p30 peptide of TT, we inserted the coding sequence for the peptide in frame between the open-reading frame for matrix and capsid of the HIV-1 structural Gag protein. Figure 1D represents the Hgpsyn-TTp30 construct. Env-VLPs-p30 particles were produced by co-transfection of 293T cells with both the BG505 NFL2P gp140-GTMCD and the Hgpsyn-TTp30 construct.

The purified Env proteins as well as the produced Env-VLP-p30 particles were analyzed by Western Blot under reducing conditions (Figure 1E). gp120 and gp140 were represented by prominent bands in the respective sizes. The pseudotyped BG505 NFL2P gp140-GTMCD protein was approximately 150 to 160 kDa in size and, therefore, runs slightly higher than the soluble trimer (Figure 1E; upper panel). In addition, three major Gag bands represent different states of the capsid maturation of the HIV-VLPs (Figure 1E; lower panel).

These characterized Env trimers were further used for the coupling onto the surface of the CaP nanoparticles.

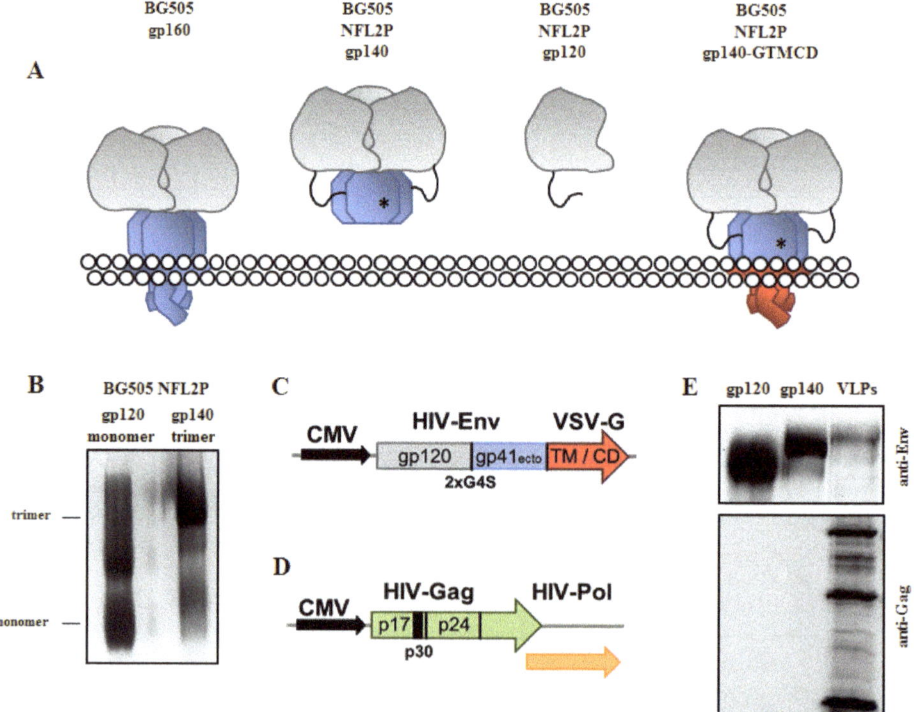

Figure 1. Production and characterization of soluble HIV-1 Env constructs and Env-p30-VLPs. (**A**) Overview of different membrane-embedded and soluble Env constructs. The native BG505 gp160 (left), that is composed of three gp120 (grey) – gp41 (blue) heterodimers, has been used previously to create a soluble stabilized recombinant trimer (BG505 NFL2P gp140; middle left) by truncation at amino acid 664 and introduction of both a point mutation (I559P) indicated with an asterisk (*) and a flexible linker (2xG4S). As a purification control, we produced a BG505 NFL2P gp120 monomer (middle right) by truncation of the protein downstream of the flexible linker. A recombinant VSV-G transmembrane and cytoplasmic domain (TM/CD; red) was introduced downstream of the gp41 ectodomain to create a membrane-bound form of the stabilized trimer for production of Env-VLP-p30 particles (BG505 NFL2P gp140-GTMCD; right). (**B**) Native PAGE of purified gp140 and gp120. Three gp140 molecules formed a globular trimer molecule represented by a major band at 700 kDa. The gp120 subunit was expressed as monomeric or dimeric proteins, but did not form trimers. (**C,D**) Schematic overview of the plasmids used for Env-VLP-p30 production. (**C**) BG505 NFL2P gp140-GTMCD construct. Colors are matched to their respective protein domains in Figure 1A. (**D**) Hgpsyn-TTp30 construct: the nucleotide sequence for the p30 peptide was introduced between the HIV-1 p17 matrix protein and the spacer protein 1 followed by the capsid protein p24. (**E**) Reducing SDS-PAGE and Western Blot of purified soluble Env proteins and Env-VLP-p30.

3.2. Design, Production and Characterization of CaP Nanoparticles Functionalized with Soluble HIV-1 Env Trimers

Stable CaP nanoparticles were synthesized and the surface modified with soluble stabilized HIV-1 Env trimer proteins (Figure 1). The obtained nanoparticles had an average solid core diameter between 38-57 nm (as determined by SEM) (Figure 2) with a positively charged surface near +27 mV, due to the stabilizing PEI polymer, and a hydrodynamic diameter between 300–400 nm. These parameters make the nanoparticles suitable for cellular uptake [26]. The complete characterization data are shown in Supplementary Table S1. For nanoparticles loaded with the T-helper peptide p30 or with CpG, the load was absorbed onto the nanoparticle core before forming the external silica shell around it. This coating provides protection to the internal loading but also permits a covalent surface modification with Env [21].

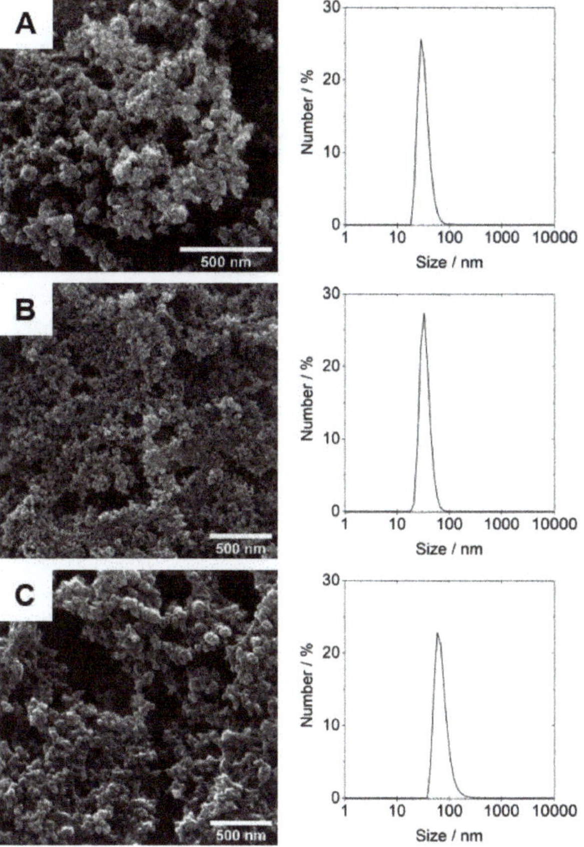

Figure 2. Scanning electron micrographs and dynamic light scattering (DLS) particle size distribution of the prepared nanoparticles: (**A**) Env-CaP, (**B**) Env-CaP-p30, (**C**) Env-CaP-CpG.

To make the protein reactive to the thiol-terminated calcium phosphate nanoparticles Env was activated with the crosslinker sulfo-SMCC by which the primary amines from the protein react with the N-hydroxysuccimide group. In the protein, 33 lysine residues are expected to have the side chain exposed to the solvent and can react with the NHS ester [27,28]. After the addition of the nanoparticles the maleimide group at the other end of the crosslinker reacts with the thiol groups on the nanoparticle surface to form a stable covalent bond between the nanoparticle and the protein.

To determine the coupling efficiency of the Env proteins to the nanoparticles, an AlexaFluor®-488 labelled Env protein was used for the surface modification. After attaching this labelled protein, the reaction yield was determined to 85% by UV-Vis spectroscopy. This factor was assumed to calculate the concentration of Env after attaching the unlabelled protein under equivalent reaction conditions. Additionally, a comparison for the Env concentration determined by nanodrop was performed. The measured concentrations are shown in Supplementary Table S1. The Env concentration determined by UV-Vis spectroscopy was used to calculate Env concentrations for all subsequent nanoparticle preparations.

3.3. Antigen-Specific Activation of Naïve B-Cells with Env-CaP Nanoparticles In Vitro

Previously we demonstrated in vitro that CaP nanoparticles functionalized with a monovalent model antigen activated naïve antigen-specific B-cells in a dose-dependent manner more efficiently than the dissolved antigen alone [12]. In contrast to monomeric proteins, each stabilized soluble Env trimer exposes up to 3 identical epitopes per molecule [29].

To investigate how the surface functionalization of CaP nanoparticle with stabilized HIV-1 Env trimers influences the activation of naïve Env-specific B cells, we incubated B-cells from transgenic mice that express the human PGT121 antibody [30] as B-cell receptors (Env B-cells) with Env-CaP and soluble Env trimers (Figure 3A). Env-VLPs served as a positive control for BCR-specific B-cell activation [31]. The stimulation with LPS showed the total polyclonal capability of the B-cells to be activated [32]. Within a concentration range of 8 ng to 200 ng of Env per 1 mL of culture medium, Env trimers arrayed on the CaP nanoparticle surface were able to activate Env-specific B cells more efficiently in a dose-dependent manner than soluble trimers in the same concentrations (Figure 3A).

Incubation of wt B-cells in the presence of Env-CaP nanoparticles, soluble Env trimers or Env-VLPs did not reveal Env-specific early B-cell activation (Figure 3B). The polyclonal activation of wt B-cells with LPS (comparable to those of Env-specific B-cells) indicated the functional activity of the wt B-cells (Figure 3) and, therefore, clearly demonstrates that all experimental Env preparations used were free of polyclonal activators. Our results are consistent with the data reported by Ingale et al., who showed that HIV-1 trimer-conjugated liposomes activated Env-specific B-cells better than soluble trimers [33].

Thus, the surface functionalization of CaP nanoparticles with soluble HIV-1 Env trimers improved the dose-dependent activation of naïve Env-specific B-cells in vitro in comparison to soluble HIV-1 Env trimers alone.

3.4. Improvement of HIV-1 Env Antibody Responses by Intrastructural Help

To analyze the induction of anti-Env antibody responses in vivo, we applied Env-CaP nanoparticles intramuscularly, as it was demonstrated to be the most appropriate way for the delivery of B- and T-cell antigens with CaP nanoparticles into draining lymphoid organs [13]. In contrast to the model antigen [13] in a highly reactive C3H mouse strain (as discussed in Reference [34]), Env-CaP nanoparticles at a dose of 10 µg of Env protein per immunization induced poor anti-Env antibody responses in Bl6 mice (Figure 4A, Env-CaP vs. naive).

Although Env-CaP nanoparticles are able to directly activate Env-specific B-cells in vitro (Figure 3), the primary anti-Env antibody response in vivo is also dependent on CD4 T-cell help [35]. The low magnitude of Env-specific IgG antibody responses induced with Env-CaP might be due to (i) a suboptimal MHC-II restricted T-cell help [14] elicited by HIV-1 BG505 Env protein in Bl6 mice [36], or (ii) suboptimal immunogenic capacities of Env-CaP (as discussed in [13]).

We have already demonstrated that the functionalization of CaP-HEL nanoparticles with the p30 peptide of TT (a universal T-helper epitope [14]) overcomes the lack of functionally active HEL-derived CD4 T-cell epitopes in Bl6 mice [13]. The functionalization of Env-CaP nanoparticles with the p30 peptide increased the magnitude of the anti-Env IgG1 antibody response (Figure 4A, Env-CaP vs. Env-CaP-p30), indicating an initially suboptimal MHC-II restricted T-cell help in mice immunized with

Env-CaP. The anti-Env IgG2c antibody response, however, remained unaffected (Figure 4B, Env-CaP vs. Env-CaP-p30).

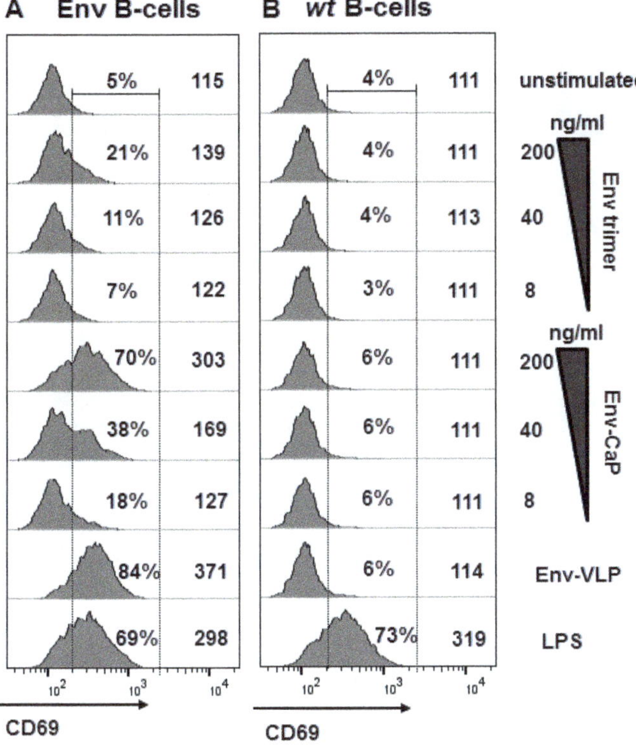

Figure 3. Activation of naïve B-cells in vitro. Naïve B-cells were isolated from PGT121 (**A**) or wt (**B**) mice and stimulated with Env trimers (8, 40 or 200 ng of Env/mL), Env-CaP nanoparticles (8, 40 or 200 ng of Env/mL), Env-VLPs (200 ng of Env/mL), or LPS (2 µg/mL). After 18h incubation, the cells were stained with a viability dye and with anti-CD19 and anti-CD69 antibodies. The histograms represent the expression of CD69 on the surface of viable CD19-positive B-cells. The numbers on the histograms indicate the percentage of the gated CD69-positive cells and medians of the total CD69 fluorescence intensity.

To further improve the Env-CaP-p30 induced anti-Env humoral immune responses, we recruited pre-existing T-helper cells generated by the licensed Tetanol pur vaccine to provide ISH for Env-specific B-cells. Mice previously vaccinated against Tetanus demonstrated significantly higher anti-Env antibody responses after Env-CaP-p30 nanoparticle immunization, than the control animals injected with DPBS instead of Tetanol pur (Figure 4A,B; Env-CaP-p30 vs. Env-CaP-p30 (ISH)). The in vitro reactivation of p30-specific CD4 T-cells from the spleens of differently immunized mice revealed an increase of potential heterologous T-cell help in the ISH group (Figure 4C–E). Intracellular cytokine staining for the Th1 cytokines [37] IFN-γ (Figure 4C) and TNF-α (Figure 4D) showed a trend of being more pronounced in the ISH group. The secretion of the Th2 cytokine [37] IL-5 after p30 reactivation was significantly higher in the ISH group (Figure 4E). CD4 T-cells that produce either Th1 or Th2 cytokines support the generation of IgG2a (IgG2c in Bl6 mice) and IgG1 antibody subclasses in mice, respectively [37,38]. The prominent Th2 cytokine profile of p30-specific CD4 T-cells is, therefore, consistent with the predominant induction of the anti-Env IgG1 antibody subclass after immunizations with Env-CaP-p30 nanoparticles (Figure 4A,B).

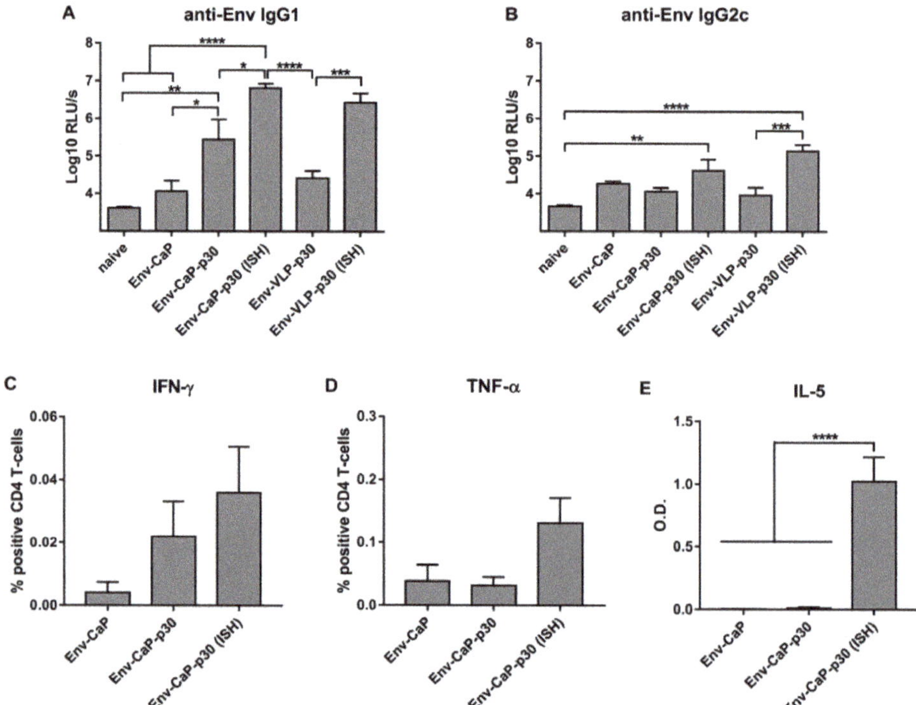

Figure 4. Improvement of Env-specific IgG subtype responses after Env-CaP-p30 immunizations by intrastructural help from Tetanus toxoid-specific T-helper cells. (**A,B**) IgG1 (**A**) and IgG2c (**B**) Env-specific antibody responses were measured in sera from wt mice primed twice with DPBS or Tetanol (ISH) and boosted 3 times with different CaP nanoparticles or Env-VLP-p30. (**C,D**) Percentages of CD4 T-cells producing IFN-γ (**C**) and TNF-α (**D**) from differently immunized wt animals after in vitro re-stimulation with p30 peptide were measured by intracellular cytokine staining. (**E**) p30-specific IL-5 cytokine secretion as determined by ELISA. The columns represent the mean values of six animals ± SEM. * $p < 0.05$; ** $p < 0.001$; *** $p < 0.0005$; **** $p < 0.0001$; one-way ANOVA with Tukey multiple comparison post-hoc test.

Previously, we demonstrated ISH in mice immunized with Tetanol pur prior to administration of HIV-1 derived VLPs containing different T helper epitopes of TT [8]. For direct comparison of the ISH effects on the anti-Env antibody induction between the CaP nanoparticle platform and the HIV-1 VLP system, we produced VLPs containing the p30 epitope of TT within Gag (Figure 1D) and presenting stabilized BG505 Env trimers in a membrane-anchored form on their surface (Figure 1A,C). The VLP antigen dosage was selected based on our previous studies [8,35] and was 0.3 µg of Env protein per immunization. For the Env-VLP-p30 particles we observed intrastructural help (Figure 4A,B; Env-VLP-p30 vs. Env-VLP-p30 (ISH)) in a magnitude comparable to the CaP nanoparticle platform (Figure 4A,B; Env-VLP-p30 (ISH) vs. Env-CaP-p30 (ISH)).

Altogether, the incorporation of T-helper cell epitopes of non-HIV proteins into CaP nanoparticles functionalized with HIV-1 Env trimers may allow the Env-specific B-cells to get T-cell help from non-HIV specific CD4 T-cells via the ISH mechanism.

3.5. Distinct Effects of ISH and CpG-Adjuvants on the Induction of Env-Specific CD4 T-Cell Responses

Along with engaging the heterologous universal p30 T-helper epitope, functionalization of the B-cell targeting CaP-nanoparticles with TLR-ligands significantly increased IgG antibody responses against a model antigen in mice [13]. The TLR9 ligand CpG demonstrated a number of advantages to other TLR-ligands tested in the study (as discussed in Reference [13]). In addition, the CpG-based adjuvant 1018 ISS [39] is already approved for clinical use.

To compare the improvement of antibody responses by CpG and ISH side-by-side, animals previously immunized against Tetanus received either Env-CaP-p30 or Env-Cap-CpG nanoparticles. Both Env-CaP-p30 and Env-Cap-CpG induced comparable total anti-Env IgG immune responses (Figure 5A), although the anti-Env IgG subtype distribution varied between the groups (Figure 5B).

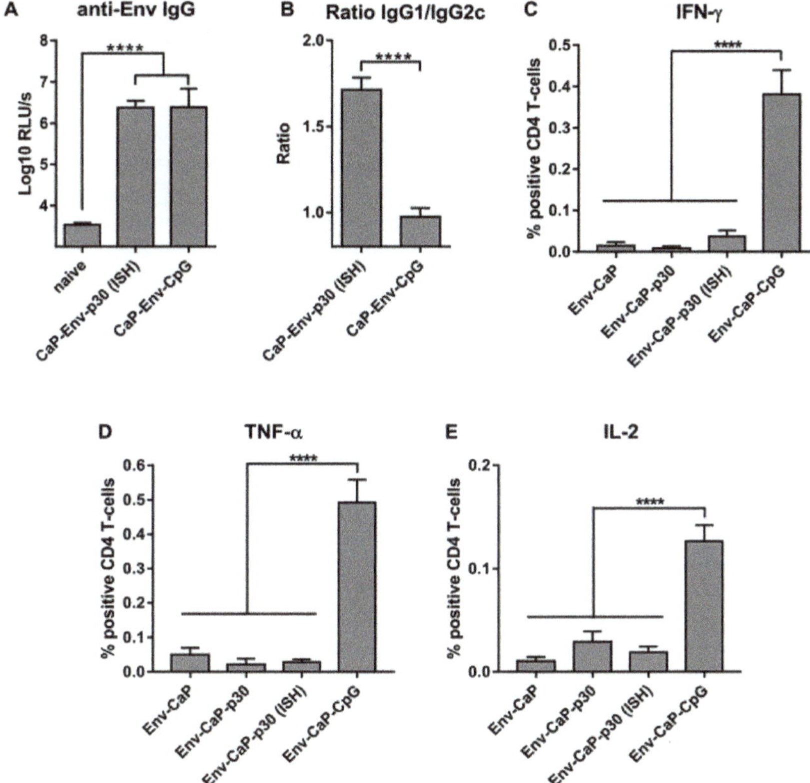

Figure 5. Characterization of anti-Env immune responses after heterologous ISH and CpG adjuvantation. (**A**) Env-specific antibody responses were measured in sera from wt mice primed twice with Tetanol and boosted 3 times with either Env-CaP-p30 or Env-CaP-CpG. The columns represent the mean values of six animals ± SEM. **** $p < 0.0001$; one-way ANOVA with Tukey multiple comparison post-hoc test. (**B**) Env-specific IgG1/IgG2c ratios of individual mice. The columns represent the mean values of six animals ± SEM. **** $p < 0.0001$; unpaired Student t-test. (**C–E**) Percentages of CD4 T-cells producing IFN-γ (**C**), TNF-α (**D**) and IL-2 (**E**) from differently immunized wt animals after in vitro re-stimulation with Env-VLPs were measured by intracellular cytokine staining. The columns represent the mean values of six animals ± SEM. **** $p < 0.0001$; one-way ANOVA with Tukey multiple comparison post-hoc test.

The efficacy of an HIV-1 vaccine, however, may not only depend on the strength of a protective antibody response induced, but also on the magnitude of vaccine-induced immune mechanisms increasing the susceptibility to infection. In particular, HIV-specific CD4 T-cell responses may increase the susceptibility to infection by expanding the number of activated CD4 T-cells as targets for the HIV-1 infection [1,6,7,40]. We therefore also compared CD4 T-cell responses in both vaccine groups.

In contrast to Env-CaP-p30 nanoparticles that recruit and support heterologous (non-HIV-1 specific) T-cell help (Figure 4), Env-CaP-CpG nanoparticles might induce increased Env-specific CD4 T-cell response as a result of TLR9 ligation in dendritic cells [41,42]. Indeed, the percentage of Env-specific CD4 T-cells in the spleens of immunized animals that can be reactivated in vitro after Env-VLP exposure was significantly higher in the Env-CaP-CpG immunized animals (Figure 5B–D). Reactivation of Env-specific CD4 T-cells producing Th1 pro-inflammatory [43] cytokines IFN-γ (Figure 5B) as well as TNF-α (Figure 5C) and IL-2 (Figure 5E), a potent mitogen and growth factor for CD4 T-cells, in mice from the Env-CaP-p30 (ISH) group was comparable to those in mice that received Env-CaP or Env-CaP-p30 after DPBS injections. These results clearly demonstrated that ISH does not increase expansion of Env-specific CD4 T-cells.

In non-human primate models of HIV/SIV infection, vaccinated animals with high numbers of vaccine-induced IFN-γ secreting T-cells were more susceptible to acquisition of challenge virus infection than poor T-cell responders [7,40], indicating that CpG functionalization of Env-CaP vaccines may potentially enhance the susceptibility to acquisition of HIV-1 infection in vaccinees.

Thus, the incorporation of T-helper epitopes of heterologous (non-HIV) proteins in Env-CaP nanoparticles was able to increase the magnitude of anti-Env specific IgG antibody responses by ISH without significant induction of Env-specific IFN-γ producing CD4 T-helper cells. This suggests that the ISH strategy for nanoparticle-based HIV-1 vaccines may allow to avoid induction of HIV-1 specific CD4 T-cell responses suspected to enhance the susceptibility to HIV-1 infection.

4. Conclusions

We demonstrated that the incorporation of T-helper cell epitopes of non-HIV proteins into CaP nanoparticles functionalized with HIV-1 Env trimers allows the Env-specific B-cells to receive T-cell help from non-HIV specific CD4 T-cells via the ISH mechanism. In a mouse model, the Env-CaP-p30 nanoparticle-based vaccine was able to improve HIV-1 Env-specific antibody responses without additional induction of HIV-specific CD4 T-cells suspected to increase the susceptibility to infection.

Supplementary Materials: The following are available online at http://www.mdpi.com/2079-4991/9/10/1389/s1, Table S1: Characterization data of the synthesized calcium phosphate nanoparticles.

Author Contributions: Conceptualization, V.T., K.Ü. and M.E.; methodology, V.T., M.E. and R.T.W.; validation, V.T. and M.E.; formal analysis, D.D., L.R.-S., V.S. and V.T.; investigation, D.D., L.R.-S. and H.T.; resources, K.Ü., R.T.W. and M.E.; writing—original draft preparation, V.T., D.D. and L.R.-S.; writing—review and editing, V.T.; visualization, D.D., L.R.-S. and V.T.; supervision, V.T., V.S., M.E. and K.Ü.; project administration, V.T. and V.S. All authors read and approved the final manuscript.

Funding: This research was funded by the European Commission (EAVI2020–European AIDS Vaccine Initiative 2020).

Acknowledgments: We thank Ghulam Nabi (Ruhr University Bochum, Germany) for his contribution to the generation of the Hgpsyn-TTp30 expression plasmid, and Michel Nussenzweig (The Rockefeller University, New York, NY, USA) for PGT-121 mice donation. We acknowledge support by Deutsche Forschungsgemeinschaft and Friedrich-Alexander-Universität Erlangen-Nürnberg (FAU) within the funding programme Open Access Publishing. L.R.-S. thanks the Ministry of Science, Technology and Telecommunications of Costa Rica (MICITT, Costa Rica) for a Ph.D. scholarship.

Conflicts of Interest: The authors declare no conflict of interest. The funders had no role in the design of the study; in the collection, analyses, or interpretation of data; in the writing of the manuscript, or in the decision to publish the results.

References

1. Temchura, V.; Tenbusch, M. The two faces of vaccine-induced immune response: Protection or increased risk of HIV infection?! *Virol. Sin.* **2014**, *29*, 7–9. [CrossRef] [PubMed]
2. Chung, A.W.; Ghebremichael, M.; Robinson, H.; Brown, E.; Choi, I.; Lane, S.; Dugast, A.S.; Schoen, M.K.; Rolland, M.; Suscovich, T.J.; et al. Polyfunctional Fc-Effector Profiles Mediated by IgG Subclass Selection Distinguish RV144 and VAX003 Vaccines. *Sci. Transl. Med.* **2014**, *6*, 228ra38. [CrossRef] [PubMed]

3. Kim, J.H.; Excler, J.-L.; Michael, N.L. Lessons from the RV144 Thai Phase III HIV-1 Vaccine Trial and the Search for Correlates of Protection. *Annu. Rev. Med.* **2015**, *66*, 423–437. [CrossRef] [PubMed]

4. Excler, J.-L.; Ake, J.; Robb, M.L.; Kim, J.H.; Plotkin, S.A. Nonneutralizing Functional Antibodies: A New "Old" Paradigm for HIV Vaccines. *Clin. Vaccine Immunol.* **2014**, *21*, 1023–1036. [CrossRef] [PubMed]

5. Ackerman, M.E.; Mikhailova, A.; Brown, E.P.; Dowell, K.G.; Walker, B.D.; Bailey-Kellogg, C.; Suscovich, T.J.; Alter, G. Polyfunctional HIV-Specific Antibody Responses Are Associated with Spontaneous HIV Control. *PLoS Pathog.* **2016**, *12*, e1005315. [CrossRef] [PubMed]

6. Lewis, G.K.; DeVico, A.L.; Gallo, R.C. Antibody persistence and T-cell balance: Two key factors confronting HIV vaccine development. *Proc. Natl. Acad. Sci. USA* **2014**, *111*, 15614–15621. [CrossRef] [PubMed]

7. Fouts, T.R.; Bagley, K.; Prado, I.J.; Bobb, K.L.; Schwartz, J.A.; Xu, R.; Zagursky, R.J.; Egan, M.A.; Eldridge, J.H.; LaBranche, C.C.; et al. Balance of cellular and humoral immunity determines the level of protection by HIV vaccines in rhesus macaque models of HIV infection. *Proc. Natl. Acad. Sci. USA* **2015**, *112*, E992–E999. [CrossRef]

8. Elsayed, H.; Nabi, G.; McKinstry, W.J.; Khoo, K.K.; Mak, J.; Salazar, A.M.; Tenbusch, M.; Temchura, V.; Überla, K. Intrastructural Help: Harnessing T Helper Cells Induced by Licensed Vaccines for Improvement of HIV Env Antibody Responses to Virus-Like Particle Vaccines. *J. Virol.* **2018**, *92*, e00141-18. [CrossRef]

9. Russell, S.M.; Liew, F.Y. T cells primed by influenza virion internal components can cooperate in the antibody response to haemagglutinin. *Nature* **1979**, *280*, 147–148. [CrossRef]

10. Temchura, V.; Überla, K. Intrastructural help. *Curr. Opin. HIV AIDS* **2017**, *12*, 272–277. [CrossRef]

11. Sokolova, V.; Westendorf, A.M.; Buer, J.; Überla, K.; Epple, M. The potential of nanoparticles for the immunization against viral infections. *J. Mater. Chem. B* **2015**, *3*, 4767–4779. [CrossRef]

12. Temchura, V.V.; Kozlova, D.; Sokolova, V.; Überla, K.; Epple, M. Targeting and activation of antigen-specific B-cells by calcium phosphate nanoparticles loaded with protein antigen. *Biomaterials* **2014**, *35*, 6098–6105. [CrossRef] [PubMed]

13. Zilker, C.; Kozlova, D.; Sokolova, V.; Yan, H.; Epple, M.; Überla, K.; Temchura, V. Nanoparticle-based B-cell targeting vaccines: Tailoring of humoral immune responses by functionalization with different TLR-ligands. *Nanomed. Nanotechnol. Biol. Med.* **2017**, *13*, 173–182. [CrossRef] [PubMed]

14. Oscherwitz, J.; Yu, F.; Cease, K.B. A heterologous helper T-cell epitope enhances the immunogenicity of a multiple-antigenic-peptide vaccine targeting the cryptic loop-neutralizing determinant of Bacillus anthracis protective antigen. *Infect. Immun.* **2009**, *77*, 5509–5518. [CrossRef] [PubMed]

15. De Taeye, S.W.; Moore, J.P.; Sanders, R.W. HIV-1 Envelope Trimer Design and Immunization Strategies to Induce Broadly Neutralizing Antibodies. *Trends Immunol.* **2016**, *37*, 221–232. [CrossRef] [PubMed]

16. Sattentau, Q.; Sattentau, J.Q. Envelope Glycoprotein Trimers as HIV-1 Vaccine Immunogens. *Vaccines* **2013**, *1*, 497–512. [CrossRef] [PubMed]

17. Escolano, A.; Steichen, J.M.; Dosenovic, P.; Kulp, D.W.; Golijanin, J.; Sok, D.; Freund, N.T.; Gitlin, A.D.; Oliveira, T.; Araki, T.; et al. Sequential Immunization Elicits Broadly Neutralizing Anti-HIV-1 Antibodies in Ig Knockin Mice. *Cell* **2016**, *166*, 1445–1458. [CrossRef] [PubMed]

18. Sharma, S.K.; deVal, N.; Bale, S.; Guenaga, J.; Tran, K.; Feng, Y.; Dubrovskaya, V.; Ward, A.B.; Wyatt, R.T. Cleavage-Independent HIV-1 Env Trimers Engineered as Soluble Native Spike Mimetics for Vaccine Design. *Cell Rep.* **2015**, *11*, 539–550. [CrossRef]

19. Wagner, R.; Graf, M.; Bieler, K.; Wolf, H.; Grunwald, T.; Foley, P.; Uberla, K. Rev-Independent Expression of Synthetic *gag-pol* Genes of Human Immunodeficiency Virus Type 1 and Simian Immunodeficiency Virus: Implications for the Safety of Lentiviral Vectors. *Hum. Gene Ther.* **2000**, *11*, 2403–2413. [CrossRef]

20. genannt Bonsmann, M.S.; Niezold, T.; Temchura, V.; Pissani, F.; Ehrhardt, K.; Brown, E.P.; Osei-Owusu, N.Y.; Hannaman, D.; Hengel, H.; Ackerman, M.E.; et al. Enhancing the Quality of Antibodies to HIV-1 Envelope by GagPol-Specific Th Cells. *J. Immunol.* **2015**, *195*, 4861–4872. [CrossRef]

21. Kozlova, D.; Chernousova, S.; Knuschke, T.; Buer, J.; Westendorf, A.M.; Epple, M. Cell targeting by antibody-functionalized calcium phosphatenanoparticles. *J. Mater. Chem.* **2012**, *22*, 396–404. [CrossRef]

22. Rojas-Sánchez, L.; Sokolova, V.; Riebe, S.; Voskuhl, J.; Epple, M. Covalent Surface Functionalization of Calcium Phosphate Nanoparticles with Fluorescent Dyes by Copper-Catalysed and by Strain-Promoted Azide-Alkyne Click Chemistry. *ChemNanoMat* **2019**, *5*, 436–446. [CrossRef]

23. Klesing, J.; Chernousova, S.; Epple, M. Freeze-dried cationic calcium phosphatenanorods as versatile carriers of nucleic acids (DNA, siRNA). *J. Mater. Chem.* **2012**, *22*, 199–204. [CrossRef]

24. Sanders, R.W.; Vesanen, M.; Schuelke, N.; Master, A.; Schiffner, L.; Kalyanaraman, R.; Paluch, M.; Berkhout, B.; Maddon, P.J.; Olson, W.C.; et al. Stabilization of the soluble, cleaved, trimeric form of the envelope glycoprotein complex of human immunodeficiency virus type 1. *J. Virol.* **2002**, *76*, 8875–8889. [CrossRef] [PubMed]

25. Finzi, A.; Pacheco, B.; Zeng, X.; Kwon, Y.D.; Kwong, P.D.; Sodroski, J. Conformational characterization of aberrant disulfide-linked HIV-1 gp120 dimers secreted from overexpressing cells. *J. Virol. Methods* **2010**, *168*, 155–161. [CrossRef] [PubMed]

26. Behzadi, S.; Serpooshan, V.; Tao, W.; Hamaly, M.A.; Alkawareek, M.Y.; Dreaden, E.C.; Brown, D.; Alkilany, A.M.; Farokhzad, O.C.; Mahmoudi, M. Cellular uptake of nanoparticles: Journey inside the cell. *Chem. Soc. Rev.* **2017**, *46*, 4218–4244. [CrossRef] [PubMed]

27. Klausen, M.S.; Jespersen, M.C.; Nielsen, H.; Jensen, K.K.; Jurtz, V.I.; Sønderby, C.K.; Sommer, M.O.A.; Winther, O.; Nielsen, M.; Petersen, B.; et al. NetSurfP-2.0, Improved prediction of protein structural features by integrated deep learning. *Proteins Struct. Funct. Bioinf.* **2019**, *87*, 520–527. [CrossRef]

28. Koniev, O.; Wagner, A. Developments and recent advancements in the field of endogenous amino acid selective bond forming reactions for bioconjugation. *Chem. Soc. Rev.* **2015**, *44*, 5495–5551. [CrossRef] [PubMed]

29. Lyumkis, D.; Julien, J.-P.; de Val, N.; Cupo, A.; Potter, C.S.; Klasse, P.-J.; Burton, D.R.; Sanders, R.W.; Moore, J.P.; Carragher, B.; et al. Cryo-EM Structure of a Fully Glycosylated Soluble Cleaved HIV-1 Envelope Trimer. *Science* **2013**, *342*, 1484–1490. [CrossRef]

30. Walker, L.M.; Huber, M.; Doores, K.J.; Falkowska, E.; Pejchal, R.; Julien, J.-P.; Wang, S.-K.; Ramos, A.; Chan-Hui, P.-Y.; Moyle, M.; et al. Broad neutralization coverage of HIV by multiple highly potent antibodies. *Nature* **2011**, *477*, 466–470. [CrossRef]

31. Gardt, O.; Grewe, B.; Tippler, B.G.; Überla, K.; Temchura, V.V. HIV-derived lentiviral particles promote T-cell independent activation and differentiation of naïve cognate conventional B2-cells in vitro. *Vaccine* **2013**, *31*, 5088–5098. [CrossRef] [PubMed]

32. Sveen, K.; Skaug, N. Comparative mitogenicity and polyclonal B cell activation capacity of eight oral or nonoral bacterial lipopolysaccharides in cultures of spleen cells from athymic (nu/nu-BALB/c) and thymic (BALB/c) mice. *Oral Microbiol. Immunol.* **1992**, *7*, 71–77. [CrossRef] [PubMed]

33. Ingale, J.; Stano, A.; Guenaga, J.; Sharma, S.K.; Nemazee, D.; Zwick, M.B.; Wyatt, R.T. High-Density Array of Well-Ordered HIV-1 Spikes on Synthetic Liposomal Nanoparticles Efficiently Activate B Cells. *Cell Rep.* **2016**, *15*, 1986–1999. [CrossRef] [PubMed]

34. Barnowski, C.; Kadzioch, N.; Damm, D.; Yan, H.; Temchura, V. Advantages and Limitations of Integrated Flagellin Adjuvants for HIV-Based Nanoparticle B-Cell Vaccines. *Pharmaceutics* **2019**, *11*, 204. [CrossRef] [PubMed]

35. Nabi, G.; Temchura, V.; Großmann, C.; Kuate, S.; Tenbusch, M.; Überla, K. T cell independent secondary antibody responses to the envelope protein of simian immunodeficiency virus. *Retrovirology* **2012**, *9*. [CrossRef] [PubMed]

36. Hu, J.K.; Crampton, J.C.; Cupo, A.; Ketas, T.; van Gils, M.J.; Sliepen, K.; de Taeye, S.W.; Sok, D.; Ozorowski, G.; Deresa, I.; et al. Murine Antibody Responses to Cleaved Soluble HIV-1 Envelope Trimers Are Highly Restricted in Specificity. *J. Virol.* **2015**, *89*, 10383–10398. [CrossRef] [PubMed]

37. Mosmann, T.R.; Coffman, R.L. TH1 and TH2 Cells: Different Patterns of Lymphokine Secretion Lead to Different Functional Properties. *Annu. Rev. Immunol.* **1989**, *7*, 145–173. [CrossRef]

38. Snapper, C.M.; Paul, W.E. Interferon-gamma and B cell stimulatory factor-1 reciprocally regulate Ig isotype production. *Science* **1987**, *236*, 944–947. [CrossRef]

39. Campbell, J.D. *Development of the CpG Adjuvant 1018, A Case Study*; Humana Press: New York, NY, USA, 2017; pp. 15–27. [CrossRef]

40. Tenbusch, M.; Ignatius, R.; Temchura, V.; Nabi, G.; Tippler, B.; Stewart-Jones, G.; Salazar, A.M.; Sauermann, U.; Stahl-Hennig, G.; Überla, K.; et al. Risk of Immunodeficiency Virus Infection May Increase with Vaccine-Induced Immune Response. *J. Virol.* **2012**, *86*, 10533–10539. [CrossRef]

41. Sokolova, V.; Knuschke, T.; Kovtun, A.; Buer, J.; Epple, M.; Westendorf, A.M. The use of calcium phosphate nanoparticles encapsulating Toll-like receptor ligands and the antigen hemagglutinin to induce dendritic cell maturation and T cell activation. *Biomaterials* **2010**, *31*, 5627–5633. [CrossRef]

42. Knuschke, T.; Sokolova, V.; Rotan, O.; Wadwa, M.; Tenbusch, M.; Hansen, W.; Staeheli, P.; Epple, M.; Buer, J.; Westendorf, A.M. Immunization with biodegradable nanoparticles efficiently induces cellular immunity and protects against influenza virus infection. *J. Immunol.* **2013**, *190*, 6221–6229. [CrossRef] [PubMed]
43. Turner, M.D.; Nedjai, B.; Hurst, T.; Pennington, D.J. Cytokines and chemokines: At the crossroads of cell signalling and inflammatory disease. *Biochim. Biophys. Acta* **2014**, *1843*, 2563–2582. [CrossRef] [PubMed]

Article

Assembly and Characterization of HBc Derived Virus-like Particles with Magnetic Core

Jakub Dalibor Rybka [1,†,*], Adam Aron Mieloch [1,2,†], Alicja Plis [1], Marcin Pyrski [3], Tomasz Pniewski [3] and Michael Giersig [1,4]

[1]   Center for Advanced Technology, Adam Mickiewicz University in Poznań, Umultowska 89C, 61-614 Poznań, Poland; amieloch@amu.edu.pl (A.A.M.); alkaplis@o2.pl (A.P.); giersig@amu.edu.pl (M.G.)
[2]   Faculty of Chemistry, Adam Mickiewicz University in Poznań, Umultowska 89B, 61-614 Poznań, Poland
[3]   Institute of Plant Genetics, Polish Academy of Sciences, Strzeszyńska 34, 60-479 Poznań, Poland; mpyr@igr.poznan.pl (M.P.); tpni@igr.poznan.pl (T.P.)
[4]   Institute of Experimental Physics, Freie Universität Berlin, Arnimallee 14, 14195 Berlin, Germany
*   Correspondence: jrybka@amu.edu.pl; Tel.: +48-61-829-1875
†   These authors contributed equally to this work.

Received: 20 December 2018; Accepted: 24 January 2019; Published: 26 January 2019

Abstract: Core-virus like particles (VLPs) assembly is a kinetically complex cascade of interactions between viral proteins, nanoparticle's surface and an ionic environment. Despite many in silico simulations regarding this process, there is still a lack of experimental data. The main goal of this study was to investigate the capsid protein of hepatitis B virus (HBc) assembly into virus-like particles with superparamagnetic iron oxide nanoparticles (SPIONs) as a magnetic core in relation to their characteristics. The native form of HBc was obtained via agroinfection of *Nicotiana benthamiana* with pEAQ-HBc plasmid. SPIONs of diameter of 15 nm were synthesized and functionalized with two ligands, providing variety in ζ-potential and hydrodynamic diameter. The antigenic potential of the assembled core-VLPs was assessed with enzyme-linked immunosorbent assay (ELISA). Morphology of SPIONs and core-VLPs was evaluated via transmission electron microscopy (TEM). The most successful core-VLPs assembly was obtained for SPIONs functionalized with dihexadecyl phosphate (DHP) at SPIONs/HBc ratio of 0.2/0.05 mg/mL. ELISA results indicate significant decrease of antigenicity concomitant with core-VLPs assembly. In summary, this study provides an experimental assessment of the crucial parameters guiding SPION-HBc VLPs assembly and evaluates the antigenicity of the obtained structures.

Keywords: virus-like particles; VLPs; hepatitis B virus capsid protein; HBc; viral self-assembly; magnetic core; HBcAg

1. Introduction

Virus-like particles (VLPs) are non-infectious and non-replicating supramolecular assemblies composed of single or multiple viral proteins, which closely resemble native virions [1]. VLPs display a unique set of immunological characteristics that render them highly potent for vaccine development such as: nanometer range size, multivalent and highly repetitive surface geometry, the ability to elicit both innate and adaptive immune response [2]. Due to favorable surface morphology and a wide range of possible modifications, VLPs have been successfully used as a platform for multivalent vaccine creation [3–5]. Several VLP-based vaccines are currently commercially available (e.g., Cervarix®, Gardasil®, Sci-B-Vac™, Mosquirix™) with more undergoing clinical trials [6]. VLPs' applicability is not limited to their immunogenic properties. Some of the use cases include: highly selective and sensitive nanobiosensor for troponin I detection, light-harvesting VLPs for use in photovoltaic or photocatalytic devices, nanofiber-like VLPs for tissue regenerating materials, nanocontainers and nanoreactors [7–11].

Hepatitis B virus (HBV) is an enveloped, icosahedral, cDNA virus that belongs to the Hepadnaviridae family. The virion has a diameter of 42 nm and is composed of a lipid envelope with hepatitis B virus surface antigen (HBsAg) and inner nucleocapsid consisting of hepatitis B virus capsid protein—HBc (named also HB core antigen, HBcAg) [12]. HBc consists of 183–185 amino acids of which 149 N-terminal amino acids form an assembly domain and 34 amino acids form C-terminal arginine-rich domain (CTAD) required for the packaging of nucleic acid [13]. HBc is a homodimeric protein that has the ability to self-assemble into icosahedral and fenestrated T = 4 (120 dimers) and T = 3 (90 dimers) capsids with respective outer diameter of 34 and 30 nm [14]. T = 4 capsid is a dominant product of a wild type HBc in vitro self-assembly (~95%) [15]. HBc capsid is highly immunogenic and has been shown to induce both B- and T-cell response [16].

Superparamagnetic iron oxide nanoparticles (SPIONs) exhibit properties, such as high magnetic susceptibility, high saturation magnetization and low toxicity [17–19]. Due to the aforementioned properties, high-yield synthesis methods and a wide array of available surface modifications, SPIONs can be utilized in: magnetic bioseparation, magnetic hyperthermia, targeted drug delivery, in diagnostics as magnetic resonance imaging (MRI) contrast agents, etc. [20–23].

Introduction of SPIONs as the core of VLPs has been performed successfully with several viral proteins of different origin [24–26]. Magnetic core adds a multitude of advantageous properties. It allows for post-assembly magnetic bioseparation, which may be crucial for large scale production [27]. It also improves cellular uptake and magnetic relaxivities resulting in higher resolution MRI images, which combined with in vivo tracking may provide essential data regarding VLPs biodistribution [28]. Functionalized core can act as a substitute for native nucleic acid, and therefore, govern the process of protein recruitment and organization during self-assembly. Rational core design can be used to facilitate the assembly and enhance such parameters as, e.g., physicochemical stability, mechanical elasticity, capacity to withstand desiccation and long-term storage. On the other hand, HBc VLPs have been shown to be potent epitope carriers [3,29].

In the study by Shen et al., HBc was genetically engineered into a truncated version, deprived of 34 C-terminal amino acids responsible for nucleic acid packaging. The removed part was replaced by six consecutive histidine residues (His-tag). Fe_3O_4 nanoparticles functionalized with nickel-nitrilotriacetic acid (nickel-NTA) chelate were used as the core. The VLPs assembly was driven by the affinity of histidine tags to the nickel-NTA chelate [28]. This study prompted us to investigate whether native HBc VLPs assembly can be successful, without resorting to genetic engineering of HBc protein.

The main objective of this study was to investigate whether provided SPION surface modification is sufficient for SPION-HBc assembly. Even though HBc subunits exhibit an ability to assemble in the absence of genetic material, electrostatic interactions between positively charged CTAD of the capsid and negatively charged nucleic acid have a major influence on the assembly process [30]. Therefore, to mimic native electrostatic interactions, negatively charged ligands were chosen for SPIONs functionalization: dihexadecyl phosphate (DHP) and PL-PEG-COOH. Both compounds were successfully used in our previous study regarding the influence of ligand charge and length on the assembly of Brome mosaic virus derived virus-like particles with magnetic core [31].

2. Materials and Methods

2.1. Reagents

Oleic acid (technical grade 90%), Iron (III) acetylacetonate (97%), Sodium chloride (99%), Dihexadecyl phosphate (90%), 1-Octadecene (90%), 2-Butanol (95.5%), Trizma®hydrochloride (99%), Calcium chloride (97%), Magnesium sulfate (99.5%), Glycine (99%), Glycerol (99%), Urea (98%), 2-(N-Morpholino)ethanesulfonic acid (99%), Sucrose (99.5%), Sigma-Aldrich (Poznan, Poland). Toluene (99.5%), n-Hexane (99%), Chloroform (98.5%) and Hydrochloric acid (30–35%), Avantor (Gliwice,

Poland). 1,2-Distearoyl-sn-glycero-3-phosphoethanolamine-N-[carboxy-(polyethyleneglycol)-2000] (ammonium salt) (PL-PEG-COOH, 2000 Da PEG (99%), Avanti, Alabaster, AL, USA). Snakeskin®Dialysis Tubing, 10K MWCO, 22 mm, Thermo Fisher Scientific (Waltham, MA, USA). All chemicals were used as received. Water was purified with Hydrolab HLP5 instrument (0.09 µS/cm, Straszyn, Poland).

2.2. Superparamagnetic Iron Oxide Nanoparticles (SPIONs) Synthesis

Spherical iron oxide nanoparticles were synthesized via thermal decomposition of iron (III) acetylacetonate Fe(acac)$_3$ [32]. Briefly, 6 mmol of Fe(acac)$_3$ and 18 mmol of oleic acid were dissolved in 40 mL of 1-octadecene. The reaction was performed with continuous stirring and nitrogen flow. Temperature of the solution was increased to 220 °C and maintained for 1 h. Subsequently, the temperature was increased further to 320 °C and maintained for 1 h. After synthesis, the solution was left to cool down to ambient temperature and 200 mL of washing solution (3:1 v/v of 2-butanol and toluene) was added. The obtained mixture was placed on a neodymium magnet and left overnight to allow nanoparticles to precipitate. Supernatant was discarded and replaced with fresh washing solution. Sonicating bath was used to resuspend nanoparticles. The washing step was performed thrice. In the final step, nanoparticles were suspended in 20 mL of chloroform. Concentration of the nanoparticles was estimated by dried sample weighing.

2.3. SPIONs Functionalization

PL-PEG-COOH functionalization was performed as per a method published elsewhere [24], with minor modifications. Briefly, 3.0 mg of PL-PEG-COOH were added to 5 mL of 1.0 mg/mL SPIONs chloroform solution. The sample was briefly sonicated in a sonic bath and left open for chloroform evaporation. The obtained waxy solid was heated for 1 min in an 80 °C water bath. The following step was adding 5 mL of miliQ water and vortexing the sample to enhance micelles formation. Subsequently, the sample was washed thrice with chloroform to remove unbound PL-PEG-COOH. Finally, water phase containing functionalized SPIONs was collected and filtered through 0.22 µm pores. Concentration of the SPION-PEG nanoparticles was measured via thermogravimetric analysis described below.

DHP functionalization was performed as per a method published elsewhere [31], with minor modifications. Briefly, 10.0 mg of dihexadecyl phosphate were added to 20 mL of hexane and dissolved with heat-assisted magnetic stirring (75 °C, ca. 10 min). After DHP dissolution, a chloroform solution containing 10.0 mg of synthesized iron oxide nanoparticles coated with oleic acid was added. The mixture was shortly sonicated and 80 mL of water were added. Subsequently, the obtained two phase solution was briefly vortexed and sonicated until the water phase became turbid. In the next step, the solution was placed in a sonicating bath for 3–4 h with no temperature control exercised. After functionalization, the solution was left overnight to allow for phase separation. The Bobtom phase was collected and placed near neodymium magnet for 24 h to separate functionalized nanoparticles from the solution. The obtained precipitate was collected, suspended in 2 mL of miliQ water and filtered through 0.22 µm pores. Concentration of the SPION-DHP nanoparticles was measured via thermogravimetric analysis described below.

2.4. Concentration Measurement via Thermogravimetric Analysis

Thermogravimetric analysis was performed to measure concentrations of the functionalized SPIONs. The analysis was performed on TGA 4000 System (Perkin Elmer apparatus, Waltham, MA, USA). Briefly, a 20 µL sample was taken for measurement. Each sample was measured in triplicate. The sample was heated from 20 to 150 °C at 10 °C/min in nitrogen atmosphere. The lowest mass was taken as fully dried sample and used for further calculations. The obtained mass was normalized for 20 mg of the initial sample mass. The mean of three measurements was calculated. Density was derived from weighing 5×15 µL of the sample and dividing the mean mass by volume. Final concentrations were: SPION-DHP = 3.52 mg/mL and SPION-PEG = 3.81 mg/mL.

2.5. HBc Production and Preparation

HBc was produced in plants via a transient expression system based on agroinfiltration. HBc expression vector was constructed on the basis of pEAQ-HT plasmid, developed by Peyret and Lomonossoff [33]. The coding sequence of HBcAg of 552 bp in length derived from HBV subtype *adw4* (GenBank: Z35717), was cloned into the vector *Age* I and *Xho* I restriction sites using sites *Age* I and compatible ends of *Sal* I, respectively, introduced by PCR using the following primers:

Forward: AACCGGTATGGACATTGACCCTTATAAAGAATTTG
Reverse: TGTCGACTGCAGTTAACATTGAGATTCCCGAGATTGAG

Complete vector pEAQ-HBc was introduced into *Agrobacterium tumefaciens* EHA105 and LBA4404 strains via electroporation.

Agroinfection was performed with *Agrobacterium* strains grown overnight on selective liquid LB medium supplemented with kanamycin (50 mg/l) and used to infiltrate leaves of 5–7 week-old *Nicotiana benthamiana* plants, cultivated in growth chamber under 5–6 klx light intensity, 16/8 h photoperiod and at a 22/16 °C temperature regime. *Agrobacterium* cells were centrifuged at 2000 g for 3 min at 4 °C and resuspended in MES buffer (10 mM 2-(N-morpholino)ethanesulphonic acid, 10 mM $MgSO_4$, pH 5.7) to optical density at a 600 nm wavelength (OD_{600}) 0.6 or 0.1 for infiltration by syringe or exsiccator, respectively. *Agrobacterium* suspension, 0.5 mL per leaf, was injected with a syringe into the bottom side of the leaves. Alternatively, whole plants were inverted and immersed in 2 L of *Agrobacterium* suspension in exsiccator (Lab Companion VDP-25G, Seoul, Korea). Pump (AGA Labor PL2, Poznań, Poland) was then applied to reach underpressure (-0.08 MPa) for approximately 1 min. The vacuum was released and applied again to ensure infiltration of the whole leaves. After 10 days following the agroinfiltration concentration of HBc in plant tissue reached approximately 1 mg/g of fresh weight (data not shown). HBc was then extracted and partially purified using sucrose density gradient as described previously [33]. The concentration of HBc directly after purification was fixed to 0.1 mg/mL. Prior to SPION encapsulation, HBc was diluted twice in a disassembly buffer.

2.6. SPION-HBc Preparation

VLPs were prepared in line with a slightly modified procedure described elsewhere [28].

HBc dissociation: 300 µL of 0.1 mg/mL HBc were diluted with 300 µL of denaturant solution (5 M urea, 300 mM NaCl, 100 mM tris-HCl) and incubated at 25 °C for 3h.

SPION-HBc assembly: The solution of dissociated HBc was divided into 100 µL aliquots (HBc conc. 0.05 mg/mL). To each aliquot, functionalized SPIONs were added to a final concentration of 0.5, 1.0 and 2.0 mg/mL. Obtained solutions were dialyzed twice against 400 mL of assembly buffer (150 mM NaCl, 10 mM $CaCl_2$, 1% *w/v* glycine, 10% *v/v* glycerol, 50 mM tris-HCl, pH = 8) for 24 h at 4 °C.

2.7. VLPs Antigenicity

Antigenicity of HBc VLPs was assessed via enzyme-linked immunosorbent assay (ELISA). HBc assembled with functionalized SPION-PEG and SPION-DHP at different concentrations (mg/mL) in comparison to the standard protein (recombined in *E. coli*, Cat No. R8A120, Meridian Life Science Inc., Memphis, TN, USA). Antigenicity defined as absorbance at 405 nm of two-fold dilution series of VLPs (from 1:160 to 1:81,920) and standard protein (from 0.5 to 0.004 µg/mL).

2.8. Statistical Analysis

Results of SPION-HBc formation were analyzed using a two-way ANOVA followed by a Duncan test; differences were considered significant at $p \leq 0.05$. Statistical analysis was performed using the Statistica 8.0 statistical software package (StatSoft Inc., Tulsa, OK, USA).

2.9. Characterization Methods

Transmission electron microscopy (TEM) images were acquired with Hitachi TEM HT7700 microscope (Tokyo, Japan). Grids were made of copper coated with a carbon film, mesh 300. Samples were prepared by placing 15 μL drop on the grid and draining the excess solution with blotting paper and left for 15 min. to dry. Subsequently, samples were negatively stained with 10 μL of 2% uranyl acetate. Particle size analysis was performed with free ImageJ software version 1.51w (NIH, Bethesda, MD, USA).

Dynamic light scattering (DLS) and ζ-potential measurements were performed on Malvern Zetasizer Nano ZS90 (Worcestershire, UK) in a Folded Capillary Zeta Cell DTS1070. Prior to measurement, samples were briefly sonicated, diluted to optimal concentration and filtered with a 0.2 μm syringe filter (Merck Millipore, Burlington, MA, USA). Measurements were repeated in triplicate.

ELISA was performed on the assembled HBc VLPs, with or without SPION core. The assay was performed in line with a procedure described previously [34]. MaxiSorp (NUNC) 96-well microplate was coated overnight at 4 °C with of HBc-specific mAb (0.5 mg/mL) (Cat. No. C31190 M, Meridian Life Science Inc., Memphis, TN, USA) in carbonate buffer pH 9.6. Each step following the coating was preceded by three washes with PBST buffer (phosphate buffered saline with additional 0.05% v/v Tween20, Sigma, Saint Louis, MO, USA). The coated wells were blocked for 1 h with 5% (w/v) fat-free milk/PBS, followed by incubation with 100 μL of antibody solution for 1 h at 25 °C. The samples were added to the PBS-filled wells and two-fold serially diluted. HBc produced in *E. coli* (Cat. No. R8A120, Meridian Life Science) was used as the reference. Rabbit polyclonal PBST antibody specific to HBc (Cat. No. LS-C67451/18649, Life Span Biosciences, Seattle, VA, USA) 0.125 mg/mL and goat anti-rabbit whole-molecule polyclonal antibody AP-conjugated (Sigma) 1:10,000 dilutions were premixed and added as the primary and secondary antibody. Finally, the substrate for alkaline phosphatase (pNPP, Sigma) was added and the reaction was developed at 25 °C for at least 30 min. The absorbance was measured at 405 nm using a microplate reader (Model 680, Bio-Rad, Hercules, CA, USA).

3. Results

3.1. SPIONs Synthesis and Functionalization

Monodispersed superparamagnetic iron oxide nanoparticles (SPIONs) of 15 nm diameter were synthesized via thermal decomposition of iron (III) acetylacetonate Fe(acac)$_3$ (Figure S1). The approximate diameter of the HBc VLP internal cavity is 25 nm for the T = 4 particles and 21 nm for the T = 3 [14]. Therefore, both structures provide sufficient marginal space to accommodate the ligands. In order to obtain negative surface charge, SPIONs were functionalized with short and long chain ligands: dihexadecyl phosphate (DHP) and 1,2-distearoyl-sn-glycero-3-phosphoethanolamine-N-[carboxy-(polyethylene glycol)-2000] (PL-PEG-COOH) (Figure 1a,b). Functionalized SPIONs will be denoted as SPION-DHP and SPION-PEG, respectively.

DHP functionalization was performed according to our previously described method [31]. PL-PEG-COOH functionalization was achieved by a slightly modified protocol by Huan et al. [24]. Both SPION-DHP and SPION-PEG were analyzed via ζ-potential and dynamic light scattering (DLS) measurements (Table 1). In both cases, functionalization is driven by hydrophobic interactions between oleic acid residues present on the surface of as-obtained SPIONs and alkyl chains of the ligands. More detailed characterization of both functionalizations can be found in our previous work [31].

Despite rather small differences in surface charge, hydrodynamic radius differs substantially. Counterintuitively, long-chain PEG ligand provided smaller hydrodynamic radius than DHP, which may be partially caused by differences in ζ-potential. Another plausible explanation is the interplay of surface charge and multilayered micelle structure formation.

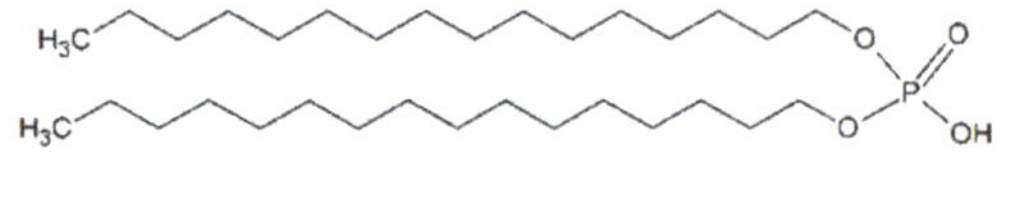

a)

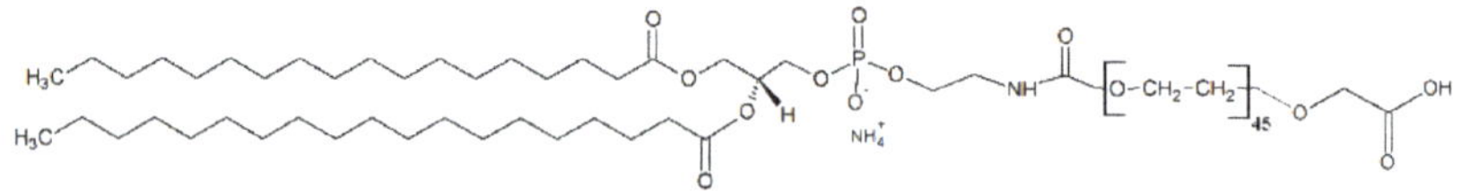

b)

Figure 1. Molecular structure of ligands used for superparamagnetic iron oxide nanoparticles (SPIONs) functionalization. (**a**) dihexadecyl phosphate (DHP); (**b**) 1,2-distearoyl-sn-glycero-3-phosphoethanolamine-N-[carboxy-(polyethylene glycol)-2000] (PL-PEG-COOH).

Table 1. ζ-potential and hydrodynamic radius of the functionalized superparamagnetic iron oxide nanoparticles (SPIONs) obtained by dynamic light scattering (DLS) (Figures S2 and S3).

	SPION-DHP	**SPION-PEG**
ζ-potential	-44.0 ± 3.4 mV	-37.3 ± 2.9 mV
Hydrodynamic diameter	53.75 ± 1.93 nm	29.69 ± 1.57 nm

3.2. VLPs-SPION Assembly

The assembly rates and core encapsidation efficiency are strictly dependent on surface charge density, capsid protein concentration and core/capsid protein stoichiometric ratio [35]. Therefore, HBc concentration was fixed at 0.05 mg/mL while SPIONs concentrations were varied between: 0.05, 0.1 and 0.2 mg/mL. It is important to note that due to differences in ligands' molecular weight and probable differences in functionalization densities, equal w/v concentrations of SPION-PEG and SPION-DHP do not represent the same amount of particles in the solution. The most successful core-VLP assembly was obtained at following concentrations: 0.2 mg/mL SPION-DHP and 0.05 mg/mL SPION-PEG (Figure 2a,b).

TEM images were used to measure the diameter of the assembled VLPs. Mean diameter of SPION-DHP-HBc was 28.4 ± 1.2 nm while SPION-PEG-HBc mean diameter was 29.9 ± 1.5 nm (Table 2). The obtained measurements indicate that in both cases capsids assembled into T = 3 symmetry (native size of T = 3 capsid is 30 nm). Nonetheless, to assess the VLPs symmetry with certainty, crystallographic studies would be required. The obtained results are concordant with thermodynamic studies of nanospheres encapsulated in virus capsids reveling, in that core surface charge and its radius determine the size of the capsid formed around the nanoparticle [36]. In this case, despite T = 4 symmetry being a predominant form of in vitro HBc self-assembly (~95%), introduction of SPION-DHP and SPION-PEG facilitated assembly into a smaller, presumably T = 3 form. This may suggest that negative surface charge density was high enough to drive the assembly into less energetically-favorable capsid morphology. Studying TEM images, SPION-DHP assembly displayed higher efficiency than SPION-PEG and resulted in minority of empty capsids and unassembled cores. In comparison, SPION-PEG assembly produced a multitude of empty capsids along with unassembled cores.

Table 2. VLPs diameter measurements obtained from transmission electron microscopy (TEM) images. Measured with ImageJ software.

	SPION-DHP-HBc	**SPION-PEG-HBc**
Diameter	28.4 ± 1.2 nm	29.9 ± 1.5 nm
Number of measured VLPs	41	19

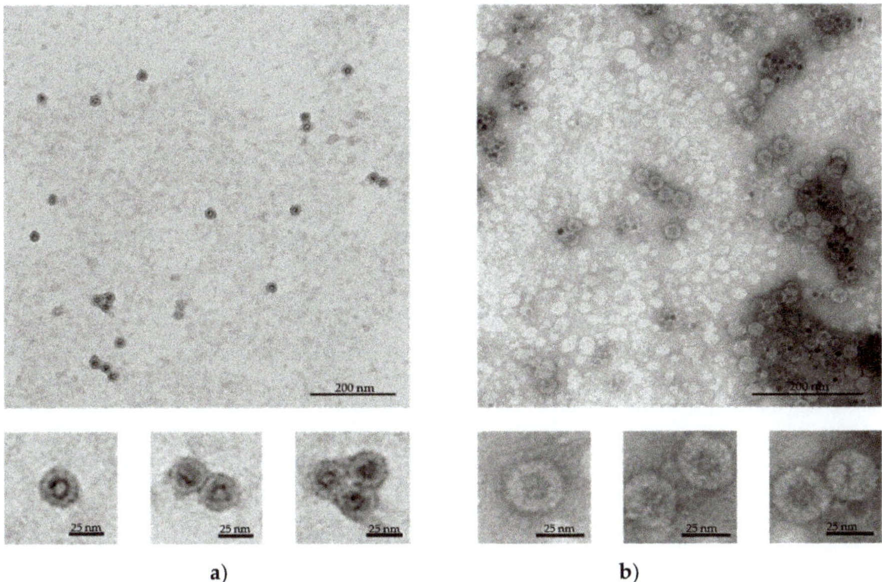

a)   b)

Figure 2. Transmission electron microscopy (TEM) images of the assembled VLPs with magnetic cores, negatively stained with 2% uranyl acetate. (**a**) SPION-DHP-HBc VLPs obtained at 0.2 mg/mL of SPION-DHP and 0.05 mg/mL of HBc; (**b**) SPION-PEG-HBc VLPs obtained at 0.05 mg/mL of SPION-PEG and 0.05 mg/mL of HBcAg.

3.3. VLPs ELISA

Antigenicity of the obtained VLPs was assessed via ELISA (Figures 3 and S4). Despite unvaried HBc protein concentrations, for all VLPs variants decreased signals of HBc detection in comparison to control (initial HBc used) were observed, as well as some significant differences in signals among VLPs variants were found. All concentrations of SPION-PEG displayed a decrease in signal intensity; however, the differences between 0.05 and 0.1 mg/mL concentrations were not statistically significant. Additionally, in comparison to analogous variants of SPION-DHP. For SPION-PEG, the highest observed amount of core-VLPs was found at 0.05 mg/mL (Figure 2B), which also resulted in decreased signal intensity, although insignificantly different from other SPION-PEG concentrations. Finally, among all variants of SPION-HBc VLPs, the significantly lowest signal, 61.6% of HBc was recorded for 0.2 mg/mL SPION-DHP, the same concentration at which the highest number of core-VLPs was observed. The obtained results indicate that core introduction into HBc derived VLPs may decrease antigenicity. This phenomenon could be explained by the proclivity of HBcAg to assemble into smaller T = 3 capsids in the presence of the SPIONs, which in turn results in higher antigen density on the VLPs surface and competitive binding of antibodies. A study by Wu et al. has shown that the amount of antibodies bound to the capsid depends on its morphology and is significantly decreased for T = 3 capsids [37]. Moreover, steric hindrance has been proven to be a crucial factor for antibody binding to surface antigens [38,39]. These results provide a great starting point for further investigations of the relationship between core properties, capsid morphology, antigenicity and biological activity of the HBc derived VLPs.

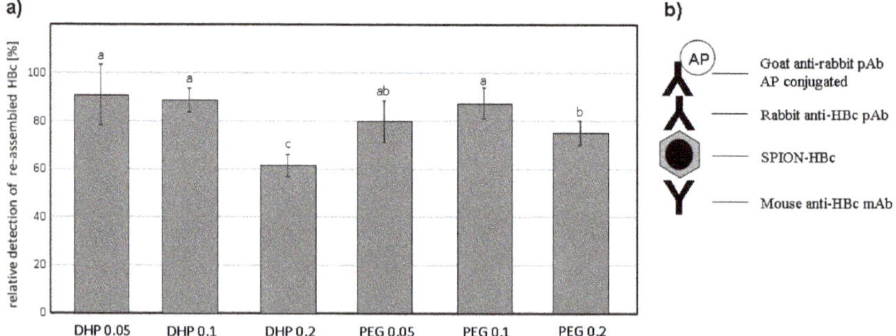

Figure 3. (**a**) HBc re-assembly on SPIONs functionalized with DHP or PEG in different concentrations (mg/mL) in comparison to the initial preparation of plant-derived antigen (100%). Statistically significant differences marked by a letter indexes; (**b**) Scheme of enzyme-linked immunosorbent assay (ELISA) test used for assay of VLP-assembled HBc and SPION-HBc VLPs. AP—alkaline phosphatase.

4. Discussion

Core-VLPs assembly, here SPION-HBc, is a kinetically complex cascade of interactions between viral proteins, nanoparticle surface and an ionic environment. In silico modeling predicts that core introduction provides a plethora of advantages such as increased assembly rates and efficiency over wider set of conditions, stimulation of the assembly below critical subunit concentration (CSC), possibility of templating VLPs morphology. Nonetheless, computational modeling results are not always confirmed in experimental studies. For example, the predicted increase of assembly efficiency driven by the increase of surface charge density has been overestimated in comparison to experimental data [35]. Moreover, simulations predominantly assume cores geometry perfectly commensurate with capsid interior. In our case, functionalized 15 nm SPIONs were not perfectly fitted into the capsid, which could elicit the existence of kinetic traps, not predicted by the computational studies [40]. The most successful core assembly was achieved at a 0.2/0.05 mg/mL core/protein ratio with SPION-DHP. Slightly lower ζ-potential concomitant with larger hydrodynamic diameter in comparison to SPION-PEG might indicate that surface charge density was higher in case of SPION-DHP. SPION-PEG assembly at 0.05/0.05 mg/mL core/protein ratio resulted in partially successful core assembly along with multitude of empty capsids and unassembled cores. This result indicates that one or several assembly parameters were suboptimal; however, due to the complexity of the process, we are unable to pinpoint the exact cause of lower efficiency. It is possible that interactions between less negatively charged SPION-PEG and positively charged domains of the HBc were insufficient to win competition over subunit-subunit attraction, resulting in the empty capsids. Additionally, it is important to note that the ligands used differed in length which could also affect assembly kinetics. In both cases, core introduction resulted in slightly smaller VLPs diameters even for T = 3 capsid morphology, which could be attributed to more compact HBc dimer-dimer spacing resulted from strong electrostatic core-HBc dimer interactions. This thesis stands in agreement with our experimental data showing smaller core-VLPs diameters for lower values of the core's ζ-potential (Table 2). VLPs morphology driven antigenicity is a crucial aspect determining its potential application, especially in the area of vaccinology. In that respect, our study demonstrated successful assembly together with substantially retained HBc antigenicity, although decreased in comparison to native HBc protein preparation containing mainly T = 4 capsids (Figure 3). This may indicate competitive binding of antibodies and/or steric clashes due to increased antigen surface density, stemming from the decreased VLPs diameter. Crucially, for many medical applications such as cell- or tissue-specific targeting, decreased immunogenicity of core antigen may be desirable. Nonetheless, this theory requires more in depth experimental investigation to be confirmed. Surface charge density is one of the most important parameters guiding core assembly. However, it is not easily measureable by standard lab equipment, which impedes rational design of physicochemical properties of the core. Therefore, we propose the use of ζ-potential and hydrodynamic diameter, as two parameters encompassing surface charge density. This approach would simplify and unify core's surface electrostatic characterization, providing more accessible tool for core-VLPs design.

5. Conclusions

This study provides an experimental assessment of the crucial parameters guiding SPION-HBc VLPs assembly and evaluates antigenicity of the obtained structures. The presented results highlight potential directions for further studies regarding the mechanism guiding HBc VLPs assembly with metallic cores as well as their antigenic properties.

Supplementary Materials: The following are available online at http://www.mdpi.com/2079-4991/9/2/155/s1, Figure S1: TEM images of as synthesized SPIONs, Figure S2: DLS results for SPION-PEG and SPION-DHP, Figure S3: ζ-potential for SPION-DHP and SPION-PEG, Figure S4: ELISA of HBcAg SPION assembly.

Author Contributions: Conceptualization, J.D.R., A.A.M. and T.P.; Data curation, J.D.R., A.A.M., A.P. and M.P.; Formal analysis, J.D.R., A.A.M. and M.P.; Funding acquisition, J.D.R. and M.G.; Investigation, J.D.R., A.A.M., A.P. and M.P.; Methodology, J.D.R., A.A.M., M.P. and T.P.; Project administration, J.D.R.; Supervision, J.D.R., T.P. and M.G.; Validation, J.D.R. and A.A.M.; Visualization, A.A.M.; Writing—original draft, J.D.R. and A.A.M.; Writing—review & editing, J.D.R., A.A.M., M.P. and T.P.

Funding: This research was funded by the National Science Center, grant number UMO-2016/23/B/NZ7/01288 and by The National Centre for Research and Development, grant number LIDER/34/0122/L-9/NCBR/2018. HBc production and ELISA was funded by statutory funds of IPG PAS.

Acknowledgments: We are grateful to: Monika Kręcisz, CAT, UAM, for her contribution; Aleksandra Gryciuk, IPG PAS, for construction of vector for plant transient expression. In addition, we are grateful to Professor George Lomonossoff, John Innes Centre, Norwich, UK, for kindly providing the pEAQ-HT vector for transient expression of HBcAg in plants and to Plant Bioscience Ltd for allowing the material to be used.

Conflicts of Interest: The authors declare no conflict of interest.

References

1. López-Sagaseta, J.; Malito, E.; Rappuoli, R.; Bottomley, M.J. Self-assembling protein nanoparticles in the design of vaccines. *Comput. Struct. Biotechnol. J.* **2016**, *14*, 58–68. [CrossRef] [PubMed]

2. Bachmann, M.F.; Jennings, G.T. Vaccine delivery: a matter of size, geometry, kinetics and molecular patterns. *Nat. Rev. Immunol.* **2010**, *10*, 787–796. [CrossRef] [PubMed]

3. Sominskaya, I.; Skrastina, D.; Dislers, A.; Vasiljev, D.; Mihailova, M.; Ose, V.; Dreilina, D.; Pumpens, P. Construction and immunological evaluation of multivalent hepatitis B virus (HBV) core virus-like particles carrying HBV and HCV epitopes. *Clin. Vaccine Immunol.* **2010**, *17*, 1027–1033. [CrossRef] [PubMed]

4. Galarza, J.M.; Latham, T.; Cupo, A. Virus-Like Particle (VLP) Vaccine Conferred Complete Protection against a Lethal Influenza Virus Challenge. *Viral Immunol.* **2005**, *18*, 244–251. [CrossRef] [PubMed]

5. Jackwood, D.J. Multivalent Virus-Like–Particle Vaccine Protects Against Classic and Variant Infectious Bursal Disease Viruses. *Avian Dis.* **2013**, *57*, 41–50. [CrossRef] [PubMed]

6. Mohsen, M.O.; Zha, L.; Cabral-Miranda, G.; Bachmann, M.F. Major findings and recent advances in virus–like particle (VLP)-based vaccines. *Semin. Immunol.* **2017**, *34*, 123–132. [CrossRef]

7. Park, J.-S.; Cho, M.K.; Lee, E.J.; Ahn, K.-Y.; Lee, K.E.; Jung, J.H.; Cho, Y.; Han, S.-S.; Kim, Y.K.; Lee, J. A highly sensitive and selective diagnostic assay based on virus nanoparticles. *Nat. Nanotechnol.* **2009**, *4*, 259–264. [CrossRef]

8. Stephanopoulos, N.; Carrico, Z.M.; Francis, M.B. Nanoscale Integration of Sensitizing Chromophores and Porphyrins with Bacteriophage MS$_2$. *Angew. Chem. Int. Ed.* **2009**, *48*, 9498–9502. [CrossRef]

9. Merzlyak, A.; Indrakanti, S.; Lee, S.-W. Genetically Engineered Nanofiber-Like Viruses For Tissue Regenerating Materials. *Nano Lett.* **2009**, *9*, 846–852. [CrossRef]

10. De la Escosura, A.; Nolte, R.J.M.; Cornelissen, J.J.L.M. Viruses and protein cages as nanocontainers and nanoreactors. *J. Mater. Chem.* **2009**, *19*, 2274. [CrossRef]

11. Uchida, M.; Klem, M.T.; Allen, M.; Suci, P.; Flenniken, M.; Gillitzer, E.; Varpness, Z.; Liepold, L.O.; Young, M.; Douglas, T. Biological Containers: Protein Cages as Multifunctional Nanoplatforms. *Adv. Mater.* **2007**, *19*, 1025–1042. [CrossRef]

12. Liang, T.J. Hepatitis B: The virus and disease. *Hepatology* **2009**, *49*, S13–S21. [CrossRef]

13. Nassal, M. The arginine-rich domain of the hepatitis B virus core protein is required for pregenome encapsidation and productive viral positive-strand DNA synthesis but not for virus assembly. *J. Virol.* **1992**, *66*, 4107–4116. [PubMed]

14. Crowther, R.A.; Kiselev, N.A.; Böttcher, B.; Berriman, J.A.; Borisova, G.P.; Ose, V.; Pumpens, P. Three-dimensional structure of hepatitis B virus core particles determined by electron cryomicroscopy. *Cell* **1994**, *77*, 943–950. [CrossRef]

15. Zlotnick, A.; Palmer, I.; Kaufman, J.D.; Stahl, S.J.; Steven, A.C.; Wingfield, P.T. Separation and crystallization of T = 3 and T = 4 icosahedral complexes of the hepatitis B virus core protein. *Acta Crystallogr. Sect. D Biol. Crystallogr.* **1999**, *55*, 717–720. [CrossRef]

16. Milich, D.R.; McLachlan, A.; Moriarty, A.; Thornton, G.B. Immune response to hepatitis B virus core antigen (HBcAg): localization of T cell recognition sites within HBcAg/HBeAg. *J. Immunol.* **1987**, *139*, 1223–1231. [PubMed]

17. Haracz, S.; Mróz, B.; Rybka, J.D.; Giersig, M.; Mrõz, B.; Rybka, J.D.; Giersig, M. Magnetic behaviour of non-interacting colloidal iron oxide nanoparticles in physiological solutions. *Cryst. Res. Technol.* **2015**, *50*, 791–796. [CrossRef]

18. Kręcisz, M.; Rybka, J.D.; Strugała, A.J.; Skalski, B.; Figlerowicz, M.; Kozak, M.; Giersig, M. Interactions between magnetic nanoparticles and model lipid bilayers—Fourier transformed infrared spectroscopy (FTIR) studies of the molecular basis of nanotoxicity. *J. Appl. Phys.* **2016**, *120*, 124701. [CrossRef]

19. Haracz, S.; Hilgendorff, M.; Rybka, J.D.; Giersig, M. Effect of surfactant for magnetic properties of iron oxide nanoparticles. *Nucl. Instruments Methods Phys. Res. Sect. B Beam Interact. Mater. Atoms* **2015**, *364*, 120–126. [CrossRef]

20. Mahmoudi, M.; Sant, S.; Wang, B.; Laurent, S.; Sen, T. Superparamagnetic iron oxide nanoparticles (SPIONs): Development, surface modification and applications in chemotherapy. *Adv. Drug Deliv. Rev.* **2011**, *63*, 24–46. [CrossRef]

21. Neuberger, T.; Schöpf, B.; Hofmann, H.; Hofmann, M.; von Rechenberg, B. Superparamagnetic nanoparticles for biomedical applications: Possibilities and limitations of a new drug delivery system. *J. Magn. Magn. Mater.* **2005**, *293*, 483–496. [CrossRef]

22. Teja, A.S.; Koh, P.-Y. Synthesis, properties, and applications of magnetic iron oxide nanoparticles. *Prog. Cryst. Growth Charact. Mater.* **2009**, *55*, 22–45. [CrossRef]

23. Malyutin, A.G.; Easterday, R.; Lozovyy, Y.; Spilotros, A.; Cheng, H.; Sanchez-Felix, O.R.; Stein, B.D.; Morgan, D.G.; Svergun, D.I.; Dragnea, B.; et al. Viruslike nanoparticles with maghemite cores allow for enhanced mri contrast agents. *Chem. Mater.* **2015**, *27*, 327–335. [CrossRef]

24. Huang, X.; Bronstein, L.M.; Retrum, J.; Dufort, C.; Tsvetkova, I.; Aniagyei, S.; Stein, B.; Stucky, G.; McKenna, B.; Remmes, N.; et al. Self-assembled virus-like particles with magnetic cores. *Nano Lett.* **2007**, *7*, 2407–2416. [CrossRef] [PubMed]

25. Okuda, M.; Eloi, J.-C.; Jones, S.E.W.; Verwegen, M.; Cornelissen, J.J.L.M.; Schwarzacher, W. Pt, Co–Pt and Fe–Pt alloy nanoclusters encapsulated in virus capsids. *Nanotechnology* **2016**, *27*, 095605. [CrossRef]

26. Gubin, S.P.; Koksharov, Y.A.; Khomutov, G.B.; Yurkov, G.Y. Magnetic nanoparticles: preparation, structure and properties. *Russ. Chem. Rev.* **2005**, *74*, 489–520. [CrossRef]

27. Fatima, H.; Kim, K.S. Magnetic nanoparticles for bioseparation. *Korean J. Chem. Eng.* **2017**, *34*, 589–599. [CrossRef]

28. Shen, L.; Zhou, J.; Wang, Y.; Kang, N.; Ke, X.; Bi, S.; Ren, L. Efficient Encapsulation of Fe_3O_4 Nanoparticles into Genetically Engineered Hepatitis B Core Virus-Like Particles Through a Specific Interaction for Potential Bioapplications. *Small* **2015**, *11*, 1190–1196. [CrossRef]

29. Roose, K.; De Baets, S.; Schepens, B.; Saelens, X. Hepatitis B core-based virus-like particles to present heterologous epitopes. *Expert Rev. Vaccines* **2013**, *12*, 183–198. [CrossRef]

30. Belyi, V.A.; Muthukumar, M. Electrostatic origin of the genome packing in viruses. *Proc. Natl. Acad. Sci. USA* **2006**, *103*, 17174–17178. [CrossRef]

31. Mieloch, A.A.; Kręcisz, M.; Rybka, J.D.; Strugała, A.; Krupiński, M.; Urbanowicz, A.; Kozak, M.; Skalski, B.; Figlerowicz, M.; Giersig, M. The influence of ligand charge and length on the assembly of *Brome mosaic virus* derived virus-like particles with magnetic core. *AIP Adv.* **2018**, *8*, 035005. [CrossRef]

32. Giersig, M.; Hilgendorff, M. Magnetic nanoparticle superstructures. *Eur. J. Inorg. Chem.* **2005**, *2005*, 3571–3583. [CrossRef]

33. Peyret, H.; Lomonossoff, G.P. The pEAQ vector series: the easy and quick way to produce recombinant proteins in plants. *Plant Mol. Biol.* **2013**, *83*, 51–58. [CrossRef] [PubMed]

34. Pyrski, M.; Rugowska, A.; Wierzbiński, K.R.; Kasprzyk, A.; Bogusiewicz, M.; Bociąg, P.; Samardakiewicz, S.; Czyż, M.; Kurpisz, M.; Pniewski, T. HBcAg produced in transgenic tobacco triggers Th1 and Th2 response when intramuscularly delivered. *Vaccine* **2017**, *35*, 5714–5721. [CrossRef] [PubMed]

35. Hagan, M.F. A theory for viral capsid assembly around electrostatic cores. *J. Chem. Phys.* **2009**, *130*, 114902. [CrossRef] [PubMed]

36. Šiber, A.; Zandi, R.; Podgornik, R. Thermodynamics of nanospheres encapsulated in virus capsids. *Phys. Rev. E Stat. Nonlinear Soft Matter Phys.* **2010**, *81*, 051919. [CrossRef]

37. Wu, W.; Chen, Z.; Cheng, N.; Watts, N.R.; Stahl, S.J.; Farci, P.; Purcell, R.H.; Wingfield, P.T.; Steven, A.C. Specificity of an anti-capsid antibody associated with Hepatitis B Virus-related acute liver failure. *J. Struct. Biol.* **2013**, *181*, 53–60. [CrossRef]

38. Kent, S.P.; Ryan, K.H.; Siegel, A.L. Steric hindrance as a factor in the reaction of labeled antibody with cell surface antigenic determinants. *J. Histochem. Cytochem.* **1978**, *26*, 618–621. [CrossRef]

39. De Michele, C.; De Los Rios, P.; Foffi, G.; Piazza, F. Simulation and Theory of Antibody Binding to Crowded Antigen—Covered Surfaces. *PLoS Comput. Biol.* **2016**, *12*, e1004752. [CrossRef]

40. Hagan, M.F. Controlling viral capsid assembly with templating. *Phys. Rev. E Stat. Nonlinear Soft Matter Phys.* **2008**, *77*, 051904. [CrossRef]

Article

Secretory Nanoparticles of *Neospora caninum* Profilin-Fused with the Transmembrane Domain of GP64 from Silkworm Hemolymph

Hamizah Suhaimi [1], Rikito Hiramatsu [2], Jian Xu [3], Tatsuya Kato [1,2,3] and Enoch Y. Park [1,2,3,*]

[1] Laboratory of Biotechnology, Department of Bioscience, Graduate School of Science and Technology, Shizuoka University, 836 Ohya, Suruga-ku, Shizuoka 422-8529, Japan; noor.hamizah.binsuhaimi.16@shizuoka.ac.jp (H.S.); kato.tatsuya@shizuoka.ac.jp (T.K.)

[2] Laboratory of Biotechnology, Department of Applied Biological Chemistry, College of Agriculture, Graduate School of Integrated Science and Technology, Shizuoka University, 836 Ohya Suruga-ku, Shizuoka 422-8529, Japan; hiramatsu.rikito.17@shizuoka.ac.jp

[3] Laboratory of Biotechnology, Research Institute of Green Science and Technology, Shizuoka University, 836 Ohya Suruga-ku, Shizuoka 422-8529, Japan; xu.jian@shizuoka.ac.jp

* Correspondence: park.enoch@shizuoka.ac.jp; Tel./Fax: +81-54-238-4887

Received: 20 February 2019; Accepted: 6 April 2019; Published: 10 April 2019

Abstract: Neosporosis, which is caused by *Neospora caninum*, is a well-known disease in the veterinary field. Infections in pregnant cattle lead to abortion via transplacental (congenitally from mother to fetus) transmission. In this study, a *N. caninum* profilin (NcPROF), was expressed in silkworm larvae by recombinant *Bombyx mori* nucleopolyhedrovirus (BmNPV) bacmid and was purified from the hemolymph. Three NcPROF constructs were investigated, native NcPROF fused with an N-terminal PA tag (PA-NcPROF), PA-NcPROF fused with the signal sequence of bombyxin from *B. mori* (bx-PA-NcPROF), and bx-PA-NcPROF with additional C-terminal transmembrane and cytoplasmic domains of GP64 from BmNPV (bx-PA-NcPROF-GP64TM). All recombinant proteins were observed extra- and intracellularly in cultured Bm5 cells and silkworm larvae. The bx-PA-NcPROF-GP64TM was partly abnormally secreted, even though it has the transmembrane domain, and only it was pelleted by ultracentrifugation, but PA-NcPROF and bx-PA-NcPROF were not. Additionally, bx-PA-NcPROF-GP64TM was successfully purified from silkworm hemolymph by anti-PA agarose beads while PA-NcPROF and bx-PA-NcPROF were not. The purified bx-PA-NcPROF-GP64TM protein bound to its receptor, mouse Toll-like receptor 11 (TLR-11), and formed unique nanoparticles. These results suggest that profilin fused with GP64TM was secreted as a nanoparticle with binding affinity to its receptor and this nanoparticle formation is advantageous for the development of vaccines to *N. caninum*.

Keywords: BmNPV bacmid; nanobiomaterials; *Neospora caninum*; *Neospora caninum* profilin; neosporosis; silkworm expression system

---

1. Introduction

Neosporosis is a disease caused by apicomplexan parasites such as *Neospora caninum* and *Neospora hugheshi* [1]. Considering its morphology, *N. caninum* is quite similar to *Toxoplasma gondii*, so neosporosis was misdiagnosed as *T. gondii* infection until 1988 [2]. Although the morphologies are quite similar, these parasites are biologically different and are considered as zoonotic compared with toxoplasmosis where it affects human, sheep and many other warm-blooded animals [3]. Neosporosis infection occurs from unsporulated oocysts in feces followed by ingestion by the cattle, transformation from the tachyzoites stage to the sporozoites stage and leading to neosporosis in the cattle [4]. Recrudescence of the infection occurs in pregnant cattle, where this parasite is transmitted from the placenta to the unborn fetus and may lead to abortion [5,6].

N. caninum, a parasite of the phylum Apicomplexa, has been known as the etiologic agent of neosporosis disease [5]. Apicomplexa contains specific organelles, micronemes, rhoptries, and dense granules. Most of the specific antigen proteins are secreted from each organelle and are located at the apical end of the *N. caninum* parasite and have been extensively studied as recombinant vaccine candidates against *N. caninum* infection [7,8]. In our previous study, we demonstrated the capability of baculoviruses displaying *N. caninum*-derived antigens, such as surface antigen 1 (NcSAG1), SAG1-related sequence 2 (NcSRS2) and microneme protein 3, as an alternative method to control neosporosis because the combination of three antigens could induce T-cell activation and interferon gamma (IFN-γ) production and suppress *N. caninum* infection in mice [9].

Profilin of *N. caninum* (NcPROF) is recognized and conserved as a vaccine candidate with high potential against neospora infections [10]. Interestingly, profilin is known as a small actin-binding protein located at the apical end of *N. caninum* tachyzoites and is essential for invasion of the host cell by regulating the polymerization and depolymerization of actin filaments [11]. Furthermore, Jenkins et al. [12] and Mansilla et al. [13] revealed that *T. gondii* profilin binds to Toll-like receptor 11 (TLR11) in mice and is responsible for activating dendritic cells and stimulating the release of cytokines such as interleukin 12 (IL-12) and IFN-γ. According to Innes et al. [14], some cytokines help in controlling the infection by inhibiting parasite multiplication.

In this study, to produce recombinant NcPROF, the *Bombyx mori* nucleopolyhedrovirus (BmNPV) bacmid-based silkworm expression system was used. The BmNPV bacmid-based silkworm expression system contributes to several advantages for recombinant protein expression, such as low cost, ease of treatment and high safety [15]. Previous studies of recombinant NcPROF have not dealt with the expression fusion of the transmembrane and cytoplasmic domains of GP64 from BmNPV. Therefore, to evaluate the expression and purification of fusion native NcPROF fused with an N-terminal PA tag, fused with the signal sequence of bombyxin from *B. mori* (bx-PA-NcPROF) fused with the transmembrane and cytoplasmic domains of GP64 from BmNPV (bx-PA-NcPROF-GP64TM), NcPROF was expressed in two other NcPROF constructs, one fused with a PA tag (PA-NcPROF) and PA-NcPROF fused with the signal sequence of bombyxin from *B. mori* (bx-PA-NcPROF). The bx-PA-NcPROF-GP64TM was successfully purified from silkworm hemolymph and the binding of bx-PA-NcPROF-GP64TM with recombinant mouse TLR11 (mTLR11), and the morphological analysis of the nanoparticles were reported.

2. Materials and Methods

2.1. Construction of the Recombinant BmNPV Bacmid Containing NcPROF Constructs

The NcPROF gene (GenBank accession no. BK006901.1) was used to express recombinant NcPROF. The bx-PA (Peptide tag, GVAMPGAEDDVV)-NcPROF-GP64TM gene was synthesized by GENEWIZ Japan (Saitama, Japan) with the bombyxin signal peptide (bx) sequence (NCBI reference sequence no. NP_001103771.1) at its N-terminus and transmembrane cytoplasmic domains of GP64 from BmNPV (GP64TM) (GenBank accession no. BAF32568.1) at its C-terminus. The synthetic sequence of bx-PA-NcPROF-GP64TM was digested using *Eco*RI (NEB, Tokyo, Japan) and *Not*I (NEB, Tokyo, Japan) and was cloned into pFastBac1 (ThermoFisher Scientific. K.K., Tokyo, Japan), resulting in the construct pFast/bx-PA-NcPROF-GP64TM. The gene coding bx-PA-NcPROF was amplified by polymerase chain reaction (PCR) using (bx-PA-NcPROF) with primer sets (Table 1). The amplified gene was digested by *Eco*RI and *Xho*I and was cloned into pFastbac1, resulting in the construct pFast/bx-PA-NcPROF. Additionally, pFast/PA-NcPROF and pFast/bx-PA-NcPROF were generated by PCR amplification with primers listed in Table 1 and then underwent self-ligation. Subsequently, the constructed pFast/bx-PA-NcPROF-GP64TM, pFast/bx-PA-NcPROF, and pFast/bx-PA-NcPROF were transformed into *Escherichia coli* BmDH10Bac, respectively, as described previously [15]. The white colonies were identified as positive transformants containing each recombinant BmNPV bacmid.

Table 1. Primers used in this study.

Name	5′ to 3′
pFastBac1	
Forward	5′-TATTCCGGATTATTCATACC-3′
Reverse	5′-ACAAATGTGGTATGGCTGATT-3′
NcPROF	
Forward	5′-GGACACAATCGGAGAGGACG-3′
Reverse	5′-GTGCACACATGGTGATGTCG-3′
pUC/M13	
Forward	5′-CCCAGTCACGACGTTGTAAAACG-3′
Reverse	5′-AGCGGATAACAATTTCACACAGG-3′
PA-NcPROF	
Forward	5′-GGCGTTGCCATGCCAGGTGC-3′
Reverse	5′-CATGAATTCCGCGCGCTTCG-3′
Bx-PA-NcPROF	
Forward	5′-GCGCGGAATTCATGAAGATACTCCTTGCT-3′
Reverse	5′-GCATGCCTCGAGTTAATAGCCAGACTGGTGAAGGTACTCG-3′

2.2. Expression of Recombinant NcPROF in Silkworm Larvae and Bm5 Cells

Each recombinant BmNPV bacmid DNA (10 or 20 μg) was mixed with 0.1% chitosan (ratio of amino group to phosphate group 2) and 2% (*w/v*) of 2-(*N*-morpholino) ethanesulfonic acid (MES) buffer [16]. The mixture of each recombinant BmNPV bacmid DNA was incubated at room temperature (RT) for approximately 45 min before injection into the silkworm larvae. Subsequently, approximately 50 μL of the mixture was injected into a fifth instar silkworm larva (Ehime Sansyu, Ehime, Japan), and the larvae were reared at 25 °C and 65 ± 5% relative humidity with the artificial diet Silkmate S2 (Nosan, Yokohama, Japan) for 6–7 day. The collected hemolymph was diluted with 10-fold of phosphate-buffered saline (PBS, pH 7.4) and was injected again into fifth-instar silkworm larva and reared for 4–5 d. The larval hemolymph was collected through cutting the caudal leg, was mixed with 5 μL of 200 mM 1-phenyl-2-thiourea, and then was centrifuged at 10,000× *g* for 10 min at 4 °C. The larval fat body was dissected in PBS mixed with 0.1% Triton-X 100 before sonication on ice with an interval time of 15 s until the solution was clear and then it was centrifuged at 10,000× *g* for 30 min at 4 °C. For the expression of NcPROF in Bm5 cells, diluted hemolymph containing recombinant BmNPVs was added to the cultured Bm5 cells in a 6-well plate at 27 °C in Sf-900II medium (ThermoFisher Scientific K.K.) supplemented with 1% antibiotic-antimycotic (ThermoFisher Scientific K.K.) and 10% FBS (Gibco, Tokyo, Japan). The culture supernatant and infected-Bm5 cells were collected at 3 d post infection. The culture supernatant (culture media) was collected after centrifugation at 10,000× *g* for 10 min at 4 °C together with the same solution as the larval hemolymph. Infected-Bm5 cells were mixed with the same solution and were subjected to the same treatment as the larval fat body, followed by separation of cell lysate soluble and insoluble fractions after centrifugation. The supernatants of the larval hemolymph, fat body, cell medium, cell lysate soluble, and insoluble were immediately frozen at −80 °C until further analysis.

2.3. Purification of bx-PA-NcPROF-GP64TM from Silkworm Larvae

bx-PA-NcPROF-GP64TM was purified from silkworm larval hemolymph using anti-PA tag affinity chromatography (WAKO Pure Chemical Industries, Osaka, Japan). Ten-times the diluted supernatant of the hemolymph was mixed with anti-PA tag affinity beads and incubated at 4 °C for 24 h with gentle agitation. Next, the beads were washed 10 times with Tris-buffered saline (TBS, pH 7.6), and the bound proteins were eluted with 0.1 M glycine-HCl (pH 3) and immediately neutralized with 1.5 M Tris-HCl (pH 8.0). The elution was conducted at RT in a stepwise manner.

2.4. Ultracentrifugation Analysis of bx-PA-NcPROF-GP64TM, bx-PA-NcPROF, and PA-NcPROF

Two milliliters of crude hemolymph was mixed with 1 mL of PBS (pH 6.2). This mixture was centrifuged at 100,000× g for 90 min at 4 °C, and the pellet was suspended in 1 mL of PBS (pH 6.2). This suspension was sonicated to dissolve the pellet, and the suspension was collected and subjected to western blotting.

2.5. Transmission Electron Microscopy Observation of bx-PA-NcPROF-GP64TM

Purified bx-PA-rNcPROF-GP64TM was analyzed through negative staining. The bx-PA-rNcPROF-GP64TM (20 µL) drop was loaded onto the surface of film 200 mesh copper grid (Nisshin Em Co. Ltd., Tokyo, Japan) within 30 s at RT. Next, the grid was washed 3 times with PBS and negatively stained with phosphotungstic acid (2% of *v/v*). To investigate the surface of the bx-PA-rNcPROF-GP64TM nanoparticle, the sample on the grid was blocked in 2% (*v/v*) bovine serum albumin (BSA) for approximately 5 min after washing with PBS 3 times. Next, the grid was incubated for 1 h at RT, and then it was loaded onto the surface drop of rat anti-PA tag monoclonal antibody (NZ-1, 1:50 in PBS) (WAKO Pure Chemical Industries). After 1 h of incubation, the grid was washed 6 times and was loaded onto the surface with goat anti-rat immunoglobulin G (IgG) (H+L)-conjugated with 12 nm gold beads (1:50 in PBS) (Jackson ImmnunoResearch Inc., West Grove, PA, USA) drops and was incubated for 1 h at RT. Finally, the grid was washed 6 times with PBS, followed negative staining with phosphotungstic acid (2% *v/v*). Images were acquired with a transmission electron microscope (TEM; JEM-2100F; JEOL, Ltd., Tokyo, Japan) operated at 100 kV.

2.6. Sodium Dodecyl Sulfate-Polyacrylamide Gel Electrophoresis (SDS-PAGE) and Western Blotting

The collected larval hemolymph and extract of a fat body were used to confirm the expression of recombinant proteins through SDS-PAGE (Bio-Rad, Hercules, CA, USA). The proteins were transferred onto a polyvinylidene fluoride (PVDF) membrane using a trans-blot SD semidry transfer cell (Bio-Rad). Next, 5% (*w/v*) skimmed milk in Tris-buffered saline containing 0.1% (*v/v*) Tween 20 (TBST) was used to block the PVDF membrane for 1 h. Thereafter, the membrane was incubated with rat anti-PA tag monoclonal antibody (NZ-1, 0.1 µg/mL) (Wako Pure Chemical Industries, Ltd.), followed by incubation with horseradish peroxidase (HRP)-conjugated goat anti-rat IgG (H+L) (1:10,000) (Bios Antibodies Inc., Woburn, MA, USA) and development for 1 min with Immobilon western chemiluminescence HRP substrate (Merck Millipore, Burlington, MA, USA). Stained proteins were detected using a molecular imager VersaDoc MP imaging systems (Bio-rad).

2.7. Binding Assay of bx-PA-NcPROF-GP64TM with Mouse TLR11

The binding assay for purified bx-PA-NcPROF-GP64TM against recombinant mTLR11 Fc chimera (R&D Systems, Minneapolis, MN, USA) was carried out by enzyme-linked immunosorbent assay (ELISA). The binding assay was performed in 96-well plates coated with 50 µL/well of mTLR11 (100 ng/well) at 4 °C overnight. The plates were blocked with 100 µL/well of 2% skimmed milk in PBS for 2 h at RT. After 2 h, the plates were washed 3 times with PBS with 0.1% Tween 20 (PBST). The purified bx-PA-NcPROF-GP64TM with different concentrations (100, 300 and 500 ng/well) with the 1 mM dithiothreitol (DTT) treatment and 100 ng without 1 mM DTT treatment was diluted in blocking buffer in triplicate, followed by incubation for 2 h at RT. The plate was washed with PBST and incubated for 2 h at RT with 50 µL/well of rat anti-PA tag monoclonal antibody (NZ-1, 1:1000 dilution). Subsequently, the plates were washed again with PBST and were incubated for 2 h at RT with 50 µL/well of HRP-conjugated goat anti-rat IgG (H+L) secondary antibody (1:5000 dilution). Next, the plates were washed with PBST and developed by incubating with 50 µL/well of 3,3′,5,5′-tetramethylbenzidine (TMBZ) (Dojindo Co. Ltd., Kanagawa, Japan) for 20 min, followed by stopping the reaction with

100 µL/well of stop solution (10% H_2SO_4). Thereafter, 1 mM DTT and BSA (100 ng/well) were used as negative controls. Finally, the absorbance of the mixture was measured using a microplate reader (Model 680, Bio-Rad, Hercules, CA, USA) at 450 nm.

3. Results and Discussion

3.1. Expression of bx-PA-NcPROF-GP64TM, bx-PA-NcPROF, and PA-NcPROF in Silkworms

In this study, three NcPROF constructs were generated. PA-NcPROF has a PA tag sequence at its N-terminus (Figure 1A), whereas bx-PA-NcPROF has the signal sequence of bombyxin from *B. mori* at its N-terminus (Figure 1B). Meanwhile, bx-PA-NcPROF-GP64TM has the transmembrane and cytoplasmic domains of GP64 from BmNPV at the C-terminus of bx-PA-NcPROF (Figure 1C).

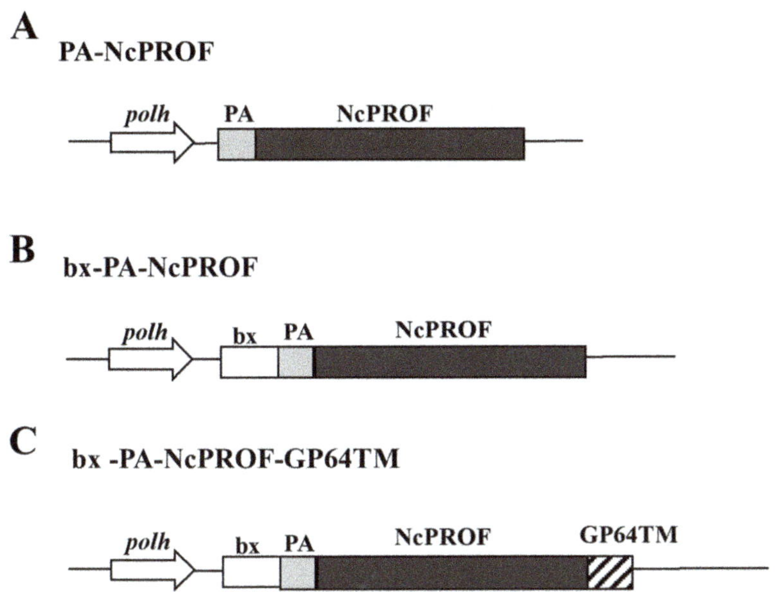

Figure 1. Constructs of native NcPROF fused with an N-terminal PA tag (PA-NcPROF) (**A**), PA-NcPROF fused with the signal sequence of bombyxin from *B. mori* (bx-PA-NcPROF) (**B**) and bx-PA-NcPROF with additional C-terminal transmembrane and cytoplasmic domains of GP64 from *Bombyx mori* nucleopolyhedrovirus (BmNPV) (bx-PA-NcPROF-GP64TM) (**C**). *polh*, polyhedrin promoter; bx, bombyxin signal sequence; GP64TM, transmembrane and cytoplasmic domains of GP64 from BmNPV; PA, PA-tag sequence.

The bombyxin signal peptide (bx) is responsible for the secretion of recombinant proteins to the hemolymph of silkworm larvae [17,18]. The GP64TM were used as a fusion partner with NcPROF as a target protein to be incorporated into the cell membrane [19]. All NcPROF constructs were observed in both the fat body and hemolymph when they were expressed in silkworm larvae (Figure 2A). PA-NcPROF, bx-PA-NcPROF, and bx-PA-NcPROF-GP64TM comprise 176, 195, and 226 amino acids, respectively, and their estimated molecular weights are 19, 21, and 25 kDa, respectively. All NcPROFs were expressed and verified intra- and extracellularly in Bm5 cells (Figure 2B). The molecular weight of PA-NcPROF was approximately 25 kDa by western blotting.

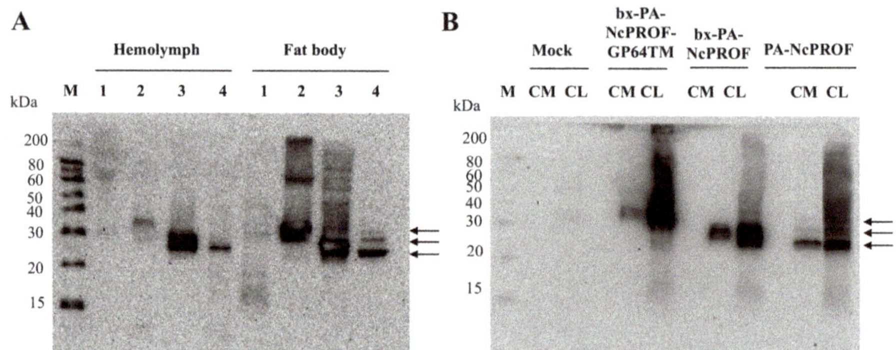

Figure 2. Western blot analysis of bx-NcPROF-GP64TM, bx-NcPROF and NcPROF expressed in silkworm larvae (**A**) and Bm5 cells (**B**). Recombinant BmNPV/bx-PA-NcPROF-GP64TM, BmNPV/bx-PA-NcPROF or BmNPV/PA-NcPROF were injected into the silkworms, followed by harvesting at 5 d post injection. Bm5 cells were harvested at 3 d post infection of the indicated recombinant BmNPV. Lanes M, 1, 2, 3 and 4 in (**A**) denote the molecular marker, mock, bx-PA-NcPROF-GP64TM, bx-PA-NcPROF, and PA-NcPROF, respectively. Lanes M, CM and CL in (**B**) denote the molecular marker, protein samples from cell medium and cell lysate, respectively. Upper, middle and lower arrows in (**A**,**B**) indicate bx-NcPROF-GP64TM, bx-NcPROF and NcPROF, respectively.

The molecular weight of native NcPROF in this study was similar to that of full-length profilin expressed in *E. coli* (22 kDa) [12]. This finding suggests that PA-NcPROF expressed in silkworm larvae may be not modified posttranslationally. bx-PA-NcPROF was observed in the fat body and hemolymph at 25–30 kDa as a large obscure band that appeared as two bands. The molecular weight of the lower band was the same as that of PA-NcPROF and that of the upper band was larger than that of PA-NcPROF. This finding suggests that bx-PA-NcPROF may be partially modified posttranslationally after the cleavage of the bombyxin signal peptide. In fact, N128 was predicted as an *N*-glycosylation site estimated by NetNGlyc 1.0 Server (http://www.cbs.dtu.dk/services/NetNGlyc/). The bx-PA-NcPROF-GP64TM product was observed at 30–40 kDa by western blotting. Interestingly, bx-PA-NcPROF-GP64TM was observed in the hemolymph and culture medium although it was fused with the transmembrane domain of GP64. In our previous report, *N. caninum*-derived antigens NcSAG1 and NcSRS2 fused with the transmembrane and cytoplasmic domains of GP64 were not observed in hemolymph [9]. In this study, bx-PA-NcPROF-GP64TM was secreted into hemolymph, in part, and the culture medium from Bm5 cells. It has been reported that profilins of *T. gondii* and *Babesia canis* were partially secreted even though no signal peptide is predicted in these genes [20,21]. In this study, NcPROF was also secreted regardless of the addition of the signal sequence. To investigate *N*-glycosylation of the *N*-glycan of three NcPROFs, PNGase F treatment was carried out (Figure S1). Molecular weight of each NcPROF was not changed between before and after the PNGase F treatment, indicating no *N*-glycan was attached into the three NcPROFs. These results suggest that the two bands of bx-PA-NcPROF did not come from its *N*-glycosylation. Hence, it showed that the bx-PA-NcPROF and bx-PA-NcPROF-GP64TM were not posttranslationally modified with an *N*-glycan in endoplasmic reticulum even though these proteins have the bx signal peptide. In addition, the discrepancy of the molecular weight of each PROF to its estimated that was not caused by the *N*-glycosylation. In nature, estimated pI of this NcPROF is around 4.89, which may have influence on the mobility of NcPROFs on the SDS-PAGE gel [22]. In addition, the negative charge of acidic residues may create repulsion and this repulsion caused the anomalous migration of protein in SDS-polyacrylamide gels [23]. This may be the reason why the molecular weight of bx-PA-NcPROF-GP64TM was increased.

The expressed NcPROFs in the Bm5 cell culture broth were further investigated by ultracentrifugation (Figure 3). It showed that PA-NcPROF and bx-PA-NcPROF were not precipitated by ultracentrifugation

(100,000× *g*), while bx-PA-NcPROF-GP64TM was pelleted. Normally, transmembrane proteins are anchored in the membrane fraction of cells and are not easily secreted. However, glycoproteins from some viruses are secreted and are partially incorporated into extracellular vesicles when they are overexpressed [24,25]. Moreover, the transmembrane and cytoplasmic domains of GP64 from BmNPV facilitated the display of recombinant proteins on BmNPV particles. These findings suggest that bx-PA-NcPROF-GP64TM was secreted as nanoparticles or by display on the envelope of BmNPV particles.

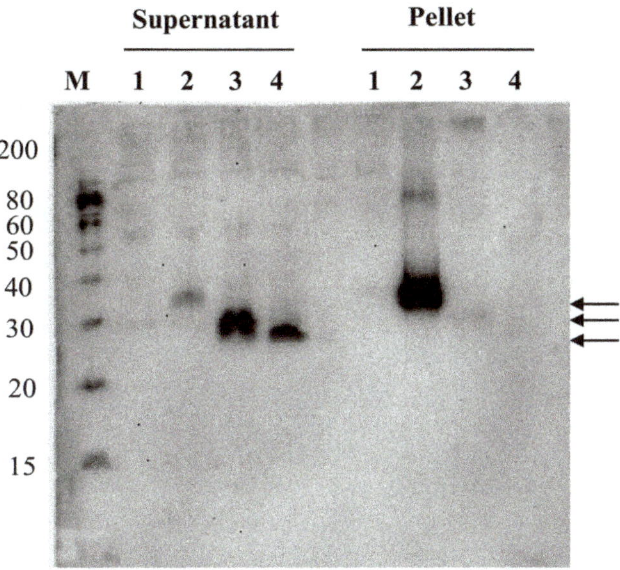

Figure 3. Western blots of soluble and insoluble fractions from the Bm5 cell culture supernatant. The culture supernatant containing bx-PA-NcPROF-GP64TM, bx-PA-NcPROF or PA-NcPROF was centrifuged at 100,000× *g*, and the pellet was resuspended in 1 mL of PBS. Lanes M, 1, 2, 3 and 4 denote the molecular marker, mock, bx-PA-NcPROF-GP64TM, bx-PA-NcPROF and PA-NcPROF, respectively. Upper, middle and lower arrows indicate bx-NcPROF-GP64TM, bx-NcPROF and NcPROF, respectively.

bx-PA-NcPROF-GP64TM was then purified from silkworm hemolymph using anti-PA tag agarose. As shown in Figure 4A, bx-PA-NcPROF-GP64TM was successfully purified as a single band in SDS-PAGE and was further verified by western blotting (Figure 4B). This finding indicates that bx-PA-NcPROF-GP64TM does not display on the surface of BmNPV particles because no band was observed in the purified sample except for bx-PA-NcPROF-GP64TM. These results suggested that bx-PA-NcPROF-GP64TM forms only some nanoparticles and can be secreted into hemolymph and the culture broth. The yield of purified NcPROF-GP64TM was approximately 14 μg from 3 mL of larval hemolymph.

Interestingly, PA-NcPROF and bx-PA-NcPROF could not be purified from hemolymph using the same protocols as bx-PA-NcPROF-GP64TM. From the results, it was demonstrated that, without modification of the transmembrane region at the C-terminus for both constructs, bx-PA-NcPROF was only slightly purified and PA-NcPROF was almost impossible to be purified from hemolymph (Figure S2) although a portion of bx-PA-NcPROF was secreted into hemolymph compared with bx-PA-NcPROF-GP64TM (Figure 2A). However, when 0.5% (*w/v*) nonyl phenoxypolyethoxylethanol (NP-40) was added, both bx-PA-NcPROF and PA-NcPROF could be purified from hemolymph and showed positive bands at 30 kDa and 25 kDa, respectively (Figure S2), suggesting that bx-PA-NcPROF and PA-NcPROF were aggregated abnormally and the aggregates were dissolved with NP-40.

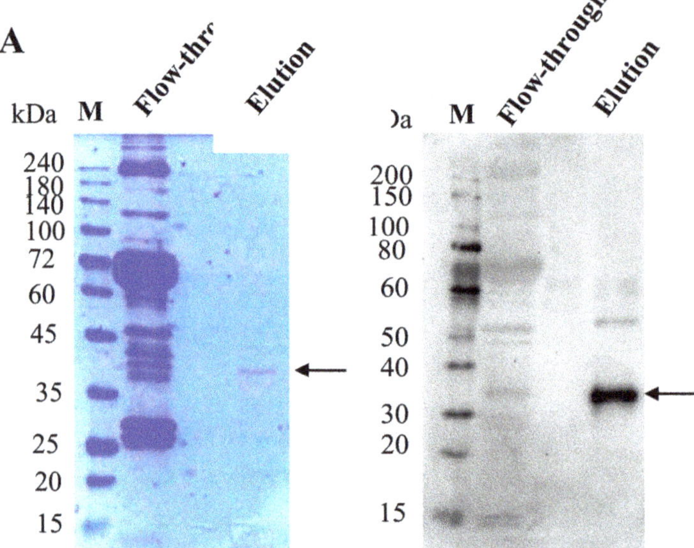

Figure 4. (**A**) Sodium dodecyl sulfate-polyacrylamide gel electrophoresis (SDS-PAGE) and western blot under reducing conditions. (**B**) SDS-PAGE of purified bx-PA-NcPROF-GP64TM from hemolymph using anti-PA tag affinity chromatography. M indicates the molecular weight marker; arrows indicate the recombinant bx-PA-NcPROF-GP64TM band.

3.2. Binding Assay of bx-NcPROF-GP64TM with mTLR11

Recently, it was reported that TLR is responsible for parasite recognition and the induction of cytokines in *T. gondii* [26]. NcPROF induces limited protection and the T-cell response in mice [13]. In this study, NcPROF expressed in silkworms was investigated to determine whether it binds to recombinant mTLR11. The binding of purified bx-NcPROF-GP64TM to mTLR11 was not clearly observed compared with that of the negative control. Because bx-PA-NcPROF-GP64TM formed multimers under nonreducing conditions, DTT was then added to purified bx-PA-NcPROF-GP64TM before ELISA (Figure 5A,B). As shown in Figure 5C, the specific binding was enhanced by an increment of almost 72.7% after DTT treatment and simultaneously increased the binding ability when different concentrations (100, 300, 500 ng/well) were challenged. Hence, it suggests that bx-PA-NcPROF-GP64TM expressed in silkworm larvae is functional. These results indicate that multimer formation might prevent bx-PA-NcPROF-GP64TM from binding to mTLR11. Similar to the results of a previous study, Hedhli et al. [27], showed that PROF of *T. gondii* secreted into *Drosophila* S2 cell broth with the signal sequence of insect binding immunoglobulin protein (BiP) enhanced the cellular and humoral responses in mice after DTT treatment. These results indicate that bx-PA-NcPROF-GP64TM was secreted through the secretory pathway in the host and irregular disulfide bonds were formed in the endoplasmic reticulum. In nature, NcPROF does not have the signal peptide at its N-terminus and does not enter the endoplasmic reticulum [20,28].

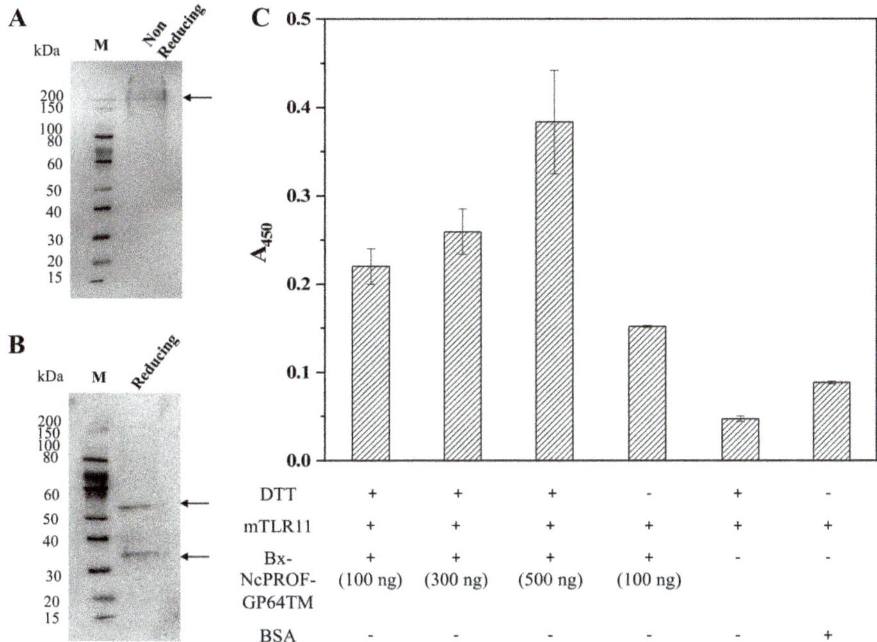

Figure 5. Western blots under reducing conditions (**A**) and nonreducing conditions (**B**) of purified bx-PA-NcPROF-GP64TM. (**C**) enzyme-linked immunosorbent assay (ELISA) analysis of the binding of purified of bx-NcPROF-GP64TM to recombinant mTLR11 Fc chimera. Purified bx-NcPROF-GP64TM (100 ng, 300 ng, 500 ng) with the 1 mM dithiothreitol (DTT) treatment and 100 ng of bx-NcPROF-GP64T without the 1 mM DTT treatment were assayed. Bovine serum albumin (BSA) was used as a negative control. The absorbance was detected at 450 nm and presented as means ±0.001~0.05 standard deviation with coefficient of variance ±0.7%~15%.

3.3. Morphology of bx-PA-NcPROF-GP64TM Nanoparticles

The morphology of purified bx-PA-NcPROF-GP64TM was observed using TEM. Nanoparticles with a diameter of 30 nm were observed in purified bx-PA-NcPROF-GP64TM (Figure 6A). To further analyze the purified bx-PA-NcPROF-GP64TM, immuno-TEM was conducted. The recombinant proteins were probed with rat anti-PA tag (NZ-1) and goat anti-rat IgG (H+L) conjugated with 12-nm gold beads. As shown in Figure 6B, bx-PA-NcPROF-GP64TM demonstrated a spherical morphology with a diameter of approximately 30 nm. These results indicate that bx-PA-NcPROF-GP64TM was secreted as nanoparticles (Figure 3).

Vesicular stomatitis virus G glycoprotein (VSV-G) could be recovered from the culture supernatant when VSV-G alone was expressed in mammalian cells [24,25]. Similarly, the S protein of severe acute respiratory syndrome-associated coronavirus was incorporated into exosomes when its transmembrane and cytoplasmic domains were replaced with those of VSV-G [29]. Extracellular vesicles containing virus-encoded glycoprotein were secreted into the culture supernatant when mammalian cells were infected with the modified *Vaccinia* virus Ankara strain [30]. Additionally, glycoprotein 64 (AcGP64) of *Autographa californica* multiple nucleopolyhedrovirus was secreted into the culture medium when AcGP64 was expressed in mammalian cells [31]. In this study, it is possible that the bx-PA-NcPROF-GP64TM protein formed nanoparticles when it was expressed in silkworm larvae and Bm5 cells. In our previous paper, antigens of *N. caninum* fused with the transmembrane and cytoplasmic domains of BmGP64 were displayed on the surface of BmNPV particles, but few of these fusion proteins were secreted into silkworm hemolymph [9]. These studies suggest that the secretion of recombinant fusion proteins with the transmembrane and cytoplasmic domains of BmGP64 depends on the properties of recombinant proteins of interest. Currently, nanoparticles were only observed in purified bx-PA-NcPROF-GP64TM but not in the other two constructs. Additionally, when purified bx-PA-NcPROF-GP64TM was treated with 1% TritonX-100, some of this protein moved to the fraction at low sucrose concentration during sucrose density gradient centrifugation (Figure S3), implying that this fusion protein forms nanoparticles. We are investigating whether these nanoparticles are aggregates or not and how these nanoparticles form.

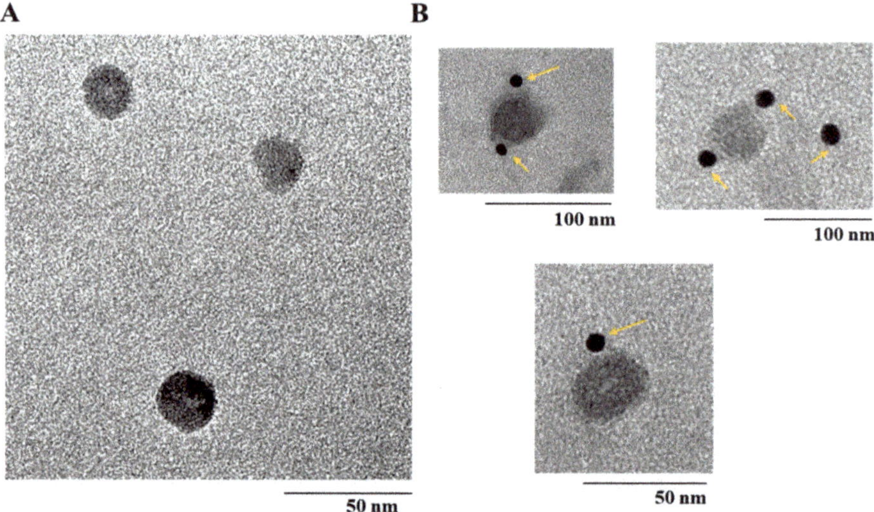

Figure 6. (**A**) Transmission electron microscope (TEM) image of bx-PA-NcPROF-GP64TM. (**B**) Immuno-TEM using gold-labeled antibody. The arrows in (**B**) anti-PA tag-conjugated-gold nanoparticles, respectively.

Apicomplexan profilin, including NcPROF, is a promising protein target with an adjuvant activity as a vaccine candidate [10]. Previous research from Gause et al. [32], stated that delivery of the subunit antigen using particle-based delivery systems can lead to significant improvement in immunogenicity because these systems have now been enhanced by engineering through the physiochemical properties of the particle to promote an immune response. Hence, it was advantageous to purify bx-PA-NcPROF-GP64TM because it naturally forms unique functional nanoparticles without modification through its physiochemical properties to deliver subunit antigens through a particle-based delivery system. Antigen-displaying nanoparticles have been normally prepared by the co-expression of viral structural proteins and antigens or chemical conjugation and genetic fusion of antigens to virus-like particles [33,34]. In this study, NcPROF-displaying nanoparticles were prepared in insect cells and silkworm larvae only by expression of bx-PA-NcPROF-GP64TM, which has the bombyxin signal sequence and transmembrane and cytoplasmic domains of BmGP64 at its N-terminus and C-terminus, respectively.

4. Conclusions

bx-PA-NcPROF-GP64TM was secreted partly into the silkworm larval hemolymph and culture broth of Bm5 cells, although this protein has the transmembrane domain of BmGP64. NcPROF was purified from silkworm hemolymph as a single band only when the bombyxin signal sequence and transmembrane and cytoplasmic domains of BmGP64 were fused with the N- and C-termini, respectively. The purified bx-PA-NcPROF-GP64TM formed unique nanoparticles and was bound to mTLR11.

Supplementary Materials: Supplementary data associated with this article can be found at http://www.mdpi.com/2079-4991/9/4/593/s1. Figure S1: Analysis of *N*-Glycan attached to crude of hemolymph NcPROF (bx-PA-NcPROF-GP64TM, bx-PA-NcPROF, PA-NcPROF). Western blotting with anti-PA tag antibody deglycosylated with PNGase F (incubation at 37 °C for overnight) in denature condition. Arrows show the molecular weight of NcPROF (bx-PA-NcPROF-GP64TM, bx-PA-NcPROF, PA-NcPROF). Figure S2: Western blots of purified bx-PA-NcPROF (A) and PA-NcPROF (B) from hemolymph with and without treatment with 0.5% of *w/v* Np40 using anti-PA tag affinity chromatography. M indicates molecular weight marker; arrows show the recombinant bx-PA-NcPROF and PA-NcPROF; Figure S3: Western blotting analysis of incorporation level of bx-PA-NcPROF-GP64TM, bx-PA-NcPROF and PA-NcPROF with and without treatment of 1% Triton X-100.

Author Contributions: H.S. performed the whole experiments and wrote the first draft of the MS. R.H. and J.X. provided a helpful discussion and helped H.S. to carry out some biological in vitro experiments. T.K., J.X. and E.Y.P. designed and wrote the critical discussion, and prepared the final version of the MS. E.Y.P. supervised the experiments. All authors have given their approval to the final version of the manuscript.

Funding: This project was financially supported by JSPS KAKENHI Grant-in-Aid for Scientific Research (A) (Grant No. 16H02544).

Acknowledgments: We extend our deep appreciation to Jeawook Lee and Kenshin Takemura for their help and guidance for acquiring the TEM and immuno-TEM images.

Conflicts of Interest: The authors declare no conflict of interest.

Abbreviations

BmNPV	*Bombyx mori* nucleopolyhedrovirus
bx	Bombyxin signal peptide from *Bombyx mori*
PA	Peptide tag (GVAMPGAEDDVV)
bx-PA-NcPROF	PA-NCPROF fused with bx
bx-PA-NcPROF-GP64TM	bx-PA-NcPROF with additional C-terminal GP64TM
GP64TM	C-terminal transmembrane and cytoplasmic domains of GP64 from BmNPV
mTLR11	Recombinant mouse TLR11
NcPROF	Neospora caninum profilin
PA-NcPROF	N-terminal PA tag with *N. caninum* profilin
TLR11	Toll-like receptor 11

References

1. Buxton, D.; McAllister, M.M.; Dubey, J.P. The comparative pathogenesis of neosporosis. *Trends Parasitol.* **2002**, *18*, 546–552. [CrossRef] [PubMed]

2. Dubey, J.P.; Schares, G.; Ortega-Mora, L.M. Epidemiology and control of neosporosis and *Neospora caninum*. *Clin. Microbiol. Rev.* **2007**, *20*, 323–367. [CrossRef] [PubMed]

3. Dubey, J.P.; Schares, G. Neosporosis in animals-the last five years. *Vet. Parasitol.* **2011**, *180*, 90–108. [CrossRef]

4. Innes, E.A.; Andrianarivo, A.G.; Björkman, C.; Williams, D.J.L.; Conrad, P.A. Immune responses to *Neospora caninum* and prospects for vaccination. *Trends Parasitol.* **2002**, *18*, 497–504. [CrossRef]

5. Donahoe, S.L.; Lindsay, S.A.; Krockenberger, M.; Phalen, D.; Šlapeta, J. A review of neosporosis and pathologic findings of *Neospora caninum* infection in wildlife. *Int. J. Parasitol. Parasit. Wildl.* **2015**, *4*, 216–238. [CrossRef] [PubMed]

6. Reichel, M.P.; Ellis, J.T. *Neospora caninum*-How close are we to development of an efficacious vaccine that prevents abortion in cattle? *Int. J. Parasitol.* **2009**, *39*, 1173–1187. [CrossRef] [PubMed]

7. Jin, C.; Yu, L.; Wang, Y.; Hu, S.; Zhang, S. Evaluation of *Neospora caninum* truncated dense granule protein 2 for serodiagnosis by enzyme-linked immunosorbent assay in dogs. *Exp. Parasitol.* **2015**, *157*, 88–91. [CrossRef]

8. Lv, Q.; Xing, S.; Gong, P.; Chang, L.; Bian, Z.; Wang, L.; Zhang, X.; Li, J. A 78 kDa host cell invasion protein of *Neospora caninum* as a potential vaccine candidate. *Exp. Parasitol.* **2015**, *148*, 56–65. [CrossRef] [PubMed]

9. Kato, T.; Otsuki, T.; Yoshimoto, M.; Itagaki, K.; Kohsaka, T.; Matsumoto, Y.; Ike, K.; Park, E.Y. Bombyx mori nucleopolyhedrovirus displaying *Neospora caninum* antigens as a vaccine candidate against *N. caninum* infection in mice. *Mol. Biotechnol.* **2015**, *57*, 145–154. [CrossRef] [PubMed]

10. Mansilla, F.C.; Capozzo, A.V. Apicomplexan profilin in vaccine development applied to bovine neosporosis. *Exp. Parasitol.* **2017**, *183*, 64–68. [CrossRef]

11. Mansilla, F.C.; Quintana, M.E.; Cardoso, N.P.; Capozzo, A.V. Fusion of foreign T-cell epitopes and addition of TLR agonists enhance immunity against *Neospora caninum* profilin in cattle. *Parasit. Immunol.* **2016**, *38*, 663–669. [CrossRef]

12. Jenkins, M.C.; Tuo, W.; Feng, X.; Cao, L.; Murphy, C.; Fetterer, R. *Neospora caninum*: cloning and expression of a gene coding for cytokine-inducing profilin. *Exp. Parasitol.* **2010**, *125*, 357–362. [CrossRef]

13. Mansilla, F.C.; Quintana, M.E.; Langellotti, C.; Wilda, M.; Martinez, A.; Fonzo, A.; Moore, D.P.; Cardosa, N.; Capozzo, A.V. Immunization with *Neospora caninum* profilin induces limited protection and a regulatory T-cell response in mice. *Exp. Parasitol.* **2016**, *160*, 1–10. [CrossRef] [PubMed]

14. Innes, E.A.; Wright, S.; Bartley, P.; Maley, S.; Macaldowie, C.; Esteban-Redondo, I.; Buxton, D. The host-parasite relationship in bovine neosporosis. *Vet. Immunol. Immunopathol.* **2005**, *108*, 29–36. [CrossRef]

15. Motohashi, T.; Shimojima, T.; Fukagawa, T.; Maenaka, K.; Park, E.Y. Efficient large-scale protein production of larvea and pupae of silkworm by *Bombyx mori* nuclear polyhedrosis virus bacmid system. *Biochem. Biophys. Res. Commun.* **2005**, *326*, 564–569. [CrossRef] [PubMed]

16. Kato, T.; Sugioka, S.; Itagaki, K.; Park, E.Y. Gene transduction in mammalian cells using *Bombyx mori* nucleopolyhedrovirus assisted by glycoprotein 64 of *Autographa californica* multiple nucleopolyhedrovirus. *Sci. Rep.* **2016**, *6*, 1–9. [CrossRef] [PubMed]

17. Otsuki, T.; Dong, J.; Kato, T.; Park, E.Y. Expression, purification and antigenicity of *Neospora caninum*-antigens using silkworm larvae targeting for subunit vaccines. *Vet. Parasitol.* **2013**, *192*, 284–287. [CrossRef] [PubMed]

18. Southey, B.R.; Sweedler, J.V.; Rodriguez-Zas, S.L. Prediction of neuropeptide cleavage sites in insects. *Bioinformatics* **2008**, *24*, 815–825. [CrossRef] [PubMed]

19. Oomens, A.G.P.; Wertz, G.W. The Baculovirus GP64 protein mediates highly stable infectivity of a human respiratory syncytial virus lacking its homologous transmembrane glycoproteins. *J. Virol.* **2004**, *78*, 124–135. [CrossRef]

20. Plattner, F.; Yarovinsky, F.; Romero, S.; Didry, D.; Carlier, M.F.; Sher, A.; Soldati-Favre, D. *Toxoplasma* profilin is essential for host cell invasion and TLR11-dependent induction of an interleukin-12 response. *Cell Host Microb.* **2008**, *3*, 77–87. [CrossRef]

21. Eichenberger, R.M.; Ramakrishnan, C.; Russo, G.; Deplazes, P.; Hehl, A.B. Genome-wide analysis of gene expression and protein secretion of *Babesia canis* during virulent infection identifies potential pathogenicity factors. *Sci. Rep.* **2017**, *7*, 3357. [CrossRef]

22. Graur, D. The evolution of electrophoretic mobility of proteins. *J. Theor. Biol.* **1986**, *118*, 443–469. [CrossRef]

23. Shi, Q.; Jackowski, G. One-Dimensional Polyacrylamide Gel Electrophoresis. In *Gel Electrophoresis of Proteins. A Practical Approach*, 3rd ed.; Hames, B.D., Ed.; Oxford University Press: New York, NY, USA, 1998; pp. 31–33.

24. Abe, A.; Miyanohara, A.; Friedmann, T. Enhanced gene transfer with fusogenic liposomes containing vesicular stomitis virus G glycoproattein. *J. Virol.* **1998**, *72*, 6159–6163.

25. Okimoto, T.; Friedmann, T.; Miyanohara, A. VSV-G envelope glycoprotein forms complexes with plasmid DNA and MLV retrovirus-like particles in cell free conditions and enhances DNA transfection. *Mol. Ther.* **2001**, *4*, 232–238. [CrossRef]

26. Yarovinsky, F. Innate immunity to *Toxoplasma gondii* infection. *Nat. Rev. Immunol.* **2014**, *14*, 109–121. [CrossRef]

27. Hedhli, D.; Moiré, N.; Akbar, H.; Laurent, F.; Héraut, B.; Dimier-Poisson, I.; Mévélec, M.N. The antigen-specific response to *Toxoplasma gondii* profilin, a TLR11/12 ligand, depends on its intrinsic adjuvant properties. *Med. Microbiol. Immunol.* **2016**, *205*, 345–352. [CrossRef]

28. Hiasa, J.; Nishimura, M.; Itamoto, K.; Xuan, X.; Inokuma, H.; Nishikawa, Y. Enzyme-linked immunosorbent assays based on *Neospora caninum* dense granule protein 7 and profilin for estimating the stage of neosporosis. *Clin. Vaccine Immunol.* **2012**, *19*, 411–417. [CrossRef]

29. Kuate, S.; Cinatl, J.; Doerr, H.W.; Uberla, K. Exosomal vaccines containing the S protein of the SARS coronavirus induce high levels of neutralizing antibodies. *Virology* **2007**, *362*, 26–37. [CrossRef]

30. Spehner, D.; Drillien, R. Extracellular vesicles containing virus-encoded membrane protein are a byproduct of infection with modified vaccinia virus Ankara. *Virus Res.* **2008**, *137*, 129–136. [CrossRef]

31. Guibinga, G.H.; Song, S.; Loring, J.; Friedmann, T. Characterization of the gene delivery properties of boculoviral-based virosomal vectors. *J. Virol. Methods* **2008**, *148*, 277–282. [CrossRef]

32. Gause, K.T.; Wheatley, A.K.; Cui, J.; Tan, Y.; Kent, S.J.; Caruso, F. Immunological principles guiding the rational design of particles for vaccine delivery. *ACS Nano* **2017**, *11*, 54–68. [CrossRef]
33. Koudelka, K.J.; Pitek, A.S.; Manchester, M.; Steinmetz, N.F. Virus-based nanoparticles as versatile nanomachines. *Annu. Rev. Virol.* **2015**, *2*, 379–401. [CrossRef]
34. Yan, D.; Wei, Y.Q.; Guo, H.C.; Sun, S.Q. The application of virus-like particles as vaccines and biological vehicles. *Appl. Microbiol. Biotechnol.* **2015**, *99*, 10415–10432. [CrossRef]

Article

Hydrophobization of Tobacco Mosaic Virus to Control the Mineralization of Organic Templates

Petia Atanasova [1,*], Vladimir Atanasov [2], Lisa Wittum [3], Alexander Southan [4], Eunjin Choi [1], Christina Wege [3], Jochen Kerres [2], Sabine Eiben [3] and Joachim Bill [1]

[1] Institute for Materials Science, University of Stuttgart, Heisenbergstr. 3, 70569 Stuttgart, Germany; choi@is.mpg.de (E.C.); bill@imw.uni-stuttgart.de (J.B.)

[2] Institute of Chemical Process Engineering, University of Stuttgart, Böblinger Straße 78, 70199 Stuttgart, Germany; vladimir.atanasov@icvt.uni-stuttgart.de (V.A.); jochen.kerres@icvt.uni-stuttgart.de (J.K.)

[3] Institute of Biomaterials and Biological Systems, University of Stuttgart, Pfaffenwaldring 57, 70569 Stuttgart, Germany; lisawittum1@yahoo.de (L.W.); christina.wege@bio.uni-stuttgart.de (C.W.); sabine.eiben@gmx.de (S.E.)

[4] Institute of Interfacial Process Engineering and Plasma Technology, University of Stuttgart, Pfaffenwaldring 31, 70569 Stuttgart, Germany; alexander.southan@igvp.uni-stuttgart.de

[*] Correspondence: atanasova@imw.uni-stuttgart.de

Received: 24 April 2019; Accepted: 20 May 2019; Published: 24 May 2019

Abstract: The robust, anisotropic tobacco mosaic virus (TMV) provides a monodisperse particle size and defined surface chemistry. Owing to these properties, it became an excellent bio-template for the synthesis of diverse nanostructured organic/inorganic functional materials. For selective mineralization of the bio-template, specific functional groups were introduced by means of different genetically encoded amino acids or peptide sequences into the polar virus surface. An alternative approach for TMV surface functionalization is chemical coupling of organic molecules. To achieve mineralization control in this work, we developed a synthetic strategy to manipulate the surface hydrophilicity of the virus through covalent coupling of polymer molecules. Three different types of polymers, namely the perfluorinated (poly(pentafluorostyrene) (PFS)), the thermo-responsive poly(propylene glycol) acrylate (PPGA), and the block-copolymer polyethylene-block-poly(ethylene glycol) were examined. We have demonstrated that covalent attachment of hydrophobic polymer molecules with proper features retains the integrity of the virus structure. In addition, it was found that the degree of the virus hydrophobicity, examined via a ZnS mineralization test, could be tuned by the polymer properties.

Keywords: tobacco mosaic virus; ZnS; bio/inorganic hybrid materials; hydrophobization; polymer coupling

---

1. Introduction

In the process of fast-growing development of new nanostructured functional materials, a huge variety of organic materials, in particular biological objects, have been used to synthesize organic/inorganic hybrids with desired properties for nanotechnological applications [1–4]. Among them, tobacco mosaic virus (TMV) is a robust tube-like plant virus, harmless for humans and animals and can be produced in scalable amounts in a green house. It has been utilized intensively as a bio-template due to its anisotropic structure, high uniformity in size and shape and defined surface chemistry. Applying 'bottom-up' approaches such as evaporative self-assembly of the capsids by convective assembly and controlled evaporation methods, homogeneous virus monolayers, aligned TMV stripes and nanowires have been produced [5–8]. TMV has been used as a template to deposit various inorganic materials (metals and semiconductors) on the exterior or the interior surface of the capsid [9–11]. The virus itself and TMV-based hybrids have found application in construction of functional devices like field-effect transistors (FETs) [12] and sensors [13–16], in Li-ion battery [17], etc. Electroconductive materials have been generated making use of the electrostatic interactions between TMV and the polymers polyaniline and polypyrrole, which enabled in-situ polymerization of the polymers on the TMV surface [18]. Most of these studies have been performed using either wild type (wt) TMV, cysteine mutants within the first 3 amino acids of the coat protein (CP) or a C-terminal lysine mutant. An additional number of TMV CP mutants exist [19–21], however especially those changing the surface charge interfere with the natural assembly process [22], forming nanotubes without incorporation of stabilizing RNA. These nanotubes vary greatly in length and are often liable to disassembly upon small changes in buffer conditions making them unavailable for mineralization studies. Although there has been a quite successful attempt to stabilize these RNA-free virus-like particles by introducing an inter CP disulfide bridge in the inner channel of the nanotubes and thus opening up the possibility to use bacterially expressed CP mutants for templating [23,24].

Another strategy to increase the possible surface properties of TMV is to use the existing stable mutants for chemical modification by coupling, which has been done to add mineralization inducing peptides, chemicals for magnetic resonance imaging, whole enzymes and polymers [25]. Recently, a DNA-controlled "stop-and-go" strategy was established to assemble two distinct, selectively addressable CPs variants with RNA into artificial TMV nanotubes with highly defined longitudinal subdomains [26]. The presence of two domains consisting of CPs with different functional groups on one artificial TMV nanoparticle [27] gives the opportunity of using, e.g., "click reactions" to specifically couple organic molecules to only one part of the particle and thereby to synthesize Janus-type TMV particles with a hydrophilic and a hydrophobic portion. The amphiphilicity of classical Janus particles provides unique chemical and physical properties, not accessible for their homogeneous counterparts. Owing to their amphiphilic, magnetic, catalytic or optical properties, they have found numerous applications in fields like drug delivery and catalysis, as surfactants and building blocks for complex 3D nanostructures, for water-repellent coatings, etc [28]. Furthermore, such particles provide additional perspectives towards exploring the influence of the genetic modification of TMV CPs on the nucleation and growth of inorganic materials within one artificial TMV particle, i.e., on the guidance of mineralization reactions. While TMV CPs can be engineered genetically or modified chemically to introduce different hydrophilic functionalities on the virus surface, TMV hydrophobization is a challenging task. An important requirement by choosing viable reaction conditions is the prevention of the hydrophilic complex protein structure from disassembly in the presence of a hydrophobic surrounding. In this regard, A. J. Patil et al. have modified the surface of cowpea mosaic virus (CPMV) with the anionic polymer-surfactant poly(ethylene glycol) 4-nonylphenyl 3-sulfopropyl ether through electrostatically directed assembly [29]. However, electrostatic interactions do not allow control over a local specific functionalization of the virus surface, which in contrast can be achieved via covalent coupling of organic molecules. Another important point concerns the degree of virus hydrophobization required to suppress subsequent mineralization, if partially coated particles are sought after. Therefore, in this work we used wt-TMV and a thiol-displaying TMV-Cys mutant [27] to examine different synthesis strategies and to establish an approach for covalent coupling of polymer molecules to the virus surface, with the aim to produce stably functionalized viruses with sufficient hydrophobicity to suppress mineral deposition on the viral protein coat. For the bioconjugation reactions, the highly hydrophobic perfluorinated (poly(pentafluorostyrene) (PFS)), the thermo-responsive poly(propylene glycol) acrylate (PPGA), which becomes hydrophobic by increasing the temperature and the block-copolymer polyethylene-block-poly(ethylene glycol) ((PE)-b-(PEG)) with a well distinguished hydrophobic part were used. We demonstrate that the integrity of the virus particles can be preserved after covalent attachment of hydrophobic polymer molecules, compatible with the structure of the virus. The achieved hydrophobicity was verified via mineralization reaction applied to the modified virus surface with ZnS, in order to examine if the degree of virus hydrophobicity can be controlled by the choice of the conjugated polymer.

2. Materials and Methods

Materials Disodium hydrogen phosphate, acetone, sodium dodecyl sulfate, tris(hydroxymethyl)aminomethane, acetic acid, sodium carbonate, ammonium persulfate, tetramethylenediamine, sodium chloride and sodium thiosulfate were obtained from Carl Roth (Karlsruhe, Germany). Triphenylphosphine, hydrazine hydrate, PE-b-PEG-OH, N-hydroxyphthalimide (97%), sodium nitrite (97%, ACS), p-toluenesulfonic acid monohydrate (ACS), p-aminoacetophenone (0.15 M in acetonitrile), diisopropyl azodicarboxylate (95%), zinc chloride ($\geq$98%) and sodium sulfide nonahydrate ($\geq$98%) were obtained from Sigma-Aldrich. Poly(propylene glycol) acrylate (M_n = 475 g mol^{-1}), silver nitrate, formaldehyde, Coomassie-Brilliant-Blue R250 and dihydrogen phosphate were obtained from Merck (Darmstadt, Germany). Acrylamide/Bis Solution, 37.5:1 was purchased from Serva (Heidelberg, Germany). Pentaflourostyrene 98% was purchased from ABCR GmbH. All solvents used in this study were in HPLC grade and used as received.

Substrate cleaning: Silicon wafers (100, p-doped polished wafers, Silchem, Germany) were used as substrates. To get a hydrophilic surface, they were thoroughly cleaned prior to use applying the following procedure: 10 min sonication in ultrapure water, 10 min sonication in ethanol/acetone (1:1, v/v), 10 min O_2 plasma treatment (30 W) and 10 min sonication in ultrapure water. After each sonication step, the substrates were washed 10 times with the corresponding solvent and dried under Ar stream. Silicon substrates with reduced hydrophilicity (not plasma treated) were cleaned only successively with ultrapure water, ethanol and acetone.

Preparation and purification of TMV: wt-TMV of the type strain U1 as well as the TMV-Cys mutant [27] were propagated in Nicotiana tabacum 'Samsun' nn plants and purified by PEG precipitation as described before [30]. The virus was stored in 10 mM sodium potassium phosphate (SPP) buffer at pH 7.4 at 4 °C. Buffer-free virus solutions were obtained by resuspension of TMV in ultrapure water after pelleting the virus in a Beckman ultracentrifuge at 35,000 rpm (corresponding to an average g force of 95,800) for 105 min at 4 °C using a 45 Ti rotor and resuspension in ultrapure water.

TMV immobilization: A droplet (3 μL, 0.2 mg mL^{-1}) buffer-free virus solution was spotted onto a substrate surface and incubated for 10 min in a closed chamber. Then, the droplet was removed and the substrate with the immobilized viruses was dried under argon stream.

Synthesis of poly(pentafluorostyrene)(PFS): The synthesis of PFS has already been reported elsewhere [31]. Briefly, emulsion polymerization reaction of pentafluorostyrene in conditions similar to styrene polymerization was used. The molecular weight was optimized to obtain high values (M_n = 52 kDa, M_w = 124 kDa and MWD = 2.4), but allowing good solubility of the polymer in solvents like tetrahydrofuran (THF) and $CHCl_3$.

Click reaction of TMV-Cys with PFS: Due to the limited solubility of PFS in N,N-dimethylacetamide (DMAc), first 1 wt% PFS in DMAc was prepared at 90 °C for 24 h. The solution was filtered through 0.2 μm filter, and silicon substrates with immobilized viruses were let to react with the filtered polymer solution for 30 min at room temperature. After the treatment, the substrates were washed with DMAc and dried. For reaction in tetrahydrofuran (THF), silicon substrates with immobilized TMV-Cys particles were placed in a vessel with 1 wt% PFS in THF/trimethylamine (TEA) solution at room temperature for 2 h. Then, the substrates were thoroughly washed with THF and dried.

Synthesis of TMV-Cys-poly(propylene glycol) acrylate (PPGA): All solutions were stored in an ice-bath for at least 30 min to prevent warming up of the PPGA. For coupling, 1–4 mg/mL of the virus solutions in 10 mM SPP buffer were incubated with the same volume of PPGA (M_n = 475) diluted in 10 mM SPP buffer to give a molecular ratio from 1:1000 to 1:10,000 of CP to PPGA. The solution was incubated in a cooled shaker at 4 °C and 300 rpm for 72 h. Before loading of the functionalized TMV-Cys on a 16/60 Sephacryl S-500 column, bulk PPGA was removed by precipitation of the virus by ultracentrifugation at 34,000 rpm for 2 h at 4°C in a pre-cooled 45 Ti rotor. As sample and running buffer 50 mM SPP containing 150 mM NaCl was used. Purification was performed on an ÄKTApurifier system in a cooling cabinet at a flowrate of 0.5 mL/min, and 1 mL fractions were collected after 75 min until the end of the purification at 250 min. The fractions containing TMV were determined by UV-Vis analysis. Fractions showing an absorbance ratio of 260 nm and 280 nm of 1.19 were pooled and concentrated using an Amicon® Ultra 4 mL centrifugal filter unit with 50,000 nominal molecular weight limit (NMWL) cut-off (Merck, Darmstadt, Germany).

Synthesis of PE-b-PEG phthalimide (**2**): A three-neck round flask was flame-dried under nitrogen stream. Triphenylphosphine (PPh$_3$, 1.377 g, 5.25 mmol), N-hydroxyphthalimide (0.895 g, 5.5 mmol), PE-b-PEG-OH (**1**) (11.25 g, 5 mmol, Mw = 2250 g/mol) and anhydrous CH$_2$Cl$_2$ (50 mL) were placed in the flask and stirred under inert atmosphere. After 15 min of stirring at room temperature, diisopropyl azodicarboxylate (DIAD, 1.08 mL, 5.5 mmol) was added in small portions (0.15–0.2 mL) into the reaction mixture. The obtained orange color of the solution was allowed to fade prior to adding the next portion DIAD. The solution was left to react under stirring in inert atmosphere at room temperature for 12 h. Then, the reaction mixture was precipitated in diethyl ether (Et$_2$O) (1 L) under vigorous stirring in an ice bath for 45 min. The white precipitate was separated from the solvent via suction filtration. The collected product was washed several times with Et$_2$O. To achieve high conversion, the reaction was repeated by using the product as a starting reagent. The obtained PE-b-PEG phthalimide (**2**) was precipitated. To improve its purity, it was dissolved in small amount of CH$_2$Cl$_2$, precipitated in Et$_2$O twice and dried under vacuum for 3.5 h. Yield = 91.6 %. FT-IR (cm^{-1}) = 3586, 2917, 2849, 1789, 1734, 1640, 1462, 1348, 1324, 1295, 1248, 1187, 1095, 1036, 995, 948, 877, 848, 808, 706, 669, 518. ^{1}H NMR (500 MHz, CHCl$_3$-d, δ ppm) = 0.88 (t, J = 6.94 Hz, 3 H), 1.25 (s, 54 H), 1.57 (quin, J = 6.94 Hz, 2 H), 3.44 (t, J = 6.94 Hz, 2 H), 3.47–3.52 (m, 1H), 3.53–3.71 (m, 180 H), 3.74–3.81 (m, 1 H), 3.82–3.90 (m, 2 H), 4.34–4.42 (m, 2 H), 7.72–7.87 (m, 4 H).

Synthesis of aminooxy PE-b-PEG (**3**): A 100 mL flame-dried two neck round flask equipped with stirrer bar was charged with anhydrous CH$_2$Cl$_2$ (10 mL) under nitrogen. Then, PE-b-PEG phthalimide (**2**) (1.0675 g, 0.446 mmoL) was added and dissolved at room temperature. After 20 min, hydrazine hydrate (58 µL, 0.468 mmoL) was added to the reaction mixture and stirred in inert atmosphere at room temperature for 1 h. The color changed from orange to yellow, and a white precipitate was monitored. Et$_2$O (0.5 L) was added to the reaction mixture, placed in an ice bath, under vigorous stirring and left to stir for 2 h. The filtered solid product was rinsed with Et$_2$O and dried under vacuum. Yield = 48.2%. FT-IR (cm^{-1}) = 2917, 2849, 1462, 1348, 1324, 1297, 1248, 1095, 1040, 995, 948, 846, 720, 532. ^{1}H NMR (500 MHz, CHCl$_3$-d, δ ppm) = 0.87 (t, J = 6.94 Hz, 3 H), 1.25 (s, 55 H), 1.57 (quin, J = 7.09 Hz, 2 H), 3.44 (t, J = 6.62 Hz, 2 H), 3.47–3.53 (m, 1H), 3.53–3.76 (m, 190 H), 3.76–3.82 (m, 2 H), 3.84–3.89 (m, 1 H), 4.00–4.04 (m, 1 H), 4.06–4.41 (m, 1 H), 4.06–4.41 (m, 1 H), 4.41–4.50 (m, 1 H)

Functionalization of wt-TMV: To introduce a ketone residue to the virus surface, diazonium salt was prepared from aqueous p-toluenesulfonic acid monohydrate (800 µL, 0.84 M), aqueous sodium nitrite (800 µL, 0.46 M) and p-aminoacetophenone (1.6 mL, 0.15 M in acetonitrile), mixed in a reaction tube and placed in an ice bath. The color of the solution was converted to yellow while stirring for 1 h. Then, virus solution (1033 µL, 4.84 mg/mL) was prepared separately by mixing wt-TMV (500 µL, 10 mg/mL) in 10 mM SPP buffer (pH 7.4) with borate buffer (533 µL, 68 mM) containing 100 mM NaCl (pH 9). 160 µL (11.97 µmoL) diazonium salt solution were mixed with the obtained virus solution and kept in a water bath in the fridge at 4°C for 2 h. The color of the solution turned from yellow to dark brown. The reaction solution was purified using a PD MiniTrap G-25 column with an exclusion limit of approximately M$_r$ = 5000 applying a gravity protocol. The column was eluted with 100 mM aqueous potassium phosphate buffer (pH 6). After the purification step, 2 mL containing 2.1 mg/mL TMV-ketone (**4**) were obtained. The synthesis of the product (**4**) was followed by SDS-PAGE analysis.

Synthesis of h-TMV (**4**): Aminooxy PE-b-PEG (**3**) (0.1614 g, 71.25 µmoL), the TMV-ketone (**4**) (2 mL, 2.1 mg/mL) and 2 mL 100 mM potassium phosphate buffer (pH 6) were mixed in a reaction tube. The mixture was gently rotated overnight at room temperature and purified with a PD MiniTrap G-25 column. Further purification was obtained by size exclusion chromatography at a flow rate of 0.5 mL min^{-1} using a 16/60 Sephacryl S-500 column and 50 mM SPP containing 150 mM NaCl as running buffer. A Pharmacia Biotech LCC-501 PLUS FPLC system with automatic fraction collector was used at room temperature. Fractions exhibiting 1.19 absorbance ratio of 260 nm and 280 nm, indicative for virus particles composed of 95 w/w % proteins and 5 w/w % RNA, were pooled and precipitated by ultracentrifugation at 34,000 rpm at 4°C in a 45 Ti rotor for 2 h. The functionalized virus particles were then resuspended in ultrapure water. The success of the reaction was proved by SDS-PAGE. The height and integrity of the hydrophobized viruses (**5**) was checked by AFM. In a control experiment, wt-TMV (10 µL, 10 mg/mL), borate buffer (10.7 µL, 68 mM) containing 100 mM NaCl (pH 9), potassium phosphate buffer (160 µL, 100 mM, pH 6) and aminooxy PE-b-PEG (**3**) (3.3 mg, 1.46 µmoL) were mixed and rotated overnight at room temperature. The reaction mixture was purified and characterized according to the procedure used for product (**5**).

Sodiumdodecylsulfate-polyacrylamide-gel electrophoresis (SDS-PAGE): Standard procedures according to Green et al. were applied [32]. Samples containing between 0.2–1 µg protein were heated for 5 min at 95 °C in sample buffer and resolved in 15% discontinuous SDS-PA gels. Fixed gels were stained using a silver staining procedure.

ZnS mineralization: Silicon substrates immobilized with the corresponding virus types were fixed in a holder and mounted perpendicular in the center of a vessel. There, the viruses were incubated in aqueous $ZnCl_2$ precursor solution (30 mL, 100 mM, pH 6.2) for 20 min. Then, equimolar aqueous Na_2S solution (pH 12.5) was added dropwise to the reaction solution applying a peristaltic pump with a constant speed (1 mL min^{-1}) under continuous stirring at room temperature. The addition of Na_2S to the $ZnCl_2$ precursor solution caused a drop in the pH to 3.5, and the reaction pH was maintained until the dropwise addition was stopped. Within the addition of the first Na_2S droplets, a white precipitate was formed in the deposition solution. After finishing the reaction (controlled according to the required reaction time (5 or 30 min)), the substrates were washed thoroughly with ultrapure water and dried under argon stream.

Sample characterization: Atomic force microscopy on a Digital Instruments MultiMode 8 from Bruker with a NanoScope 5 controller operated in tapping mode was used to image the immobilized and the mineralized viruses. Silicon cantilevers and PPP-NCHR-W (Nanosensors) n^{+} doped tips with resistivity 0.01–0.02 Ω cm were used. The virus height of wt-TMV and h-TMV was evaluated with the manufacturer's software Nanoscope. The height of a single virus was averaged over the whole virus length with the software. The average height of 10 virus particles from each virus type was used to obtain the virus height distribution of the bare virus. The virus height of wt-TMV and h-TMV after treatment with ZnS deposition solution for 5 min, 20 min and 30 min was determined applying the same procedure, taking 5 to 10 virus particles of each virus type for the evaluation and comparing only wt-TMV and h-TMV particles mineralized simultaneously in the same deposition solution. ^{1}H-NMR measurements were conducted on Bruker Avance 500 in 500 MHz field and analyzed by ACD/Spectrus Processor. Fourier transform infrared spectroscopy (FT-IR) analysis of the samples was performed at a Bruker FT-IR spectrometer. The samples were investigated in the wavelength range: 500–4000/cm^{-1} and the spectra were analyzed by ACD/Spectrus Processor.

3. Results and Discussion

The wt-TMV particle consists of a genomic RNA helix, embedded into about 2130 identical CP subunits. It has a tube-like structure, which is 300 nm long. The inner diameter of the longitudinal central channel is 4 nm, while the outer nanotube diameter is 18 nm when the capsid is dispersed in solution. It may be flattened to a height of around 14 nm upon immobilization on a hydrophilic silicon substrate surface [12,33]. The viral CPs can be engineered genetically to introduce amino acid residues on the virus surface with certain terminal groups appropriate for bioconjugation reactions. In this study, wt-TMV and a TMV-Cys mutant, presenting a thiol group for coupling on each CP were used to find appropriate synthetic pathways for covalent attachment of organic molecules making the virus surface hydrophobic without destroying the integrity of the capsid, in order to suppress the virus mineralization by inorganic deposits as shown in Scheme 1.

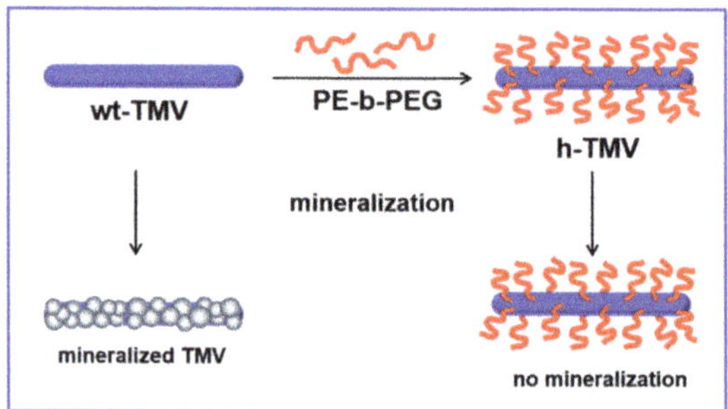

Scheme 1. Schematic representation of tobacco mosaic virus (TMV) functionalization making the virus surface hydrophobic, which restricts subsequent mineralization.

Recently, we reported on selective mineralization of wt-TMV with ZnS at room temperature in aqueous solution and at various reaction pH [10]. $ZnCl_2$ and Na_2S were used as ZnS precursors without the need of any further additives as structure directing or complexing agents. The growth mechanism of the inorganic material on the virus surface was studied comparing the optical properties, band gap (Eg) and particle size of solution-grown ZnS nanoparticles and ZnS nanoparticles mineralized onto the virus template. It was shown that the virus particles played templating function, and heterogeneous nucleation triggered by the virus surface was confirmed. The XRD analysis confirmed formation of ZnS with cubic lattice structure in solution and on the virus surface. Since well-studied, the mineralization of wt-TMV with ZnS applying this reaction was used as a reference in this work and compared to the mineralization behavior of the functionalized virus particles in order to identify, if the change of virus surface hydrophilicity influenced the mineralization process.

3.1. Reaction of TMV with Highly Hydrophobic Polymers

A covalent attachment of a highly hydrophobic polymer such as poly(pentafluorostyrene) (PFS) to the virus surface could provide both specific coupling to the virus surface and high surface hydrophobicity. Therefore, first a click reaction of PFS to a TMV-Cys was examined. Our previous study revealed that the reactivity of the para-fluorine function in the PFS with nucleophiles such as sulfides is enhanced by the reduced electron density of the perfluorinated phenyl ring, allowing the use of relatively mild reaction conditions [31]. The reaction is conducted in *N,N*-dimethylacetamide (DMAc) or tetrahydrofuran/ triethylamine (THF/TEA) at room temperature. For the click reaction, TMV-Cys particles were selected, because each of their CPs provides a single surface-accessible thiol group, introduced by genetic modification of the third CP amino acid from serine to cysteine (S3C) [27]. The bioconjugation was conducted according to the reaction pathway presented in Scheme 2A.

Before starting with the coupling reaction, the stability of TMV-Cys in DMAc and in THF with or without TEA was studied at room temperature. To do this, viruses immobilized on a plasma cleaned silicon substrate (Figure 1a) were incubated in DMAc for 30 min. This led to partial loss of the virus integrity especially visible in Figure 1b, where the viruses became roughened and the virus sharp ends are lost. The mean virus height is reduced from 14.5 nm to around 13 nm. Finally, reaction with PFS in DMAc for the same reaction time led to even further disassembly of the viruses, where the virus coat appeared irregular, with an uneven coverage of the RNA with CPs (Figure 1c).

Scheme 2. Click reaction of TMV-Cys mutant with poly(pentafluorostyrene) (PFS) and poly(propylene glycol) acrylate (PPGA).

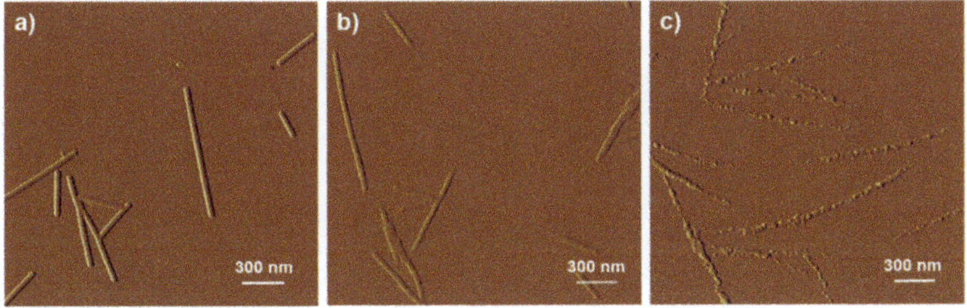

Figure 1. AFM amplitude images of: (**a**) TMV-Cys virus particles, (**b**) TMV-Cys incubated in N,N-dimethylacetamide (DMAc) for 30 min and (**c**) TMV-Cys reacted with poly(pentafluorostyrene) (PFS) in DMAc for 30 min.

In comparison, the virus structure seemed stable in both THF and THF/TEA for a few hours (Figure 2a,b). This is in contradiction to our previous experiments on the stability of TMV dispersed in THF/H$_2$O mixtures, where TMVs appeared to be unstable. The reason for the increased stability in this case might be explained by the immobilization of the viruses on the substrate. Based on the enhanced overall structural stability and the need of an extended reaction time for better educts conversion, the TMV-Cys particles were let to react with PFS in THF/TEA for 2 h at room temperature. The AFM analysis indicated that the capsids stayed intact and retained their integrity after this treatment without change in virus height (Figure 2c). However, subsequent mineralization resulted in deposition of ZnS on the virus surface, and the virus height increased to 16.5 nm (Figure 2d,e). The latter suggests that although stable in this reaction solution, the viruses could not be functionalized with PFS. This may be attributed to the lower reactivity of PFS in THF/TEA compared to that in DMAc, which not only requires longer reaction time but usually also higher temperatures not suitable for TMV.

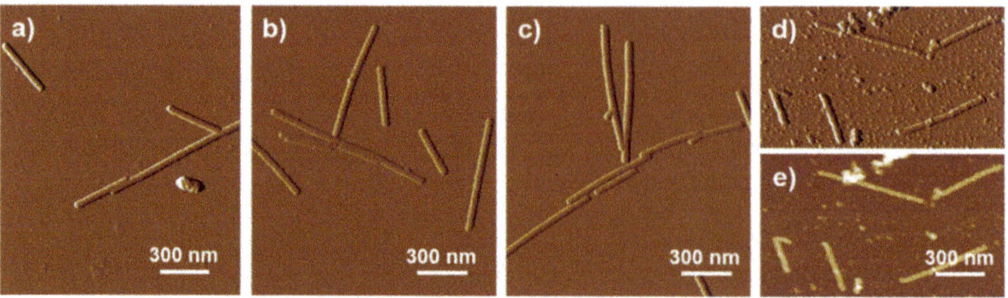

Figure 2. AFM amplitude images of TMV-Cys virus particles treated: (**a**) 1 h in tetrahydrofuran (THF), (**b**) 1 h in THF/TEA, (**c**) 2 h with PFS in THF/TEA at room temperature. (**d**) AFM amplitude image of viruses from (**c**) mineralized 5 min with ZnS and (**e**) AFM height image of (**d**).

3.2. TMV Hydrophobization with Aliphatic Polyethers

As established in the previous section, the highly hydrophobic environment and the aggressive reaction solvent necessary for dissolving the polymer caused strain on the virus structure and subsequent disintegration. As an alternative, so-called thermo-responsive polymers e.g., poly (propylene glycol)methacrylate can be applied [34]. This polymer can change its hydrophobicity with temperature. It is water-soluble at temperatures below ~10 °C and neutral pH, and becomes insoluble at higher temperatures due to conformational changes inducing hydrophobization. This would allow a chemical coupling of the polymer at 4 °C in its hydrophilic form to the virus surface with expectation that the virus will retain its stability at these conditions. Then, a mineralization test at room temperature or at elevated temperatures, when the polymer becomes hydrophobic, will be conducted to examine the obtained hydrophobicity. Due to the higher reactivity of the acrylate group over the methacrylate group as a Michael addition donor, poly(propylene glycol)acrylate (PPGA) (M_n = 475 g mol^{-1}) was chosen for our experiments (Scheme 2B).

To determine the coupling efficiency and specificity, TMV-Cys and wt-TMV were incubated at 4 °C where PPGA is completely soluble (see Supplementary Materials) for 72 h, with PPGA added at different ratios with regard to the number of CPs. SDS-PAGE analysis of the CPs revealed for a ratio of 1000:1 PPGA:CP-Cys only a slight smear above the CP band, which increased substantially at a ratio of 3300:1, but not significantly more at a further excess of PPGA:CP-Cys of 10,000:1 (see Figure 3). The coupling of PPGA to CP-Cys did not result in a defined band due to the polydispersity of PPGA. The absence of any smear above wt-CPs of wt-TMV particles treated in parallel shows on the one hand that the reaction was specific for TMV-Cys, and on the other hand that the TMV structure stayed intact during the reaction, as during TMV particle disassembly an otherwise hidden cysteine residue would become accessible to the reaction with PPGA.

As PPGA was used in high excess, residual unreacted polymer was removed prior to further investigations by a combined ultracentrifugation/fast protein liquid chromatography (UC/FPLC) approach. All purification steps were performed below 5 °C to prevent switching of PPGA into the hydrophobic state. The first ultracentrifugation step was necessary to get rid of the bulk amount of PPGA, which may result in a broadening of the peaks due to overloading size-exclusion chromatography. About 25 % of the TMV initially used for coupling could be obtained in pure form. Figure 4a,b show AFM images of PPGA-functionalized TMV-Cys before and after purification, where residual excess PPGA appears as large spots around the TMV rods in the first image and had disappeared in samples after the cleaning procedure.

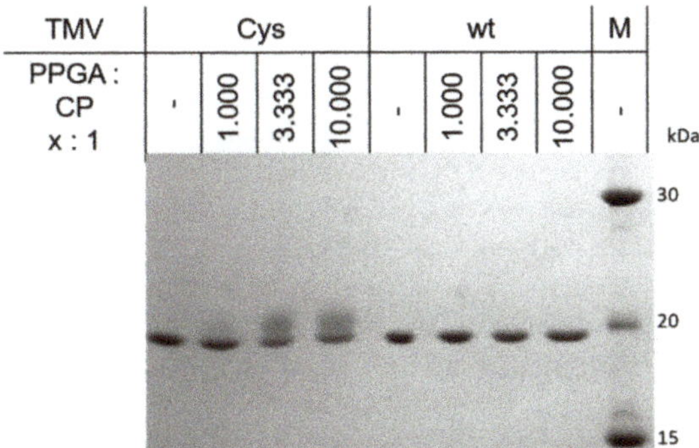

Figure 3. Silver-stained 15%-SDS-PAGE showing the coupling efficiency at different ratios of PPGA to CP using TMV-Cys and no coupling of PPGA to wt-TMV. The product of the conjugation reaction can be seen as a smear above the CP band at about 20 kDa.

After having shown that the thermo-responsive PPGA was successfully coupled to TMV-Cys, the degree of hydrophobicity achieved with PPGA was investigated by subjecting the modified nanotubes to mineralization reaction with ZnS [10]. PPGA-functionalized TMV-Cys viruses were immobilized on a plasma cleaned silicon substrate, which was placed perpendicularly in $ZnCl_2$ aqueous solution in order to avoid attachment of big ZnS agglomerates formed in solution on the substrate surface. The mineralization reaction was allowed to proceed for 5 min, which based on previous experiments is enough to observe differences in the mineralization behavior between untreated TMV-Cys and hydrophobized particles, respectively. The amplitude AFM image in Figure 4c clearly shows that the PPGA-functionalized TMV-Cys surface is covered with small ZnS nanoparticles, and the virus height calculated from the corresponding AFM height images was increased by around 1.6 nm. This leads to the conclusion that, although successfully attached to the virus surface preserving the capsid structure, the thermo-responsive PPGA residues could not provide sufficient hydrophobicity to fully suppress virus mineralization.

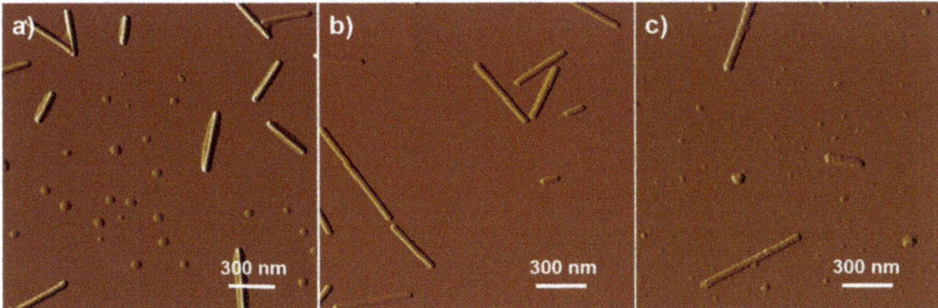

Figure 4. AFM amplitude images of TMV-Cys-PPGA: (**a**) before and (**b**) after purification. (**c**) TMV-Cys-PPGA mineralized with ZnS.

3.3. TMV Hydrophobization with Block-Co-Polymers

Covalent attachment of PPGA to the virus CPs was possible without affecting the overall capsid structure, however, the obtained viruses were not hydrophobic enough to suppress mineralization. A possible reason for this could be the length of the PPGA molecules, and, as obvious from the SDS-PAGE analysis presented in Figure 3, only part of the viral CPs was covalently coupled with PPGA. Another possible reason may be connected to the surface charge of the functionalized viruses. The overall surface charge of organic templates (in our case functionalized TMV) plays an important role for mineralization in solution. The presence of even partial charges like from the free electron pairs of the oxygen in the PPGA polyether can induce mineralization of the PPGA-functionalized viruses. Therefore in the next step, a synthetic pathway developed by Schlick et al. [35]. was used. In their work, the TMV capsid exterior was covered with polyethylene glycol (PEG) to make TMV soluble in various hydrophobic organic solvents and to increase its thermal stability. Although PEG introduced sufficient hydrophobicity on the virus surface to transfer the PEG-TMV conjugate from aqueous solution into organic solvents, the presence of oxygen in the PEG chains would support mineralization of the virus, similar to the case of TMV-Cys-PPGA. Therefore, in order to introduce sufficient hydrophobicity without affecting the integrity of the virus structure, we covalently coupled the block-co-polymer $(PE)_{11}$-$(PEG)_{34}$-OH (M_w 2250 g/moL) to the wt-TMV surface. The non-polar polyethylene (PE) part is expected to ensure sufficient hydrophobicity of the virus surface and to suppress the mineralization of the organic template. The PEG chain, compatible with the virus, should separate the virus from the hydrophobic PE chain. Additionally, the PEG spacer unit is responsible to attach the block-co-polymer to the virus CP [36]. To do this, we used the already reported electrophilic substitution reaction with diazonium salts at the *ortho* position of the phenyl ring of the tyrosine residue (Y139) in TMV CPs [35]. The polymer coupling was done in a three-step procedure as can be seen in Scheme 3.

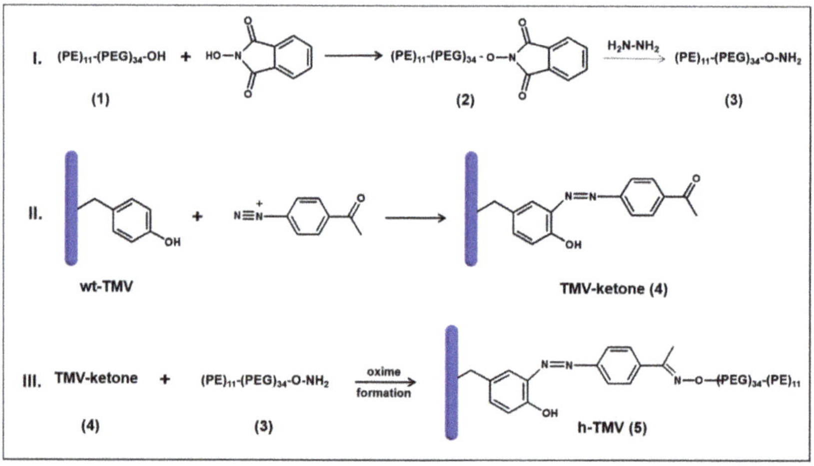

Scheme 3. Synthetic pathway for functionalization of $(PE)_{11}$-$(PEG)_{34}$-OH block-co-polymer **1** and formation of hydrophobized h-TMV.

To this end, the block-co-polymer **1** converted into the alkoxyamine **3** was reacted with the ketone-functionalized wt-TMV **4** to form the ketoxime **5**. In step I, the terminal hydroxyl group of the block-co-polymer **1** was activated with N-hydroxyphtalimide. As a good leaving group, N-hydroxyphtalimide facilitated the subsequent reaction with hydrazine resulting in the formation of alkoxyamine **3**. The success of the reactions was verified by ^{1}H-NMR combined with FT-IR spectroscopy [37]. The appearance of protons with a chemical shift in the low field region around 7.8 ppm and an IR signal in 1730–1790 cm^{-1} corresponding to aromatic and carbonyl residues by product **2** (Figure 5) and their disappearance by product **3** confirmed the attachment and the subsequent replacement of N-hydroxyphtalimide from the polymer.

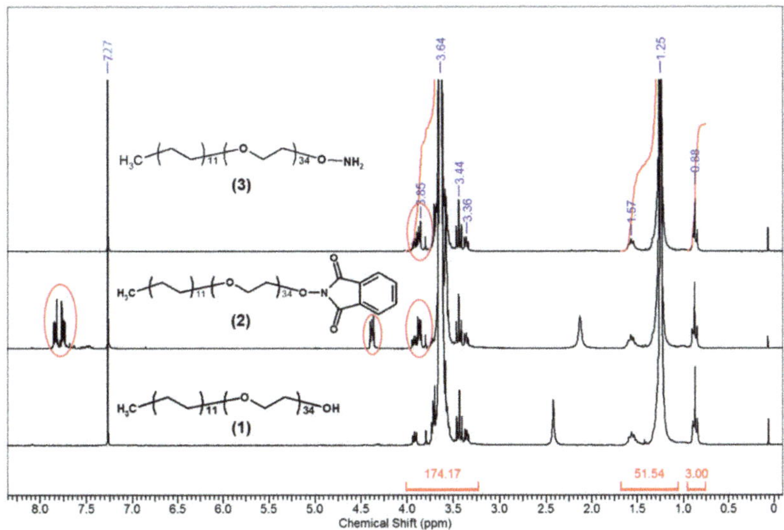

Figure 5. ^{1}H-NMR of **(1)** (PE)$_{11}$-(PEG)$_{34}$-OH block-co-polymer, **(2)** activated with the N-hydroxyphtalimide derivative and **(3)** the corresponding alkoxyamine.

In step II, the virus CPs were functionalized and a terminal ketone group was introduced, which was done through the accessible tyrosine groups of CPs. It was assumed that only one of the four tyrosine residues, present in each TMV CP, should be modified since the other three are hidden between the CPs after assembly to the helical TMV particle structure [35]. The ketone group was included by treatment of TMV with diazonium salt, prepared prior to use from sodium nitrite and p-aminoacetophenone. The virus functionalization was performed at 4 °C, pH 9 for 2 h. PD MiniTrap G-25 columns were used to separate the obtained TMV-ketone **4** from the added in surplus diazonium salt. SDS-PAGE analysis confirmed the CP modification (see Figure S1) while AFM examination showed preserved virus particles. The hydrophobization of wt-TMV (Scheme 3, step III) was done by reaction of 250 eq. aminooxy PE-b-PEG **3** with TMV-ketone **4** in potassium phosphate buffer (pH 6) at room temperature for 15 h. The hydrophobized TMVs (h-TMV) were purified through PD MiniTrap G-25 columns. Then, the h-TMVs were further separated from the not reacted polymer residue via FPLC and concentrated via UC. The shift of the CP band in the SDS-PA gel (Figure S1) to one with higher molecular weight clearly indicated the successful attachment of polymer to the virus CPs compared to the control experiment, where the aminooxy PE-b-PEG **3** was reacted with a not functionalized wt-TMV and no change in the position of wt-TMV CPs was observed. The blurring of the band corresponding to the CPs of h-TMV can be explained by the polydispersity of the block-co-polymer. The AFM height images in Figure 6 confirmed that the h-TMV particles remained intact. The AFM cross-sections have shown that compared to wt-TMV (Figure 6a,c), the polymer attachment on the virus surface led to an increase in virus height of about 3 nm on hydrophilic (O$_2$-plasma treated) (Figure 6b) but also on hydrophobic (not plasma treated) (Figure 6d) silicon substrates.

As stated in previous reports, very polar and protic solvents lead to precipitation of PEG functionalized TMV [36]. Therefore, the stability of the h-TMV in the precursor solution was examined before starting with the mineralization test with ZnS. A silicon substrate with immobilized h-TMV was placed in ZnCl$_2$ aqueous solution. After 30 min stirring at room temperature, the viruses could still be detected as intact particles via AFM (see Figure S2a). Then, a substrate with h-TMV and one substrate with immobilized wt-TMV, used as a reference, were mineralized simultaneously with ZnS for different reaction times. In Figure 7 are shown the corresponding AFM height, amplitude and phase images of wt-TMV and h-TMV after 5 and 30 min treatment. It can be seen that after 5 min reaction time, the surface of wt-TMV became rough due to attachment of ZnS particles, while the surface of h-TMV remained smooth. This is especially visible comparing their AFM amplitude and phase images. Increase of the reaction time to 30 min maintained the trend. The amount of the ZnS particles on wt-TMV surface was increased, which is well detectable also on the AFM height image. Opposed to this, no specific ZnS deposition on h-TMV was observed.

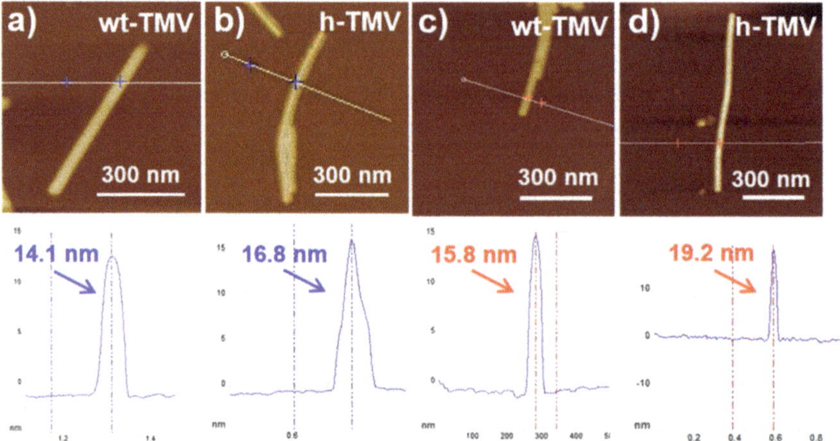

Figure 6. AFM height images with the corresponding cross-sections showing the virus height of: (**a**) wt-TMV and (**b**) h-TMV immobilized on hydrophilic (O_2-plasma treated) silicon substrates. (**c**) wt-TMV and (**d**) h-TMV immobilized on hydrophobic (not plasma treated) silicon substrates.

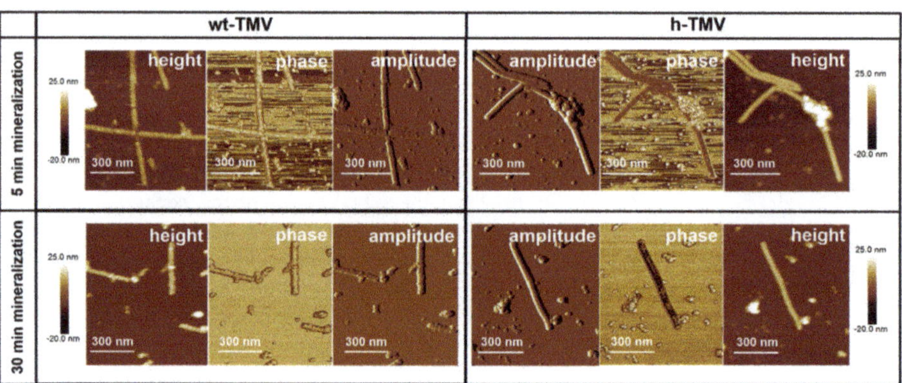

Figure 7. AFM height, amplitude and phase images of wt-TMV and h-TMV mineralized simultaneously with ZnS for 5 and 30 min, respectively.

The observation of a selective deposition of ZnS nanoparticles only on the surface of wt-TMV, but not on the surface of the hydrophobized counterpart was further supported by evaluation and comparison of the virus height of wt-TMV and h-TMV treated with ZnS deposition solution for a certain reaction time (Figure 8). Since small fluctuations in the mineralization conditions might cause deviations in the deposition rate and hence might influence the resulting virus height, for better assessment, the virus heights of wt-TMV and h-TMV mineralized in the same deposition solution and within the same experiment were compared. The chart diagram in Figure 8 shows that in comparison to the hydrophobized viruses, whose surface remain smooth with the reaction time, the wt-TMV particles increase their height linearly with the mineralization time. The latter is a result of selective deposition of ZnS on the virus surface with a constant rate, which is in a good agreement with our previous observations for mineralization of wt-TMV with ZnS applying this reaction [10].

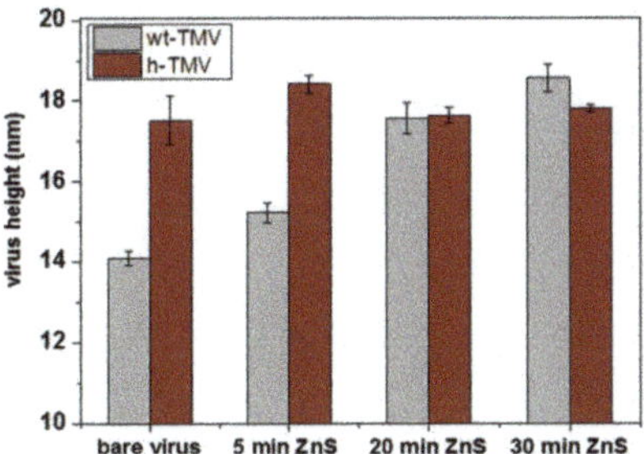

Figure 8. Virus height of not mineralized wt-TMV and h-TMV and the height of both virus types when mineralized with ZnS after 5, 20 and 30 min.

4. Conclusions

In this work, we have demonstrated that the surface hydrophilicity of TMV particles can be manipulated via covalent attachment of polymer molecules. Experiments with highly hydrophobic PFS, the thermo-responsive PPGA, and the block-copolymer PE-b-PEG led to the conclusion that the success of the reaction depends on the polymer features. The use of perfluorinated polymers showed that highly hydrophobic polymers are not appropriate candidates for virus functionalization. Although the virus particles displayed significant stability enhancement in organic solvents when immobilized on solid supports, the reaction failed due to the need of either aggressive organic solvent (DMAc) or elevated reaction temperature both not compatible with the virus. In contrast, the thermo-responsive PPGA polyether, insoluble in water above 8 °C, was successfully attached to the TMV-Cys surface, and also PE-b-PEG was successfully bonded to wt-TMV CPs. As a result, stable and intact functionalized particles were synthesized. The covalent coupling to virus CPs was confirmed via SDS-PAGE analysis. While PPGA only partially coupled to TMV-Cys CPs, complete coupling efficiency was reached with PE-b-PEG and wt-TMV CPs. The height of the hydrophobized with PE-b-PEG virus (h-TMV), detected by AFM, showed a significant increase by around 2.5 nm. The degree of virus hydrophobicity, achieved with both polymers, was examined by mineralization with ZnS. The incomplete coverage of the virus surface with PPGA and the presence of partial charges in the polymer chains could not suppress virus mineralization. However, combination of a virus-compatible PEG block and a pure hydrophobic PE block in PE-b-PEG together with bioconjugation of the polymer to most CPs in TMV led to the formation of intact h-TMV, fully covered with polymer molecules exposing their hydrophobic parts outwards, which completely hindered subsequent mineralization of the virus surface. This synthetic approach could be used for the synthesis of Janus-type TMV particles with a hydrophobic and a hydrophilic part in future experiments, to study the mineralization behavior of TMV mutants within a single artificial TMV particle with an intrinsic hydrophobic reference. Furthermore, the self-organization of pure and mineralized Janus-type TMV particles in complex highly organized structures as well as their potential integration in interfacial catalysis and as drug cargo is envisioned.

Supplementary Materials: The following are available online at http://www.mdpi.com/2079-4991/9/5/800/s1, Figure S1: SDS-PAGE of samples containing: (**1**) wt-TMV, (**2**) TMV-ketone, (**3**) TMV-ketone after purification with a PD MiniTrap G-25 column, (**4**) h-TMV, (**5**) h-TMV after purification with a PD MiniTrap G-25 column and (**6**) control experiment including reaction of wt-TMV with PE-b-PEG. Figure S2: Lower magnification AFM height images showing the stability of the h-TMV particles: (**a**) after 30 min stirring in 100 mM ZnCl$_2$ precursor solution and (**b**) after 5 min mineralization with ZnS.

Author Contributions: Conceptualization, P.A., S.E., V.A. and A.S.; Investigation, P.A., S.E., V.A., L.W. and E.C.; Validation, E.C. and L.W.; Writing—Original Draft, P.A., S.E. and V.A.; Writing—Review & Editing, J.B., C.W., P.A., S.E., V.A., E.C. and A.S.; Supervision, J.B., J.K., P.A. and S.E.

Funding: This research was funded by the priority program SPP 1569 of the DFG (BI 469/19-3, EI 901/1 3), Bundesministerium für Bildung und Forschung (BMBF) on account of "HT Linked" project with FKZ: 03SF0531C, the Carl Zeiss Foundation and the University of Stuttgart within the Projekthaus NanoBioMater.

Conflicts of Interest: The authors declare no conflict of interest. The funders had no role in the design of the study; in the collection, analyses, or interpretation of data; in the writing of the manuscript, or in the decision to publish the results.

References

1. Yan, H.; Park, S.H.; Finkelstein, G.; Reif, J.H.; LaBean, T.H. DNA-templated self-assembly of protein arrays and highly conductive nanowires. *Science* **2003**, *301*, 1882–1884. [CrossRef]
2. Tseng, R.J.; Tsai, C.L.; Ma, L.P.; Ouyang, J.Y. Digital memory device based on tobacco mosaic virus conjugated with nanoparticles. *Nat. Nanotechnol.* **2006**, *1*, 72–77. [CrossRef]
3. Mao, C.B.; Solis, D.J.; Reiss, B.D.; Kottmann, S.T.; Sweeney, R.Y.; Hayhurst, A.; Georgiou, G.; Iverson, B.; Belcher, A.M. Virus-based toolkit for the directed synthesis of magnetic and semiconducting nanowires. *Science* **2004**, *303*, 213–217. [CrossRef]
4. Rothemund, P.W.K. Folding DNA to create nanoscale shapes and patterns. *Nature* **2006**, *440*, 297–302. [CrossRef] [PubMed]
5. Atanasova, P.; Stitz, N.; Sanctis, S.; Maurer, J.H.; Hoffmann, R.C.; Eiben, S.; Jeske, H.; Schneider, J.J.; Bill, J. Genetically improved monolayer-forming tobacco mosaic viruses to generate nanostructured semiconducting bio/inorganic hybrids. *Langmuir* **2015**, *31*, 3897–3903. [CrossRef] [PubMed]
6. Wargacki, S.P.; Pate, B.; Vaia, R.A. Fabrication of 2D ordered films of tobacco mosaic virus (TMV): Processing morphology correlations for convective assembly. *Langmuir* **2008**, *24*, 5439–5444. [CrossRef] [PubMed]
7. Lin, Z. *Evaporative Self-Assembly of Ordered Complex Structures*; World Scientific: Singapore; Hackensack, NJ, USA, 2012; 380p.
8. Lomonossoff, G.P.; Wege, C. Chapter Six—TMV Particles: The Journey From Fundamental Studies to Bionanotechnology Applications. In *Advances in Virus Research*; Palukaitis, P., Roossinck, M.J., Eds.; Academic Press: Cambridge, MA, USA, 2018; pp. 149–176.
9. Fan, X.Z.; Pomerantseva, E.; Gnerlich, M.; Brown, A.; Gerasopoulos, K.; McCarthy, M.; Culver, J.; Ghodssi, R. Tobacco mosaic virus: A biological building block for micro/nano/bio systems. *J. Vac. Sci. Technol. A* **2013**, *31*. [CrossRef]
10. Atanasova, P.; Kim, I.; Chen, B.; Eiben, S.; Bill, J. Controllable virus-directed synthesis of nanostructured hybrids induced by organic/inorganic interactions. *Adv. Biosys.* **2017**, *1*, 1700106. [CrossRef]
11. Bittner, A.M.; Alonso, J.M.; Górzny, M.L.; Wege, C. Nanoscale science and technology with plant viruses and bacteriophages. In *Structure and Physics of Viruses: An Integrated Textbook*; Mateu, M.G., Ed.; Springer Science+Business Media: Dordrecht, The Netherlands, 2013; pp. 667–702.
12. Atanasova, P.; Rothenstein, D.; Schneider, J.J.; Hoffmann, R.C.; Dilfer, S.; Eiben, S.; Wege, C.; Jeske, H.; Bill, J. Virus-templated synthesis of ZnO nanostructures and formation of field-effect transistors. *Adv. Mater.* **2011**, *23*, 4918–4922. [CrossRef]
13. Zang, F.H.; Gerasopoulos, K.; Fan, X.Z.; Brown, A.D.; Culver, J.N.; Ghodssi, R. An electrochemical sensor for selective TNT sensing based on tobacco mosaic virus-like particle binding agents. *Chem. Commun.* **2014**, *50*, 12977–12980. [CrossRef] [PubMed]
14. Bruckman, M.A.; Liu, J.; Koley, G.; Li, Y.; Benicewicz, B.; Niu, Z.W.; Wang, Q.A. Tobacco mosaic virus based thin film sensor for detection of volatile organic compounds. *J. Mater. Chem.* **2010**, *20*, 5715–5719. [CrossRef]
15. Koch, C.; Eber, F.J.; Azucena, C.; Forste, A.; Walheim, S.; Schimmel, T.; Bittner, A.M.; Jeske, H.; Gliemann, H.; Eiben, S.; et al. Novel roles for well-known players: From tobacco mosaic virus pests to enzymatically active assemblies. *Beilstein J. Nanotech.* **2016**, *7*. [CrossRef] [PubMed]
16. Eiben, S.; Koch, C.; Altintoprak, K.; Southan, A.; Tovar, G.; Laschat, S.; Weiss, I.; Wege, C. Plant virus-based materials for biomedical applications: Trends and prospects. *Adv. Drug Delivery Rev.* **2018**. [CrossRef]
17. Chen, X.L.; Gerasopoulos, K.; Guo, J.C.; Brown, A.; Wang, C.S.; Ghodssi, R.; Culver, J.N. Virus-enabled silicon anode for lithium-ion batteries. *ACS Nano* **2010**, *4*, 5366–5372. [CrossRef] [PubMed]

18. Niu, Z.; Liu, J.; Lee, L.A.; Bruckman, M.A.; Zhao, D.; Koley, G.; Wang, Q. Biological templated synthesis of water-soluble conductive polymeric nanowires. *Nano Lett.* **2007**, *7*, 3729–3733. [CrossRef]

19. Turpen, T.H.; Reinl, S.J.; Charoenvit, Y.; Hoffman, S.L.; Fallarme, V.; Grill, L.K. Malarial epitopes expressed on the surface of recombinant tobacco mosaic-virus. *Biotechnology* **1995**, *13*, 53–57. [PubMed]

20. Lee, L.A.; Nguyen, Q.L.; Wu, L.; Horvath, G.; Nelson, R.S.; Wang, Q. Mutant plant viruses with cell binding motifs provide differential adhesion strengths and morphologies. *Biomacromolecules* **2012**, *13*, 422–431. [CrossRef]

21. Lu, B.; Taraporewala, Z.F.; Stubbs, G.; Culver, J.N. Intersubunit interactions allowing a carboxylate mutant coat protein to inhibit tobamovirus disassembly. *Virology* **1998**, *244*, 13–19. [CrossRef] [PubMed]

22. Bendahmane, M.; Koo, M.; Karrer, E.; Beachy, R.N. Display of epitopes on the surface of tobacco mosaic virus: Impact of charge and isoelectric point of the epitope on virus-host interactions. *J. Mol. Biol.* **1999**, *290*, 9–20. [CrossRef]

23. Zhou, K.; Li, F.; Dai, G.; Meng, C.; Wang, Q. Disulfide bond: Dramatically enhanced assembly capability and structural stability of tobacco mosaic virus nanorods. *Biomacromolecules* **2013**, *14*, 2593–2600. [CrossRef]

24. Zhang, J.; Zhou, K.; Wang, Q. Tailoring the self-assembly behaviors of recombinant tobacco mosaic virus by rationally introducing covalent bonding at the protein–protein interface. *Small* **2016**, *12*, 4955–4959. [CrossRef]

25. Chu, S.; Brown, A.D.; Culver, J.N.; Ghodssi, R. Tobacco mosaic virus as a versatile platform for molecular assembly and device fabrication. *Biotechnol. J.* **2018**, *13*, 1800147. [CrossRef]

26. Schneider, A.; Eber, F.J.; Wenz, N.L.; Altintoprak, K.; Jeske, H.; Eiben, S.; Wege, C. Dynamic DNA-controlled "stop-and-go" assembly of well-defined protein domains on RNA-scaffolded TMV-like nanotubes. *Nanoscale* **2016**, *8*, 19853–19866. [CrossRef] [PubMed]

27. Geiger, F.C.; Eber, F.J.; Eiben, S.; Mueller, A.; Jeske, H.; Spatz, J.P.; Wege, C. TMV nanorods with programmed longitudinal domains of differently addressable coat proteins. *Nanoscale* **2013**, *5*, 3808–3816. [CrossRef]

28. Hu, J.; Zhou, S.X.; Sun, Y.Y.; Fang, X.S.; Wu, L.M. Fabrication, properties and applications of Janus particles. *Chem. Soc. Rev.* **2012**, *41*, 4356–4378. [CrossRef]

29. Patil, A.J.; McGrath, N.; Barclay, J.E.; Evans, D.J.; Colfen, H.; Manners, I.; Perriman, A.W.; Mann, S. Liquid viruses by nanoscale engineering of capsid surfaces. *Adv. Mater.* **2012**, *24*, 4557–4563. [CrossRef]

30. Gooding, G.V.; Hebert, T.T. A simple technique for purification of tobacco mosaic virus in large quantities. *Phytopathology* **1967**, *57*, 1285.

31. Atanasov, V.; Burger, M.; Lyonnard, S.; Porcar, L.; Kerres, J. Sulfonated poly(pentafluorostyrene): Synthesis & characterization. *Solid State Ionics* **2013**, *252*, 75–83. [CrossRef]

32. Green, M.R.; Sambrook, J. *Molecular Cloning: A Laboratory Manual*, 4th ed.; Cold Spring Harbor Laboratory Press: Cold Spring Harbor, NY, USA, 2012.

33. Knez, M.; Sumser, M.P.; Bittner, A.M.; Wege, C.; Jeske, H.; Hoffmann, D.M.P.; Kuhnke, K.; Kern, K. Binding the tobacco mosaic virus to inorganic surfaces. *Langmuir* **2004**, *20*, 441–447. [CrossRef]

34. Suljovrujic, E.; Micic, M. Smart poly(oligo(propylene glycol) methacrylate) hydrogel prepared by gamma radiation. *Nucl. Instrum. Meth. B* **2015**, *342*, 206–214. [CrossRef]

35. Schlick, T.L.; Ding, Z.B.; Kovacs, E.W.; Francis, M.B. Dual-surface modification of the tobacco mosaic virus. *J. Am. Chem. Soc.* **2005**, *127*, 3718–3723. [CrossRef] [PubMed]

36. Holder, P.G.; Finley, D.T.; Stephanopoulos, N.; Walton, R.; Clark, D.S.; Francis, M.B. Dramatic Thermal stability of virus-polymer conjugates in hydrophobic solvents. *Langmuir* **2010**, *26*, 17383–17388. [CrossRef] [PubMed]

37. Iha, R.K.; Van Horn, B.A.; Wooley, K.L. Complex, degradable polyester materials via ketoxime ether-based functionalization: Amphiphilic, multifunctional graft copolymers and their resulting solution-state aggregates. *J. Polym. Sci. Pol. Chem.* **2010**, *48*, 3553–3563. [CrossRef]

Article

Elongated Flexuous Plant Virus-Derived Nanoparticles Functionalized for Autoantibody Detection

Carmen Yuste-Calvo [1], Mercedes López-Santalla [2,3], Lucía Zurita [1], César F. Cruz-Fernández [1], Flora Sánchez [1], Marina I. Garín [2,3] and Fernando Ponz [1,*]

[1] Centro de Biotecnología y Genómica de Plantas, Universidad Politécnica de Madrid-Instituto Nacional de Investigación y Tecnología Agraria y Alimentaria (CBGP, UPM-INIA), Campus Montegancedo, Autopista M-40, km 38. Pozuelo de Alarcón, 28223 Madrid, Spain; carmen.yus.cal@gmail.com (C.Y.-C.); lucia.zurita@inia.es (L.Z.); cesarfcruzf@gmail.com (C.F.C.-F.); florasanchez@telefonica.net (F.S.)
[2] Division of Hematopoietic Innovative Therapies, Centro de Investigaciones Energéticas, Medioambientales y Tecnológicas (CIEMAT) and Centro de Investigación Biomédica en Red de Enfermedades Raras (CIBER-ER), 28040 Madrid, Spain; Mercedes.LopezSantalla@externos.ciemat.es (M.L.-S.); marina.garin@ciemat.es (M.I.G.)
[3] Advanced Therapy Unit, Instituto de Investigación Sanitaria Fundación Jiménez Díaz (IIS-FJD/UAM), 28040 Madrid, Spain
* Correspondence: fponz@inia.es; Tel.: +34-910-679-187

Received: 16 September 2019; Accepted: 3 October 2019; Published: 10 October 2019

Abstract: Nanoparticles derived from the elongated flexuous capsids of *Turnip mosaic virus* (TuMV) have been shown to be efficient tools for antibody sensing with a very high sensitivity if adequately functionalized with the corresponding epitopes. Taking advantage of this possibility, TuMV virus-like particles (VLPs) have been genetically derivatized with a peptide from the chaperonin Hsp60, a protein described to be involved in inflammation processes and autoimmune diseases. Antibodies against the peptide have been previously shown to have a diagnostic value in at least one autoimmune disease, multiple sclerosis. The functionalized Hsp60-VLPs showed their significant increase in sensing potency when compared to monoclonal antibody detection of the peptide in a conventional immunoassay. Additionally, the developed Hsp60-VLPs allowed the detection of autoantibodies against the Hsp60 peptide in an in vivo mouse model of dextran sodium sulfate (DSS)-induced colitis. The detection of minute amounts of the autoantibodies allowed us to perform the analysis of their evolution during the progression of the disease. The anti-Hsp60 autoantibody levels in the sera of the inflamed mice went down during the induction phase of the disease. Increased levels of the anti-HSP60 autoantibodies were detected during the resolution phase of the disease. An extension of a previously proposed model for the involvement of Hsp60 in inflammatory processes is considered, incorporating a role for Hsp60 autoantibodies. This, and related models, can now be experimentally tested thanks to the autoantibody detection hypersensitivity provided by the functionalized VLPs.

Keywords: VNPs; Hsp60; IBD; autoantibody; inflammation; diagnosis

---

1. Introduction

The use of viral nanoparticles (VNPs) for biomedical applications has become a new tool in theranostics. Specifically, VNPs derived from plant viruses offer advantages in terms of biosafety since they are only plant pathogens and also because of plant virus variability in nature, size, and structure, that allows a specific design of VNPs depending on the application [1–13]. We work with *Turnip mosaic virus* (TuMV), a virion with an elongated and flexuous structure, 700 nm long and 12 nm wide, with *ca.* 2000 copies of the coat protein in each particle, from which multifunctional VNPs can be obtained by genetic fusion and/or chemical conjugation to the coat protein (CP) [14–17]. One potential application is their use as tools for the detection of circulating antibody levels that frequently are too low to be detected by conventional detection methodologies, thus contributing to improve the diagnosis, progression, and/or prognosis of immunity-mediated pathologies, especially at the early stages of the disease when no clear symptoms of the inflammation are evident.

Inflammatory bowel disease (IBD) is a pathology encompassing two chronic inflammatory disorders; Crohn's disease, characterized by transmural inflammation usually affecting the terminal ileus and/or the large intestine; and ulcerative colitis, where inflammation occurs in the lining of the colon mucosa [18,19]. In these pathologies, our research focuses on diagnosis improvement and the development of personalized therapy, in order to achieve improvement in the patients' quality of life. On the other hand, we seek to gain new knowledge about the multifactorial pathogenesis of these disorders, for which there are several representative animal models, capable of mimicking the symptomatology and pathology of the disease. For IBD, a murine model induced by dextran sodium sulfate (DSS) has been developed [20,21]. This compound, when administered in drinking water, induces inflammation of the colon mucosa, as well as ulceration, which leads to severe weight loss and, in extreme cases, lethality.

We have explored the deployment of Heat Shock Protein 60 (Hsp60) as a possible autoantigen in IBD, due to the implication of Hsp60 in inflammatory processes [22–31], including IBD [32–35]. Heat shock proteins are a broad family of molecular chaperons capable of reacting to cellular stress, especially in thermal changes. However, they also respond in other stress situations, such as tissue damage, cellular injury, or heavy metal poisoning. Hsp60, despite being a protein whose main function occurs in the mitochondria, is able to act in the cytoplasm modulating the immune response associated with inflammation. Both the complete protein and the peptides derived therefrom can act as natural regulators of the inflammatory reaction together with other cytokines, chemokines, and autoantibodies, regulating the immune system [30,32,36–39]. The activity of Hsp60 has been associated with various autoimmune and inflammatory pathologies, such as atherosclerosis, diabetes, rheumatoid arthritis, or multiple sclerosis [23,25,29,40–42].

Alterations in the levels of anti-Hsp60 autoantibodies have been described in several inflammatory pathologies [22,23,26,27,37], making the autoantibody detection a novel strategy with potential application for the diagnosis, progression, and/or prognosis of the disease. In these studies, exhaustive analyses by peptide arrays have allowed the identification of certain epitopes for the design of specific VNPs where the election of a peptide with diagnostic value is the key to success. Based on these premises, we have designed functionalized TuMV VNPs with an epitope, described previously in multiple sclerosis [40]. The multimeric presentation of this epitope on TuMV VNPs (Hsp60-virus-like particles (VLPs)) allowed us the quantitative detection of anti-Hsp60 autoantibodies in an in vivo model of intestinal inflammation induced by DSS. The high detection levels of the developed Hsp60-VLPs may represent a novel tool that can be used for the diagnosis, progression, and/or prognosis in inflammation-mediated disorders.

2. Materials and Methods

2.1. Construction of CP and Hsp60-CP Expression Plasmids

Constructs corresponding to infectious clones were obtained by designing synthetic genes (GeneArt), based on the p35Tunos-Vec01-Nat1 sequence, as shown in Figure S1, derived from an original TuMV infectious clone [43]. Synthetic DNA consisted of a partial NIb gene, starting with the sequence corresponding to the restriction site *Mlu* I, followed by alanine codon (first CP amino acid) to maintain the protease recognition sequence, followed by the sequence encoding the Hsp60 peptide, which comprises amino acids 301–320 of human Hsp60 (KAPGFGDNRKNQLKDMAIAT, sequence conserved in mice), and the restriction site *Nae* I, corresponding to the first two CP amino acids. The DNA was digested with *Mlu* I and *Nae* I. The resulting fragment was purified. Vector p35Tunos-vec01-Nat1 was also digested with these two restriction enzymes, and the two fragments ligated together to obtain the recombinant vector with the sequence encoding Hsp60 peptide.

For VLP expression plasmids, PCR (GeneAmp®PCR System 9700, Applied Biosystems, CA, United States) was performed, amplifying the whole modified CP with two extra codons—one corresponding to the initial methionine, and a STOP codon. The nucleotide sequence CACC was also added at the 5' end to allow directional cloning into a pENTR-D-TOPO vector, as shown in Figure S2. Then the fragment was cloned into a pEAQ-HT vector, as shown in Figure S3, by Gateway cloning with LR clonase enzyme.

2.2. Production and Purification of VLPs

For VLP production, the pEAQ construct was transformed into *Agrobacterium tumefaciens* LBA4404 for agroinfiltration mediated transitory expression in *Nicotiana benthamiana* plants [44,45]. The same procedure was followed for non-modified VLPs. Plant growth, *Agrobacterium* culture preparation, agroinfiltration, tissue harvesting, and VLP purification were performed as previously described by us [16,17].

2.3. VLP Characterization

Characterization of assembled purified VLPs was performed by SDS-PAGE, western blot and transmission electron microscopy (TEM, ICTS-CNME, Madrid, Spain).

The conditions for SDS-PAGE and western blot were as described [16]. Anti-Hsp60 D307 antibody was from Invitrogen.

To check structural integrity, TEM was performed. Electron microscopy grids (400 mesh copper, carbon coated) were coated at room temperature for 15 min with a 10 µL drop of VLPs diluted at 0.02 mg/mL final concentration (50 mM borate buffer, pH 8.1), and washed with buffer. Finally, the grids were rinsed with distilled water and stained with 2% uranyl acetate for 2 min. Samples were examined on a transmission electron microscope (JEM JEOL 1010, Tokyo, Japan).

2.4. IBD Murine Model

Adult C57BL/6J (8 weeks-old) mice used in these studies were obtained from the Jackson Laboratory (ref. 000664). The regulations concerning experimental animal welfare (RD 223/1998 and Directive 2010/63/EU protocols) were followed. The ethics committee for animal research of the CIEMAT (Proex. 414/15) and Comunidad de Madrid (based on the RD 53/2013) reviewed and approved all protocols.

The IBD was induced in mice by dextran sodium sulfate (DSS, MP Biochemicals), a sulphated polymer cytotoxic on intestinal epithelial cells and macrophages. In addition, enteral DSS favors Gram-negative anaerobic bacteria increases, which together with the erosive potential on the intestinal barrier and the macrophages' inappropriate response, would lead to the appearance of intestinal lesions [20,21]. This model reproduces the clinical, histopathological, and immune characteristics observed in humans, inducing chronic colitis associated with diarrhea and weight loss.

The chemical compound was administered in drinking water at 1.35% (*w/v*) over 7 days *ad libitum* and disease progress was determined by monitoring mouse weight and the number of granulocytes in peripheral blood by an automated blood cell-counter (Abacus, Diatron, Budapest, Hungary). Normal distribution was analyzed by the Shapiro–Wilks test.

Non-parametric techniques (Mann–Whitney U test) were used. Autoantibody levels were determined in mouse sera, obtained at different times after disease induction. All trials were of a "simple blind" type.

2.5. Immunoassays

Indirect enzyme-linked immunosorbent assays (ELISA) were performed in order to evaluate the sensitivity provided by the VLPs, in comparison with the whole Hsp60 protein and free peptide at equal peptide amounts, and also to measure autoantibody levels in peripheral blood. Plates (Nunc

MaxiSorp) were coated with 1 µg purified VLPwt or Hsp60-VLPs, resuspended in 50 mM sodium carbonate buffer, pH 9.6; or equivalent amounts of Hsp60 protein or peptide, resuspended in the same buffer and incubated overnight at 4 °C. Plates were washed intensively, and incubated for 1 h with commercial antibodies [Hsp60 (D307) Antibody 4870S, CellSignal and Anti-Hsp60 antibody (ab46798), Abcam; different dilutions in phosphate-buffered saline, pH 7.4, 0.05% Tween 20 (*v/v*), 2% PVP-40 (*w/v*)] or overnight at 4 °C for mice sera (diluted 1:100 in the same buffer). Then, secondary antibodies, diluted 1:1500 in the same buffer, were added and incubated for 1 h at room temperature. Color was developed by alkaline phosphatase reaction and detected after addition of *p*-nitrophenylphosphate. Absorbance was measured at 405 nm (TECAN Genios Pro). Differences between both coating-VLPs were calculated, since this would be the measure attributable to autoantibodies against Hsp60. The normalization process was done as follows. A serum pool was made using the sera from healthy mice and this pool was included in all the plates in order to calculate the background of each plate. The normalized values of autoantibodies directed against Hsp60 in the DSS serum samples were obtained by taking into account the mean value of the healthy serum pool in each plate and the mean value of the healthy serum pool from all the plates, in order to minimize the effects of the variation in the background signal. The data obtained were analyzed based on the non-parametric Mann–Whitney U test.

3. Results

3.1. Production and Characterization of Hsp60-VLPs

Our first Hsp60-VNP derivatization approach was the production of virions genetically modified with the Hsp60-peptide fused to the CP N-terminus. However, the insertion of the peptide interfered with the infectivity of the modified virus, so this strategy was discarded. Starting out from the virion construct, a second construction strategy was followed to generate now modified VLPs.

VLPs modified by genetic fusion of the Hsp60-peptide to the CP were made by transient expression of the fusion protein CP-Hsp60 peptide in plants. As shown before [16], the fusion protein will self-assemble into VLPs within plant cells. Once purified nanoparticles were obtained, they were characterized by SDS-PAGE, western-blot assays, and transmission electron microscopy, as shown in Figure 1.

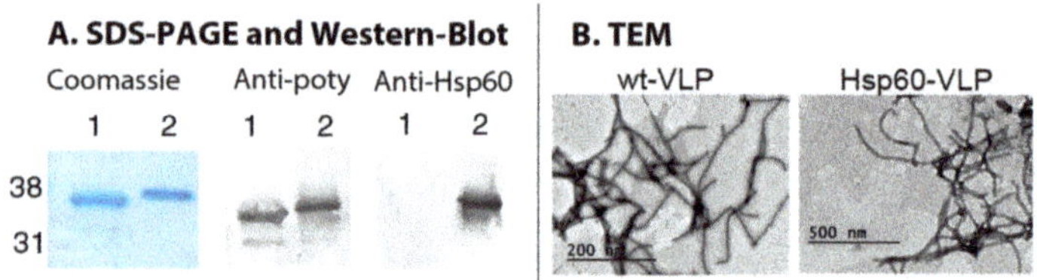

Figure 1. Hsp60-virus-like particles (VLPs) characterization. (**A**) SDS-PAGE and western blot of wild type VLPs (1) and Hsp60-VLPs (2). (**B**) Micrographs of wt-VLPs and Hsp60-VLPs, where particles of about 700 nm are shown.

The results obtained showed a modified electrophoretic mobility for Hsp60-VLPs CP, consistent with a molecular weight increase of approximately 2.5 kDa. This fusion protein was detected specifically not only by antibodies directed against the virus, but also by those directed against the Hsp60 protein. Finally, TEM structural characterization showed elongated and flexuous particles with similar length and thickness than non-modified particles, so no detectable significant changes in VLP structure were found after peptide insertion.

3.2. Increased Sensitivity of Hsp60 Peptide Detection by Specific Hsp60 Antibodies

Multimeric peptide presentation on the surface of TuMV VNPs increases its detectability by specific antibodies in conventional immunoassays like ELISA [15,17,46]. To evaluate this in the Hsp60-VLPs construct, ELISA and dot-blot assays were performed using three types of presentations: Hsp60-VLPs, free Hsp60 peptide, and complete Hsp60 protein, as shown in Figure 2. Assays were performed using equal peptide amounts presented in the three forms. Two different antibodies were used to assess the response, a monoclonal antibody that specifically recognizes an epitope within the peptide 301–320, and a polyclonal antibody against complete Hsp60.

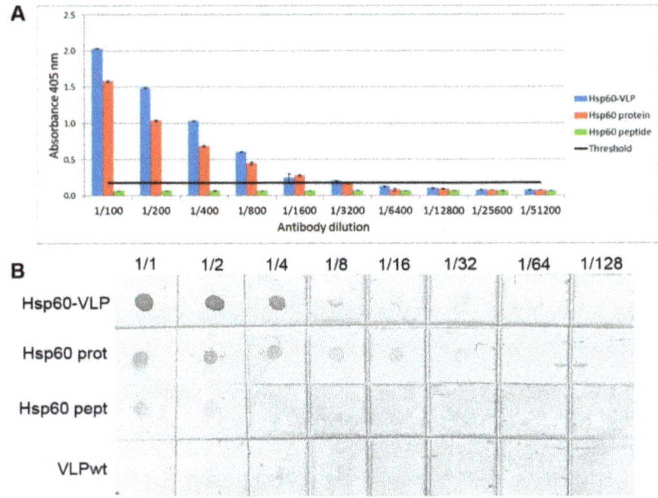

Figure 2. Specificity and sensitivity assays. (**A**) ELISA using monoclonal Hsp60 antibody dilutions, comparing Hsp60-VLPs with free Hsp60 peptide and Hsp60 protein. The black line indicates the positive signal threshold and the standard error is also shown (three replicates). (**B**) Dot blot using the same components, and also VLPs-wt.

Using the monoclonal antibody, it was found that free Hsp60 peptide was hardly recognized, while both the complete Hsp60 protein and the Hsp60-VLPs were easily detected, with Hsp60-VLPs being the most sensitive system.

Hsp60-VLP detectability, compared with the complete protein using polyclonal antibodies, was evaluated, as shown in Figure S4. In this case, complete Hsp60 protein presents greater sensitivity, possibly due to the higher number of epitopes present on its complete surface.

The results obtained show that, although the complete HSP60 protein contains various epitopes, the multimeric presentation in VNPs of a defined epitope allows differential detection by measurement of specific autoantibodies in biological fluids, such as peripheral blood serum. These results highlight the Hsp60-VLPs potential for specific antibody detection, with the epitope election being the key to develop TuMV-VNPs as a diagnostic tool.

3.3. Hsp60-VLPs Autoantibody Detection

Once sensitivity and specificity were shown through different immunoassays, the potential of Hsp60-VLPs as a diagnostic tool of autoimmune pathologies, exemplified by IBD, was tested, measuring autoantibodies against Hsp60 in a pool of control sera, as shown in Figure 3. In these pathologies, autoantibody levels are usually very low at early stages, but early diagnosis by detecting these low levels would improve patients' prognosis.

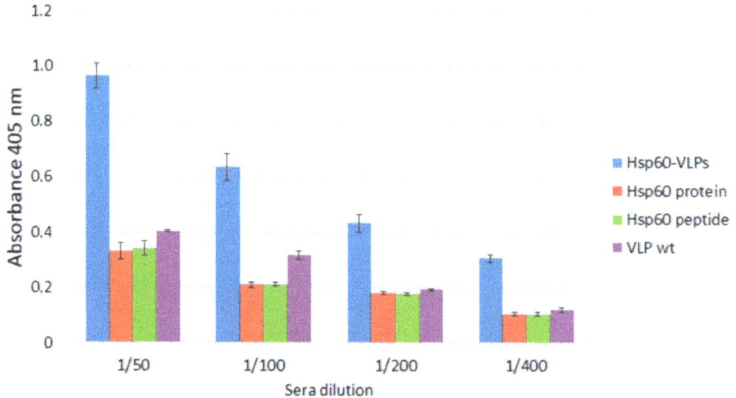

Figure 3. Anti-HSP60 autoantibody detection in the sera of healthy mice by ELISA, using different antigen-presentation platforms; Hsp60-VLP (blue), complete protein Hsp60 (red), Hsp60 peptide (green), and VLPwt (purple) and at different sera dilutions. Standard error bars shown (three replicates).

To compare the data obtained with all detection systems (Hsp60-VLPs, complete Hsp60 protein, and free peptide), an ELISA was developed with a pool of sera from healthy, non-inflamed mice, in order to determine physiological levels of anti-Hsp60 autoantibodies. Unmodified VLPs (VLPwt, not displaying the Hsp60 peptide) were used as a negative control.

The results showed that only with the Hsp60-VLPs could the anti-Hsp60 autoantibodies be detected above the background of the plate and negative controls in healthy mice sera. Therefore, the high ability of Hsp60-VLPs to detect autoantibody levels, undetectable by conventional procedures such as ELISA, was demonstrated. This result indicates that VLPs can be used as a tool to detect variations in the levels of anti-Hsp60 autoantibodies in mice sera when the traditional methodologies based on the complete Hsp60 protein cannot be detected, even at basal autoantibody levels in non-inflamed mice.

3.4. Hsp60-VLPs for Autoantibody Detection in DSS-Induced Colitis

Due to the proven ability of Hsp60-VLPs to detect low levels of autoantibodies in the sera samples, potential diagnostic applications were assessed. Since Hsp60 is deeply involved in inflammatory processes [23,25,47], a murine model of the inflammatory bowel disease (IBD) induced by DSS was selected for further studies [20,21]. DSS-induced colitis is a favorite model for chemically induced colitis because it is a simple one and highly similar to human IBD. DSS induces intestinal inflammation through a yet non-deciphered mechanism, although it probably results from damage to the epithelial monolayer lining the large intestine. This lesion would allow dissemination of proinflammatory intestinal contents (e.g., bacteria and their products) into the tissue underneath [20].

The evolution of Hsp60 autoantibody levels in the inflammatory model induced by DSS was studied. The progress of the pathology, as well as the recovery after removing the chemical agent, was determined by monitoring the body weight and granulocytes levels in peripheral blood, as shown in Figure 4.

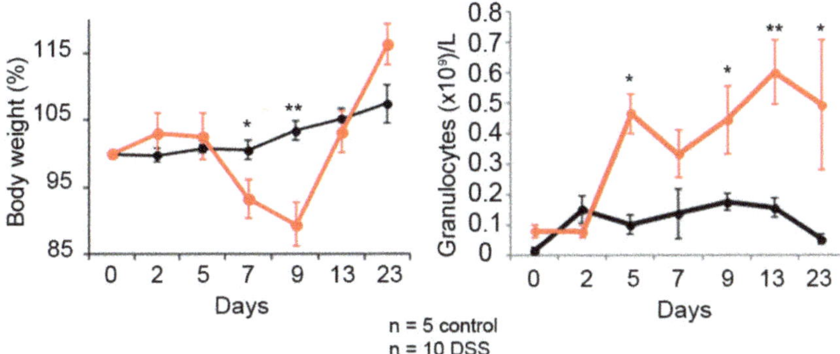

Figure 4. Colitis status of mice treated with dextran sodium sulfate (DSS) in the drinking water. Assessment of the pathology progression by determination of body weight loss and granulocyte levels in peripheral blood. Healthy mice are shown in black and the group of mice treated with DSS is shown in red. The mean and the standard error of the mean are shown, as well as those significant results according to the nonparametric Mann–Whitney U test (* $p \leq 0.05$ and ** $p \leq 0.01$).

The results show a significant decrease in body weight of mice treated with DSS, significant at day 7 and 9. This decrease continues until the DSS was withdrawn, after which the mice recovered weight and reverted to a non-pathological state. Regarding granulocyte levels, there was an increase in blood, in this case significantly at 5 days. In this case the reversal begins later, several days after withdrawing DSS.

To evaluate the potential of Hsp60-VLPs as a novel tool in the knowledge of the inflammatory pathology, an ELISA was performed. Autoantibody levels measured with the complete Hsp60 protein were not detectable, as shown in Figure S5, so only data obtained with Hsp60-VLPs and VLPs-wt were used to assess autoantibody levels in sera of the mice diluted 100-fold. Data were normalized, eliminating the signal attributable to the background which could be originated by the viral component of nanoparticles, and the results were analyzed, as shown in Figure 5.

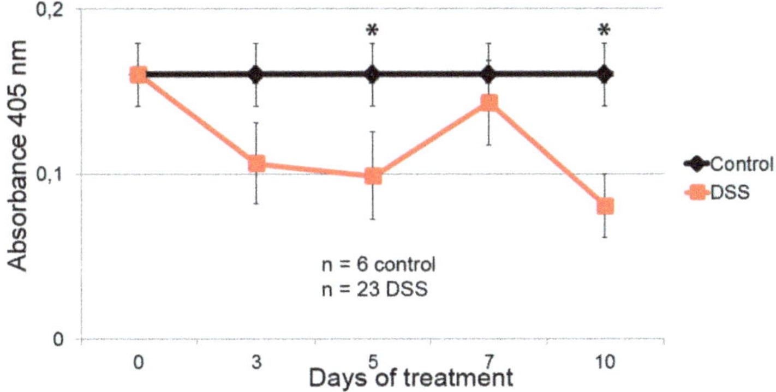

Figure 5. Detection of anti-HSP60 autoantibody in the sera of the healthy and DSS-colitic mice. Healthy mice are represented in black, and in red are mice treated with DSS in the drinking water for 7 days. The levels of anti-HSP60 autoantibodies in the control group has been defined as the average value of all the mice in the control group, including the values of the DSS-treated colitic mice at time 0. Mean and standard error of the mean are shown, as well as those significant results according to the Mann–Whitney U nonparametric test (* $p \leq 0.05$).

The results showed a marked decrease in anti-Hsp60 autoantibody levels following the progress of inflammation at 5 and 10 days (significant according to the Mann–Whitney U nonparametric test), maintaining higher levels in healthy individuals and the basal levels at day 0 in the DSS group. These results demonstrate TuMV functionalized VNPs as a hypersensitive tool in the progression analysis of the inflammation, being able to detect levels of autoantibodies in cases where conventional methodology is not capable.

4. Discussion

In the diagnosis of autoimmune pathologies, current techniques face two main problems, especially when diagnosis is made by detecting specific autoantibodies in serum—background and low autoantibody levels. The high background is due to numerous substances present in serum, and low autoantibody levels occur at early stages of pathology development, a time in which a proper diagnosis would allow early treatment, and a better patient prognosis [48,49]. This is in addition to problems associated with the use of biological samples, such as peripheral blood sera. There is also an added problem when specific epitopes are needed as the basis of a detection system. These epitopes, small peptides, are unable to adhere efficiently to test plates [50], thus leading to very low sensitivity compared to systems deploying high molecular weight proteins in which the handicap is the presence of numerous epitopes, not all of them with diagnostic value. All these inconveniences in conventional methodologies make it increasingly convenient to develop new approaches for a highly sensitive technique, capable of detecting very low levels of autoantibody, or rather small changes, in diluted sera.

To address these challenges, the use of nanoparticles as a multimeric presentation system is an alternative approach, which should provide high sensitivity and solve the epitope–adhesion problem [6,9,51,52]. We have previously shown that functionalized nanoparticles derived from TuMV significantly increased antibody detection with respect to conventional systems [16,17,46]. Although this system has not been used in the diagnosis, progression, and/or prognosis of specific pathologies yet, its potential makes it a good candidate for the design of a hypersensitive analysis system by the detection of serum autoantibodies.

To test TuMV VNPs as a new tool for autoantibody detection, Hsp60 was chosen as an antigen. This protein is involved in numerous inflammatory mechanisms, also related to autoimmune pathologies [27,30,32,33,35,38,40]. Regarding Hsp60 specific commercial antibodies, sensitivity tests were made by comparing the Hsp60-VLPs' detection capacity with respect to the complete recombinant Hsp60 protein. The results showed that, despite not showing a greater sensitivity in polyclonal antibody detection as shown in Figure S1, VLPs sensed epitope-specific monoclonal antibodies with greater sensitivity, as shown in Figure 2. This may be due not only to multimeric presentation in the nanoparticle, but also to a better epitope-exposure on the VLP surface with respect to the complete Hsp60 protein.

According to these results, and because this peptide was chosen due to its diagnostic potential in autoimmune pathologies [49], an evaluation of its efficacy in IBD diagnosis in a DSS-induced colitis murine model [20,21] was decided. In the sera of these mice, nanoparticles exhibited greater detection power not only compared to the free peptide, but also to the complete Hsp60, which was not able to detect autoantibodies under the same conditions. As mentioned above, the enhanced sensitivity of the Hsp60-VLPs could be provided not only because of a multimeric presentation of the chosen epitope, but also because of its exposure on the particle surface since the epitope may have a more internal location in the complete Hsp60 but be exposed on the surface in VLPs. This approach should allow the development of sets of designed nanoparticles for each disease, selecting peptides with diagnostic value present in various autoantigens, and presenting them on the same multifunctionalized nanoparticle [15], improving diagnosis, progression, and prognosis of autoimmune diseases.

Thanks to the high sensitivity shown by Hsp60-VLPs, it was possible to evaluate the evolution of autoantibody levels with disease progress in a murine model. In this case, and against expectations, autoantibody levels decreased concomitantly with the inflammatory process, as shown in Figure 5, suggesting a new view regarding the mechanism of action of Hsp60 (and its autoantibodies) in the inflammatory process. According to these results, anti-Hsp60 autoantibodies could be considered as biomarkers of a non-pathological inflammation state, participating in immune homeostasis as "immunomodulators", recently called "immunculus" ("immune homunculus"), as it was already described for other molecules and autoantibodies [36,39,53]. According to this model, and taking into account studies relating a decrease in autoantibody levels with autoimmune disease development [25,26,32,41,54,55], it seems that autoantibodies in healthy individuals may play a role as a mechanism to regulate inflammation, avoiding a pathological inflammatory response caused by Hsp60. This protein, and peptides derived thereof, have been implicated in pro- and anti-inflammatory processes, and a model has been previously proposed [30]. The results presented in this study allow us to extend tentatively this model, involving now a role for Hsp60 autoantibodies in the progression of the inflammation. A general view of such an extended model is presented in Figure 6. The presence of large amounts of Hsp60 protein (and/or other similar chaperonins present in intestinal microbiota) would be related to an activation of the inflammatory response, triggering the production of autoantibodies. These autoantibodies would act as neutralizers, regulating protein levels, and thereby deactivating the inflammatory response and even activating new anti-inflammatory pathways. The regulatory balance of these pathways in both directions would be essential to maintain a non-pathological physiological state. This proposed process would imply that, when an inflammatory process is induced by external agents, as occurs in experimental colitis induced by DSS, inflammation would be accompanied by a decrease in the levels of regulatory autoantibodies.

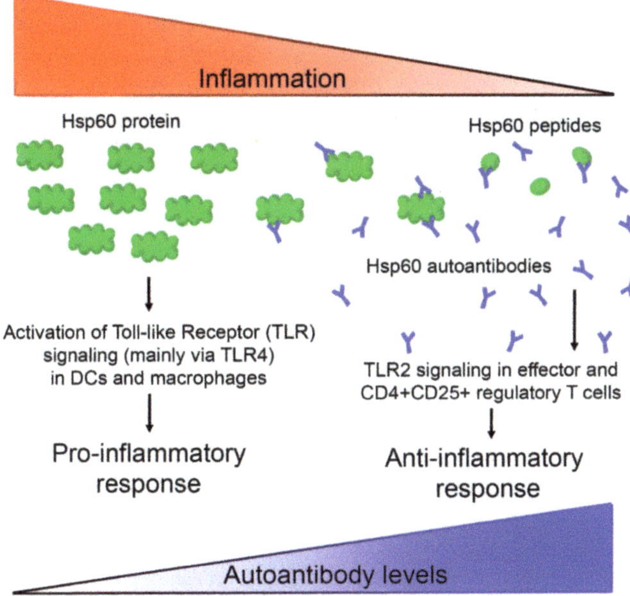

Figure 6. A possible Hsp60 autoantibody regulation system based on a previously described model [30]. High Hsp60 protein levels, related with endogenous proteins and/or similar chaperonins present in microbiota or other pathogens, would activate inflammation pathways, which in turn would induce the production of autoantibodies against Hsp60, neutralizing the antigen, and generating different peptides that activate an anti-inflammatory response.

All the ideas raised would place Hsp60 as a study target. In this case, TuMV VLPs allowed the identification of possible markers in inflammatory pathologies, being able to generate diagnostic systems of high sensitivity and specificity, useful to identify new diagnostic and therapeutic targets. In addition, considering cases such as Hsp60, where autoantibodies can act as regulatory agents, functionalized VLPs could be used not only as a diagnostic tool, but also in therapy for immunization, where the great immunogenic power of VNPs from TuMV has been demonstrated [17]. This would be related to other studies connecting autoantibodies and inflammation, in which immunization with the protein in patients induced symptomatic amelioration [25,41,56]. This work demonstrates that VNPs are a good alternative in the biomedical field, acting as a functionalizable platform for different applications, from diagnosis to treatment. In this context, exploring the possibility of using several epitopes from Hsp60 (or even from other new autoantigens) in multifunctional nanoparticles, would open the door to new theranostic tools to improve prognosis of patients with autoimmune pathologies.

5. Conclusions

This study lays the groundwork for the development of new theranostic tools, based on viral nanoparticles functionalized with specific epitopes involved in autoimmune pathologies, allowing the creation of new systems for autoantibody-based specialized diagnosis and for targeted treatments.

Supplementary Materials: The following are available online at http://www.mdpi.com/2079-4991/9/10/1438/s1, Figure S1. Vector p35Tunos-Vec01-NAT1. Figure S2. Vector pENTR/D-TOPO. Figure S3. Vector pEAQ-HT-Dest1. Figure S4. Sensitivity comparison between different detection platforms using commercial antibodies. Figure S5. Autoantibody levels in sera measured with Hsp60-VLPs (blue) and Hsp60 protein (red).

Author Contributions: Conceptualization, F.P., F.S., and M.I.G.; methodology, C.Y.-C., L.Z., C.F.C.-F., M.L.-S., and F.S.; software, C.Y.-C. and M.L.-S.; validation, C.Y.-C. and M.L.-S; formal analysis, C.Y.-C., M.L.-S., F.S., M.I.G., and F.P.; writing—original draft preparation, C.Y.-C.; writing—review and editing, M.L.-S, M.I.G., and F.P.

Funding: This work was funded by INIA (grant RTA2015-00017). Support from the Spanish Ministry of Science, Innovation and Universities in Severo Ochoa Excellence Accreditations (SEV-2016-0672 to CBGP) is also acknowledged.

Acknowledgments: We thank George Lomonossoff for providing the pEAQ vectors.

Conflicts of Interest: The authors declare no conflict of interest.

References

1. Steele, J.F.C.; Peyret, H.; Saunders, K.; Castells-Graells, R.; Marsian, J.; Meshcheriakova, Y.; Lomonossoff, G.P. Synthetic plant virology for nanobiotechnology and nanomedicine. *Wiley Interdiscip. Rev. Nanomed. Nanobiotechnol.* **2017**, *9*, e1447. [CrossRef] [PubMed]
2. Rohovie, M.J.; Nagasawa, M.; Swartz, J.R. Virus-like particles: Next-generation nanoparticles for targeted therapeutic delivery. *Bioeng. Transl. Med.* **2017**, *2*, 43–57. [CrossRef] [PubMed]
3. Lee, K.L.; Twyman, R.M.; Fiering, S.; Steinmetz, N.F. Virus-based nanoparticles as platform technologies for modern vaccines. *Wiley Interdiscip. Rev. Nanomed. Nanobiotechnol.* **2016**, *8*, 554–578. [CrossRef] [PubMed]
4. Schwarz, B.; Douglas, T. Development of virus-like particles for diagnostic and prophylactic biomedical applications. *Wiley Interdiscip. Rev. Nanomed. Nanobiotechnol.* **2015**, *7*, 722–735. [CrossRef] [PubMed]
5. Lomonossoff, G.P.; Evans, D.J. Applications of plant viruses in bionanotechnology. *Curr. Top. Microbiol. Immunol.* **2014**, *375*, 61–87. [CrossRef] [PubMed]
6. Glasgow, J.; Tullman-Ercek, D. Production and applications of engineered viral capsids. *Appl. Microbiol. Biotechnol.* **2014**, *98*, 5847–5858. [CrossRef] [PubMed]
7. Chen, Q.; Lai, H. Plant-derived virus-like particles as vaccines. *Hum. Vaccines Immunother.* **2013**, *9*, 26–49. [CrossRef] [PubMed]
8. Alonso, J.M.; Gorzny, M.L.; Bittner, A.M. The physics of tobacco mosaic virus and virus-based devices in biotechnology. *Trends Biotechnol.* **2013**, *31*, 530–538. [CrossRef]
9. Yildiz, I.; Shukla, S.; Steinmetz, N.F. Applications of viral nanoparticles in medicine. *Curr. Opin. Biotechnol.* **2011**, *22*, 901–908. [CrossRef]

10. Pokorski, J.K.; Steinmetz, N.F. The art of engineering viral nanoparticles. *Mol. Pharm.* **2011**, *8*, 29–43. [CrossRef]

11. Plummer, E.M.; Manchester, M. Viral nanoparticles and virus-like particles: Platforms for contemporary vaccine design. *Wiley Interdiscip. Rev. Nanomed. Nanobiotechnol.* **2011**, *3*, 174–196. [CrossRef] [PubMed]

12. Ludwig, C.; Wagner, R. Virus-like particles-universal molecular toolboxes. *Curr. Opin. Biotechnol.* **2007**, *18*, 537–545. [CrossRef] [PubMed]

13. Grgacic, E.V.; Anderson, D.A. Virus-like particles: Passport to immune recognition. *Methods* **2006**, *40*, 60–65. [CrossRef] [PubMed]

14. Cuenca, S.; Mansilla, C.; Aguado, M.; Yuste-Calvo, C.; Sánchez, F.; Sánchez-Montero, J.M.; Ponz, F. Nanonets derived from *Turnip mosaic virus* as scaffolds for increased enzymatic activity of immobilized *Candida antarctica* lipase B. *Front. Plant Sci.* **2016**, *7*, 464. [CrossRef] [PubMed]

15. Yuste-Calvo, C.; González-Gamboa, I.; Pacios, L.F.; Sánchez, F.; Ponz, F. Structure-based multifunctionalization of flexuous elongated viral nanoparticles. *ACS Omega* **2019**, *4*, 5019–5028. [CrossRef]

16. González-Gamboa, I.; Manrique, P.; Sánchez, F.; Ponz, F. Plant-made potyvirus-like particles used for log-increasing antibody sensing capacity. *J. Biotechnol.* **2017**, *254*, 17–24. [CrossRef] [PubMed]

17. Sánchez, F.; Sáez, M.; Lunello, P.; Ponz, F. Plant viral elongated nanoparticles modified for log-increases of foreign peptide immunogenicity and specific antibody detection. *J. Biotechnol.* **2013**, *168*, 409–415. [CrossRef] [PubMed]

18. Lemmens, B.; De Hertogh, G.; Sagaert, X. Inflammatory Bowel Diseases. In *Pathobiology of Human Disease*; McManus, L.M., Mitchell, R.N., Eds.; Academic Press: San Diego, CA, USA, 2014; Volume 19, pp. 1297–1304.

19. Denmark, V.K.; Mayer, L. Chapter 60—Inflammatory Bowel Diseases. In *The Autoimmune Diseases*, 5th ed.; Rose, N.R., Mackay, I.R., Eds.; Academic Press: Boston, MA, USA, 2014; pp. 873–888.

20. Eichele, D.D.; Kharbanda, K.K. Dextran sodium sulfate colitis murine model: An indispensable tool for advancing our understanding of inflammatory bowel diseases pathogenesis. *World J. Gastroenterol.* **2017**, *23*, 6016–6029. [CrossRef]

21. Chassaing, B.; Aitken, J.D.; Malleshappa, M.; Vijay-Kumar, M. Dextran sulfate sodium (DSS)-induced colitis in mice. *Curr. Protoc. Immunol.* **2014**, *104*, 15–25. [CrossRef]

22. Van Eden, W.; Jansen, M.A.A.; Ludwig, I.S.; Leufkens, P.; van der Goes, M.C.; van Laar, J.M.; Broere, F. Heat shock proteins can be surrogate autoantigens for induction of antigen specific therapeutic tolerance in rheumatoid arthritis. *Front. Immunol.* **2019**, *10*, 279. [CrossRef]

23. Mantej, J.; Polasik, K.; Piotrowska, E.; Tukaj, S. Autoantibodies to heat shock proteins 60, 70, and 90 in patients with rheumatoid arthritis. *Cell Stress Chaperones* **2019**, *24*, 283–287. [CrossRef] [PubMed]

24. Gonciarz, W.; Matusiak, A.; Rudnicka, K.; Rechcinski, T.; Chalubinski, M.; Czkwianianc, E.; Broncel, M.; Gajewski, A.; Chmiela, M. Autoantibodies to a specific peptide epitope of human Hsp60 (ATVLA) with homology to *Helicobacter pylori* HspB in *H. pylori*-infected patients. *APMIS* **2019**, *127*, 139–149. [CrossRef] [PubMed]

25. Ulmansky, R.; Naparstek, Y. Protective antibodies against HSP60 for autoimmune inflammatory diseases. *Clin. Immunol.* **2018**, *186*, 63. [CrossRef] [PubMed]

26. Ponnusamy, T.; Venkatachala, S.K.; Ramanjappa, M.; Kakkar, V.V.; Mundkur, L.A. Inverse association of ApoB and HSP60 antibodies with coronary artery disease in Indian population. *Heart Asia* **2018**, *10*, e011018. [CrossRef] [PubMed]

27. Selli, M.E.; Wick, G.; Wraith, D.C.; Newby, A.C. Autoimmunity to HSP60 during diet induced obesity in mice. *Int. J. Obes. (Lond.)* **2017**, *41*, 348–351. [CrossRef] [PubMed]

28. Jang, E.J.; Jung, K.Y.; Hwang, E.; Jang, Y.J. Characterization of human anti-heat shock protein 60 monoclonal autoantibody Fab fragments in atherosclerosis: Genetic and functional analysis. *Mol. Immunol.* **2013**, *54*, 338–346. [CrossRef] [PubMed]

29. Kimura, A.; Sakurai, T.; Yamada, M.; Koumura, A.; Hayashi, Y.; Tanaka, Y.; Hozumi, I.; Takemura, M.; Seishima, M.; Inuzuka, T. Elevated anti-heat shock protein 60 antibody titer is related to white matter hyperintensities. *J. Stroke Cerebrovasc. Dis.* **2012**, *21*, 305–309. [CrossRef]

30. Quintana, F.J.; Cohen, I.R. The HSP60 immune system network. *Trends Immunol.* **2011**, *32*, 89–95. [CrossRef]

31. Khalil, A.A.; Kabapy, N.F.; Deraz, S.F.; Smith, C. Heat shock proteins in oncology: Diagnostic biomarkers or therapeutic targets? *Biochim. Biophys. Acta* **2011**, *1816*, 89–104. [CrossRef]

32. Cappello, F.; Mazzola, M.; Jurjus, A.; Zeenny, M.N.; Jurjus, R.; Carini, F.; Leone, A.; Bonaventura, G.; Tomasello, G.; Bucchieri, F.; et al. Hsp60 as a novel target in IBD management: A prospect. *Front. Pharmacol.* **2019**, *10*, 26. [CrossRef]

33. Tomasello, G.; Sciume, C.; Rappa, F.; Rodolico, V.; Zerilli, M.; Martorana, A.; Cicero, G.; De Luca, R.; Damiani, P.; Accardo, F.M.; et al. Hsp10, Hsp70, and Hsp90 immunohistochemical levels change in ulcerative colitis after therapy. *Eur. J. Histochem.* **2011**, *55*, e38. [CrossRef] [PubMed]

34. Tomasello, G.; Rodolico, V.; Zerilli, M.; Martorana, A.; Bucchieri, F.; Pitruzzella, A.; Marino Gammazza, A.; David, S.; Rappa, F.; Zummo, G.; et al. Changes in immunohistochemical levels and subcellular localization after therapy and correlation and colocalization with CD68 suggest a pathogenetic role of Hsp60 in ulcerative colitis. *Appl. Immunohistochem. Mol. Morphol.* **2011**, *19*, 552–561. [CrossRef] [PubMed]

35. Rodolico, V.; Tomasello, G.; Zerilli, M.; Martorana, A.; Pitruzzella, A.; Gammazza, A.M.; David, S.; Zummo, G.; Damiani, P.; Accomando, S.; et al. Hsp60 and Hsp10 increase in colon mucosa of Crohn's disease and ulcerative colitis. *Cell Stress Chaperones* **2010**, *15*, 877–884. [CrossRef] [PubMed]

36. Kolinski, T.; Marek-Trzonkowska, N.; Trzonkowski, P.; Siebert, J. Heat shock proteins (HSPs) in the homeostasis of regulatory T cells (Tregs). *Cent. Eur. J. Immunol.* **2016**, *41*, 317–323. [CrossRef] [PubMed]

37. Alard, J.E.; Pers, J.O.; Youinou, P.; Jamin, C. Chapter 41—Heat shock protein autoantibodies. In *Autoantibodies*, 3rd ed.; Shoenfeld, Y., Meroni, P.L., Gershwin, M.E., Eds.; Elsevier: San Diego, CA, USA, 2014; pp. 343–348.

38. Grundtman, C.; Kreutmayer, S.B.; Almanzar, G.; Wick, M.C.; Wick, G. Heat shock protein 60 and immune inflammatory responses in atherosclerosis. *Arterioscler. Thromb. Vasc. Biol.* **2011**, *31*, 960–968. [CrossRef] [PubMed]

39. Habich, C.; Burkart, V. Heat shock protein 60: Regulatory role on innate immune cells. *Cell. Mol. Life Sci.* **2007**, *64*, 742–751. [CrossRef] [PubMed]

40. Pinar, O.; Ozden, Y.A.; Omur, E.; Muhtesem, G. Heat shock proteins in multiple sclerosis. In *Multiple Sclerosis: Bench to Bedside: Global Perspectives on a Silent Killer*; Asea, A.A.A., Geraci, F., Kaur, P., Eds.; Springer International Publishing Ag: Champaign, IL, USA, 2017; Volume 958, pp. 29–42.

41. Ulmansky, R.; Landstein, D.; Moallem, E.; Loeb, V.; Levin, A.; Meyuhas, R.; Katzavian, G.; Yair, S.; Naparstek, Y. A humanized monoclonal antibody against Heat Shock Protein 60 suppresses murine arthritis and colitis and skews the cytokine balance toward an anti-inflammatory response. *J. Immunol.* **2015**, *194*, 5103–5109. [CrossRef] [PubMed]

42. Wu, C.T.; Ou, L.S.; Yeh, K.W.; Lee, W.I.; Huang, J.L. Serum heat shock protein 60 can predict remission of flare-up in juvenile idiopathic arthritis. *Clin. Rheumatol.* **2011**, *30*, 959–965. [CrossRef] [PubMed]

43. Touriño, A.; Sánchez, F.; Fereres, A.; Ponz, F. High expression of foreign proteins from a biosafe viral vector derived from *Turnip mosaic virus*. *Span. J. Agric. Res.* **2008**, *6*, 48–58. [CrossRef]

44. D'Aoust, M.A.; Lavoie, P.O.; Couture, M.M.; Trepanier, S.; Guay, J.M.; Dargis, M.; Mongrand, S.; Landry, N.; Ward, B.J.; Vezina, L.P. Influenza virus-like particles produced by transient expression in *Nicotiana benthamiana* induce a protective immune response against a lethal viral challenge in mice. *Plant Biotechnol. J.* **2008**, *6*, 930–940. [CrossRef] [PubMed]

45. Sainsbury, F.; Liu, L.; Lomonossoff, G.P. Cowpea mosaic virus-based systems for the expression of antigens and antibodies in plants. *Methods Mol. Biol.* **2009**, *483*, 25–39. [CrossRef] [PubMed]

46. Duval, F.; Cruz-Vega, D.E.; González-Gamboa, I.; González-Garza, M.T.; Ponz, F.; Sánchez, F.; Alarcon-Galvan, G.; Moreno-Cuevas, J.E. Detection of autoantibodies to vascular endothelial growth factor receptor-3 in bile duct ligated rats and correlations with a panel of traditional markers of liver diseases. *Dis. Markers* **2016**, *2016*. [CrossRef] [PubMed]

47. Van Eden, W.; van Herwijnen, M.; Wagenaar, J.; van Kooten, P.; Broere, F.; van der Zee, R. Stress proteins are used by the immune system for cognate interactions with anti-inflammatory regulatory T cells. *FEBS Lett.* **2013**, *587*, 1951–1958. [CrossRef] [PubMed]

48. Gecse, K.B.; Vermeire, S. Differential diagnosis of inflammatory bowel disease: Imitations and complications. *Lancet Gastroenterol. Hepatol.* **2018**, *3*, 644–653. [CrossRef]

49. Quintana, F.J.; Farez, M.F.; Viglietta, V.; Iglesias, A.H.; Merbl, Y.; Izquierdo, G.; Lucas, M.; Basso, A.S.; Khoury, S.J.; Lucchinetti, C.F.; et al. Antigen microarrays identify unique serum autoantibody signatures in clinical and pathologic subtypes of multiple sclerosis. *Proc. Natl. Acad. Sci. USA* **2008**, *105*, 18889–18894. [CrossRef] [PubMed]

50. Antimisiaris, G.S. Protein/peptide drug delivery: Advanced drug delivery systems (DDSs). *Biotech Int.* **2010**, 22, 10–13.
51. Lico, C.; Benvenuto, E.; Baschieri, S. The two-faced *Potato virus X*: From plant pathogen to smart nanoparticle. *Front. Plant Sci.* **2015**, *6*, 1009. [CrossRef]
52. Higa, A.M.; Mambrini, G.P.; Ierich, J.C.M.; Garcia, P.S.; Scramin, J.A.; Peroni, L.A.; Okuda-Shinagawa, N.M.; Machini, M.T.; Trivinho-Strixino, F.; Leite, F.L. Peptide-conjugated silver nanoparticle for autoantibody recognition. *J. Nanosci. Nanotechnol.* **2019**, *19*, 7564–7573. [CrossRef]
53. Poletaev, A.; Osipenko, L. General network of natural autoantibodies as immunological homunculus (Immunculus). *Autoimmun. Rev.* **2003**, *2*, 264–271. [CrossRef]
54. Cohen, I.R. Autoantibody repertoires, natural biomarkers, and system controllers. *Trends Immunol.* **2013**, *34*, 620–625. [CrossRef]
55. Papuc, E.; Krupski, W.; Kurys-Denis, E.; Rejdak, K. Antibodies against small heat-shock proteins in Alzheimer's disease as a part of natural human immune repertoire or activation of humoral response? *J. Neural Transm. (Vienna)* **2016**, *123*, 455–461. [CrossRef] [PubMed]
56. Zhong, Y.; Tang, H.; Wang, X.; Zeng, Q.; Liu, Y.; Zhao, X.I.; Yu, K.; Shi, H.; Zhu, R.; Mao, X. Intranasal immunization with heat shock protein 60 induces CD4(+) CD25(+) GARP(+) and type 1 regulatory T cells and inhibits early atherosclerosis. *Clin. Exp. Immunol.* **2016**, *183*, 452–468. [CrossRef] [PubMed]

Communication

Hierarchical Cluster Analysis of Medical Chemicals Detected by a Bacteriophage-Based Colorimetric Sensor Array

Chuntae Kim [1,†], Hansong Lee [2,3,†], Vasanthan Devaraj [4,†], Won-Geun Kim [1], Yujin Lee [1], Yeji Kim [1], Na-Na Jeong [5], Eun Jung Choi [4], Sang Hong Baek [6], Dong-Wook Han [7,*], Hokeun Sun [2,3,*] and Jin-Woo Oh [1,4,8,*]

[1] Department of Nanofusion Technology, Pusan National University, Busan 46241, Korea; chuntae1122@gmail.com (C.K.); kim1guen@gmail.com (W.-G.K.); pinky204@hanmail.net (Y.L.); kkyeaj0153@naver.com (Y.K.)

[2] Interdisciplinary Program of Genomic Data Science, Pusan National University, Busan 46241, Korea; hansong798@naver.com

[3] Department of Statistics, Pusan National University, Busan 46241, Korea

[4] Research Center for Energy Convergence and Technology, Pusan National University, Busan 46241, Korea; vasanth@pusan.ac.kr (V.D.); eunjung721203@gmail.com (E.J.C.)

[5] BK21PLUS Program in Embodiment: Health-Society Interaction, Department of Public Health Sciences, Graduate School, Korea University, Seoul 02841, Korea; nana8931@naver.com

[6] Laboratory of Cardiovascular Disease, Division of Cardiology, School of Medicine, The Catholic University of Korea, Seoul 06591, Korea; whitesh@catholic.ac.kr

[7] Department of Cogno-Mechatronics Engineering, College of Nanoscience & Nanotechnology, Pusan National University, Busan 46241, Korea

[8] Department of Nanoenergy Engineering, Pusan National University, Busan 46241, Korea

* Correspondence: nanohan@pusan.ac.kr (D.-W.H.); hsun@pusan.ac.kr (H.S.); ojw@pusan.ac.kr (J.-W.O.)

† These authors contributed equally to this work.

Received: 29 November 2019; Accepted: 6 January 2020; Published: 9 January 2020

Abstract: M13 bacteriophage-based colorimetric sensors, especially multi-array sensors, have been successfully demonstrated to be a powerful platform for detecting extremely small amounts of target molecules. Colorimetric sensors can be fabricated easily using self-assembly of genetically engineered M13 bacteriophage which incorporates peptide libraries on its surface. However, the ability to discriminate many types of target molecules is still required. In this work, we introduce a statistical method to efficiently analyze a huge amount of numerical results in order to classify various types of target molecules. To enhance the selectivity of M13 bacteriophage-based colorimetric sensors, a multi-array sensor system can be an appropriate platform. On this basis, a pattern-recognizing multi-array biosensor platform was fabricated by integrating three types of sensors in which genetically engineered M13 bacteriophages (wild-, RGD-, and EEEE-type) were utilized as a primary building block. This sensor system was used to analyze a pattern of color change caused by a reaction between the sensor array and external substances, followed by separating the specific target substances by means of hierarchical cluster analysis. The biosensor platform could detect drug contaminants such as hormone drugs (estrogen) and antibiotics. We expect that the proposed biosensor system could be used for the development of a first-analysis kit, which would be inexpensive and easy to supply and could be applied in monitoring the environment and health care.

Keywords: M13 bacteriophage; multi-array sensors; hierarchical cluster analysis; high selectivity

---

1. Introduction

M13 bacteriophage, one kind of filamentous bacteriophages, have been utilized as receptors in biosensors [1,2]. By means of genetic engineering techniques, it is possible to modify and achieve a better binding affinity of M13 bacteriophage towards desired target molecules [3–5]. Among various types of M13 bacteriophage-based biosensors, colorimetric sensor systems have been intensively investigated due to their facile fabrication process and sensing method [6–8]. Colorimetric sensors fabricated by self-assembly of M13 bacteriophage result in nanostructures with varying size and periodicity [9]. When white light is illuminated onto the nanostructure, specific wavelengths determined by Bragg's law are scattered more dominantly from the nanostructures. After the penetration of external chemicals, the self-assembled nanostructures swell, resulting in a change of the wavelengths that are scattered [10]. The observed color change is detected by a complementary metal–oxide–semiconductor detector, and this is followed by image analysis. The image analysis results consist of numerical values that can be used to determine the type and concentration of the external chemicals. Using genetic engineering techniques, M13 bacteriophage-based colorimetric sensors can display sensitive color changes toward desired target materials. Furthermore, by integrating various types of genetically engineered M13 bacteriophage-based colorimetric sensors on a single chip to fabricate a sensor array, a number of target molecules can be classified by analyzing the pattern of a color change. Even though M13 bacteriophage-based multi-array sensor systems have been successfully demonstrated as a powerful platform for the detection of extremely small amounts of target molecules, the discrimination of many types of target molecules is still both necessary and challenging. However, as the types of genetically engineered M13 bacteriophage integrated on sensor arrays increase, the analysis of the sensing results becomes complicated due to the huge amount of numerical data. In this work, we introduce a statistical method, henceforth called hierarchical cluster analysis, to classify many types of medical chemicals. Antibiotics are used widely in the livestock industry on a daily basis, but the excessive use of antibiotics raises the problem of increased antibiotic resistance in animals and humans [11]. If estrogen compounds, such as estrone (E1), 17β-estradiol (E2), and estriol (E3), and oral contraceptives released from humans and animals such as endocrine disrupters (similar to 17α-ethinylestradiol (EE2)) contaminate the environment through sewage and the excrement of animals, they can adversely affect the ecology in the water [12]. Estrogenic and antibiotic substances are found in trace amounts (~ng/L) in effluent, fresh water, river water, and even in drinking water [13,14], as a consequence of their ineffective removal at wastewater treatment plants (WWTP). In general, as any particular substance is not always present by itself, an effective method to comprehensively classify unknown compounds is necessary [15]. In this regard, M13 bacteriophage-based sensor arrays and hierarchical cluster analysis are a suitable sensor system to detect and classify various types of unknown compounds. Figure 1 is a schematic illustration describing our sensor system.

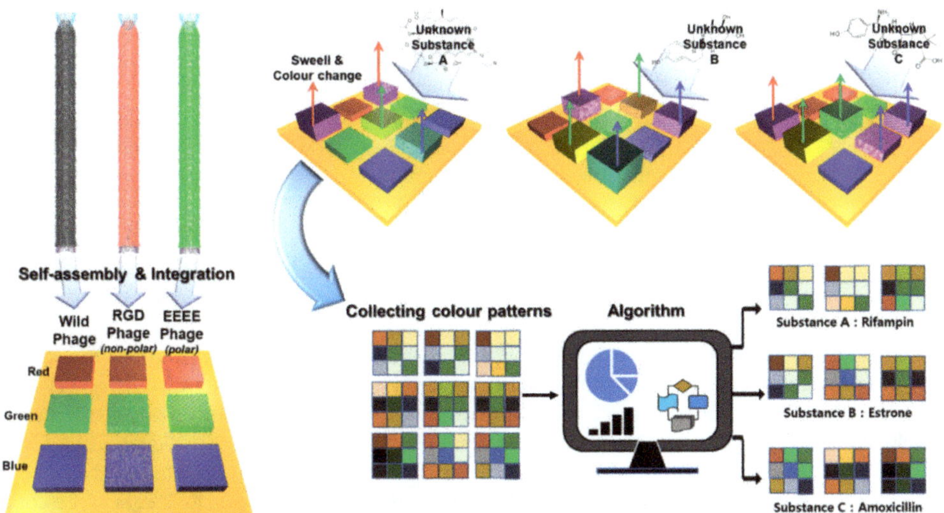

Figure 1. Schematic illustration of the M13 bacteriophage-based colorimetric sensor array. The sensor array consists of a functional M13 bacteriophage with a modified major coat protein (pVIII). When a color band is reacting with a target analyte, each type of sensor chip shows its own color change value according to the individual M13 bacteriophage's characteristics. A color pattern is formed as a unique response value, and it is possible to construct a sensor platform which can discriminate unknown textures through pattern analysis.

2. Materials and Methods

2.1. Sensor Analysis Analytes

Four types of estrogen drugs and four types of antibiotics were purchased from Sigma-Aldrich (Seoul, South Korea) and a local pharmacy (Medipharm, Busan, South Korea), ground, and used directly. In real-world situations, all such analytes exist in various physical conditions that cause molecules to vibrate or diffuse. To mimic these kinds of physical conditions, sensing experiments were carried out at four different temperatures (30, 60, 90, and 120 °C) for each material.

2.2. Development of Functional Bioreporter Materials

The wild types of filamentous phage particles were based on M13KE, which was purchased from New England Biolabs (Ipswich, MA, USA). The major coat protein (pVIII) of an M13 bacteriophage with an arginine-, glycine-, and aspartic acid-based functional peptide sequence (-RGD-) [16] and a glutamic acid-based peptide sequence (-EEEE-) were fabricated on M13KE phage [17].

2.3. Phage Colorimetric Sensor-Based Multi-Array Chip

A silicon wafer with a 100 nm gold film on a 5 nm platinum adhesion layer was selected as a substrate to deposit the M13 bacteriophage. Simple pulling methods [18] were utilized to prepare three kinds of colorimetric sensors consisting of wild-, RGD-, and EEEE-type M13 bacteriophages. The three types of colorimetric sensors consisted of red-, green-, and blue-colored M13 bacteriophage films that were determined by pulling speeds of 30 μm/min, 40 μm/min, and 50 μm/min, respectively. The fabricated colorimetric sensors were integrated into a single chip.

2.4. Colorimetric Signal Analysis and Data Processing

The color change was observed using a handheld digital microscope (Celestron DELUXE HANDHELD DIGITAL MICROSCOPE, Torrance, CA, USA) in the chamber of a color sensor array set made from phage chips. The colorimetric signal of the unit cell constituting each sensor chip was measured as the average value of a ~1000 pixel square. A method of setting the values of ΔR, ΔG, and ΔB was used, and the changes in each variable were used as the criteria for the amount of color change [7].

2.5. Analytical Data Statistics

Essentially, model-less clustering analyses, such as principal component analysis or hierarchical clustering analysis, use hierarchical analysis because they are unsuitable for predictive (i.e., classificatory) use. Hierarchical clustering analysis is a statistical analysis that classifies nearby objects into the same group using the distance that indicates the similarity of each object [19]. The method is called hierarchical cluster analysis because it forms a tree-like hierarchical structure by starting from objects at the closest distance and combining them. Generally, Euclidean distances are commonly used for distance calculations. On the other hand, the color distance * is used so that the color characteristics of RGB can be utilized, as the data used for the analysis are RGB data, representing a color change value that occurs when one sample reacts with another sample.

$$\sum_{i=1}^{n} \sqrt{\left(\left(R_i^a - R_i^b\right)^2 + \left(B_i^a - B_i^b\right)^2 + \left(G_i^a - G_i^b\right)^2\right)}$$

* Color distance: The color distance metric calculates the Euclidean distance in color space between each pair of clusters, ignoring their size. The distance between images a and b (using RGB) was calculated as in the above expression, where n is the number of bins.

3. Results and Discussion

Figure 1 depicts a schematic illustration of the M13 bacteriophage-based multi-array sensor chip and color-pattern analysis process. The multi-array sensor chip shows specific color patterns according to the applied materials due to the different chemical groups expressed on the M13 bacteriophage. Furthermore, microstructures of colorimetric sensors determined by the pulling speed contribute to the color change, owing to their different surface-to-volume ratios [7,8,10,18]. After collecting the color patterns, a statistical analysis process determines the types of unknown substances. Figure 2 displays the specific color patterns toward each target chemical at a temperature of 120 °C. The specific color patterns were generated by calculating pixel differentiation between initial color and color after exposure to the target chemical. As shown in Figure 2, the M13 bacteriophage-based sensor array displays specific color change according to the exposed chemicals. However, because this was a qualitative analysis, there were limitations in the exact classification of the target chemicals. The M13 bacteriophage-based sensor array, different from specialized sensors such as litmus paper or a pregnancy tester, is a non-specific sensor system, which means small amounts of color changes should be analyzed and classified exactly. Signal detection and data analysis by the sensor of the drug component are based on the inherent chemical properties of the material. The analysis of female hormone-related drugs, antibiotic components, and their clustering tendencies were analyzed through various features. Through the RGB distance values, the Euclidean distance was set to the RGB distance, and a hierarchical classification was performed using the Ward Linkage (Ward.D) method [20] (see Figure 3). The similarity between two clusters was measured on the basis of the increment of the error sum of squares (ESS) when two clusters were combined. The increase of ESS was measured as the distance between two clusters when the distance matrix was obtained. This method was less sensitive to noise or outlier data than the Single Linkage Method [20]. The clustering results showed a close relationship with each chemical composition. They confirmed that estrogen-based drugs and antibiotics-based drugs can be classified within a large frame and partially. In particular, the similar chemical structures of antibiotic substances made their distances very close to each other. Their vapor pressure values are similar to each-other at various temperatures, and these compounds react comparably with sensor array chips. The similar composition of the pharmaceutical forms containing each compound also had an impact.

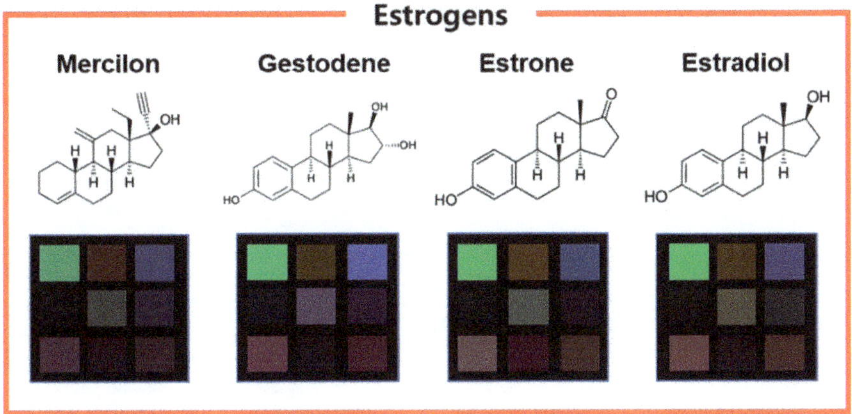

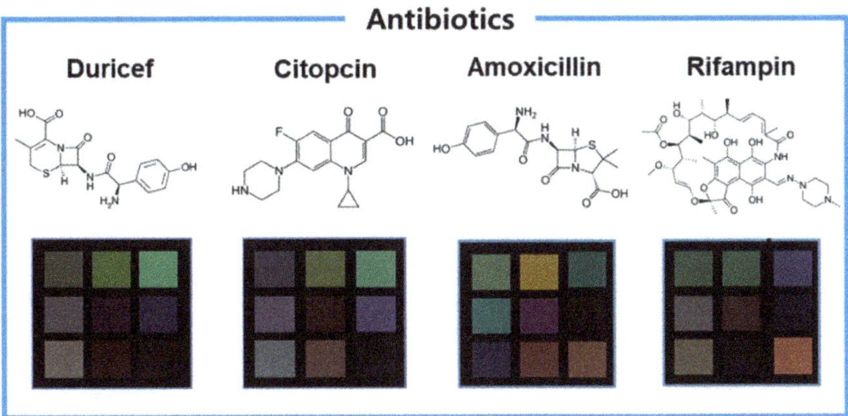

Figure 2. Image of the colorimetric sensor array chip after exposure to medical chemicals. The color pixels represent the mean value of the variation of the RGB values (in 8 bit) as a function of the M13 bacteriophage bundle's structural change.

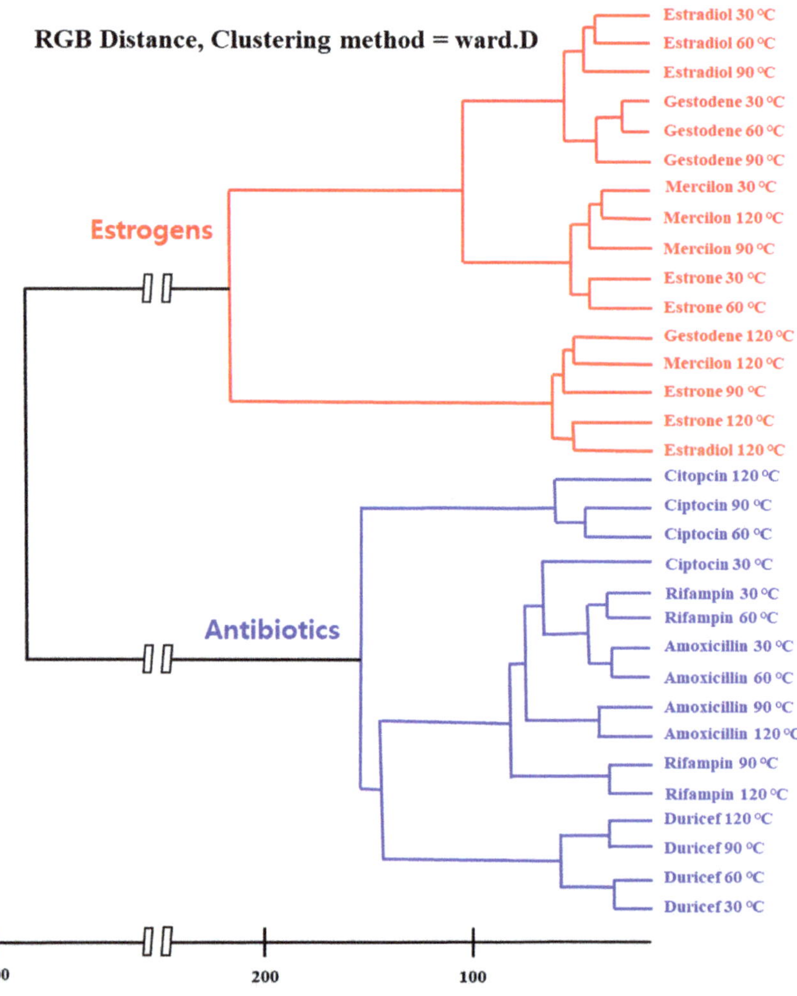

Figure 3. Hierarchical cluster analysis dendrogram for eight types of medical chemicals using the Ward.D linkage method, which is based on the linear model criterion of least squares. The Euclidean distance was set to the ΔRGB matrix which is made of the colorimetric shift values.

4. Conclusions

We fabricated a functional M13 bacteriophage-based photonic crystal structure and set a complex detection sensor device with three kinds of sensor chips with chemically inherent functional groups. We propose a sensor model that can classify antibiotics and hormone drugs by analyzing the color change amount of a 3 × 3 squared band, i.e., nine sections by setting three color bands per chip. In particular, commercial chemicals can be classified according to the product brand, so that the clustering results can slightly overlap. We see the possibility of integrating this method in a simple sensor kit for quality assurance and quality control of commercial drugs. This colorimetric sensor array model could be used to fabricate a very simple sensor platform consisting of a sensor chip 1 square centimeter wide, a chamber of about 30 cc capacity, and a small webcam for color change observation. The multi-array sensor system proposed in this work consists of nine different types of sensor chips, but in the future, by increasing the types of sensors, the selectivity of the multi-array sensor system could significantly ameliorate. Furthermore, our multi-array sensor platform could be used for user-friendly pre-analyses that could be simple and performed in real time, before instrument-based chromatographic analyses.

Author Contributions: Sensor fabrication, C.K. and Y.L.; Experimental set-up construction, Y.K. and V.D.; Preparation of genetically engineered M13 bacteriophage, C.K. and E.J.C.; Mass amplification of M13 bacteriophage, C.K. and N.-N.J.; Preparation of manuscript, C.K. and S.H.B.; Schematic illustrations, W.-G.K.; Hierarchical cluster analysis, H.L. and H.S.; Supervision and editing manuscript, D.-W.H. and J.-W.O. All authors have read and agreed to the published version of the manuscript.

Funding: This work was supported by a grant of the Korea Health Technology R&D Project through the Korea Health Industry Development Institute (KHIDI) funded by the Ministry of Health & Welfare, Republic of Korea (HI17C1662) and by Korea Institute of Planning and Evaluation for Technology in Food, Agriculture and Forestry(IPET) through Advanced Production Technology Development Program, funded by Ministry of Agriculture, Food and Rural Affairs(MAFRA)(318104-3) and by National Research Foundation of Korea (2019R1A4A1024116).

Conflicts of Interest: The authors declare no conflict of interest.

References

1. Petrenko, V.A.; Vodyanoy, V.J. Phage display for detection of biological threat agents. *J. Microbiol. Methods* **2003**, *53*, 253. [CrossRef]

2. Goldman, E.R.; Pazirandeh, M.P.; Mauro, J.M.; King, K.D.; Frey, J.C.; Anderson, G.P. Phage-displayed peptides as biosensor reagents. *J. Mol. Recognit.* **2000**, *13*, 382. [CrossRef]

3. Smith, G.P. Filamentous fusion phage: Novel expression vectors that display cloned antigens on the virion surface. *Science* **1985**, *228*, 1315. [CrossRef] [PubMed]

4. Smith, G.P.; Petrenko, V.A. Phage display. *Chem. Rev.* **1997**, *97*, 391. [CrossRef] [PubMed]

5. Petrenko, V.A.; Smith, G.P. Phages from landscape libraries as substitute antibodies. *Protein Eng. Des. Sel.* **1996**, *9*, 797. [CrossRef] [PubMed]

6. Kim, C.; Lee, S.-Y.; Kim, W.-G.; Oh, J.-W. High sensitive and selective virus based structural colorimetric sensor. *Mol. Cryst. Liq. Cryst.* **2014**, *598*, 171. [CrossRef]

7. Moon, J.-S.; Lee, Y.; Shin, D.M.; Kim, C.; Kim, W.G.; Park, M.; Han, J.; Song, H.; Kim, K.; Oh, J.W. Identification of endocrine disrupting chemicals using a virus-based colorimetric sensor. *Chem. Asian. J.* **2016**, *11*, 3097. [CrossRef] [PubMed]

8. Moon, J.-S.; Kim, W.-G.; Shin, D.-M.; Lee, S.-Y.; Kim, C.; Lee, Y.; Han, J.; Kim, K.; Yoo, S.Y.; Oh, J.-W. Bioinspired M-13 bacteriophage-based photonic nose for differential cell recognition. *Chem. Sci.* **2017**, *8*, 921. [CrossRef] [PubMed]

9. Chung, W.-J.; Oh, J.-W.; Kwak, K.; Lee, B.Y.; Meyer, J.; Wang, E.; Hexemer, A.; Lee, S.-W. Biomimetic self-templating supramolecular structures. *Nature* **2011**, *478*, 364. [CrossRef] [PubMed]

10. Oh, J.-W.; Chung, W.-J.; Heo, K.; Jin, H.-E.; Lee, B.Y.; Wang, E.; Zueger, C.; Wong, W.; Meyer, J.; Kim, C.; et al. Biomimetic virus-based colourimetric sensors. *Nat. Commun.* **2014**, *5*, 3043. [CrossRef] [PubMed]

11. Mathew, A.G.; Cissell, R.; Liamthong, S. Antibiotic resistance in bacteria associated with food animals: A united states perspective of livestock production. *Foodborne Pathog. Dis.* **2007**, *4*, 115. [CrossRef] [PubMed]

12. Mills, L.J.; Chichester, C. Review of evidence: Are endocrine-disrupting chemicals in the aquatic environment impacting fish populations? *Sic. Total Environ.* **2005**, *343*, 1. [CrossRef] [PubMed]

13. Hamid, H.; Eskicioglu, C. Fate of estrogenic hormones in wastewater and sludge treatment: A review of properties and analytical detection techniques in sludge matrix. *Water Res.* **2012**, *46*, 5813. [CrossRef] [PubMed]

14. Kuch, H.M.; Ballschmiter, K. Determination of endocrine-disrupting phenolic compounds and estrogens in surface and drinking water by HRGC-(NCI)-MS in the picogram per liter range. *Environ. Sci. Technol.* **2001**, *35*, 3201. [CrossRef] [PubMed]

15. Rakow, N.A.; Sen, A.; Janzen, M.C.; Ponder, J.B.; Suslick, K.S. Molecular recognition and discrimination of amines with a colorimetric array. *Angew. Chem. Int. Ed.* **2005**, *44*, 4528. [CrossRef] [PubMed]

16. Choi, D.S.; Jin, H.-E.; Yoo, S.Y.; Lee, S.-W. Cyclic RGD peptide incorporation on phage major coat proteins for improved internalization by HeLa Cells. *Bioconjug. Chem.* **2014**, *25*, 216. [CrossRef] [PubMed]

17. Lee, B.Y.; Zhang, J.; Zueger, C.; Chung, W.-J.; Yoo, S.Y.; Wang, E.; Meyer, J.; Ramesh, R.; Lee, S.-W. Virus-based piezoelectric energy generation. *Nat. Nanotechnol.* **2012**, *7*, 351. [CrossRef] [PubMed]

18. Kim, W.-G.; Kim, K.; Ha, S.-H.; Song, H.; Yu, H.-W.; Kim, C.; Kim, J.-M.; Oh, J.-W. Virus-based full colour pixels using a microheater. *Sci. Rep.* **2015**, *5*, 13757. [CrossRef] [PubMed]
19. Anton, H. *Elementary Linear Algebra, Binder Ready Version*; John Wiley & Sons: Hoboken, NJ, USA, 2013.
20. Ward, J.H., Jr. Hierarchical grouping to optimize an objective function. *J. Am. Stat. Assoc.* **1963**, *58*, 236. [CrossRef]

 *nanomaterials* 

Review

Research Progress of M13 Bacteriophage-Based Biosensors

Jong-Sik Moon [1,*,†], Eun Jung Choi [2,†], Na-Na Jeong [3,†], Jong-Ryeul Sohn [3], Dong-Wook Han [4] and Jin-Woo Oh [2,5,*]

[1] National Core Research Center for Hybrid Materials Solution, Pusan National University, Busan 46241, Korea
[2] Department of BIT Fusion Technology Center, Pusan National University, Busan 46241, Korea; eunjung721203@gmail.com
[3] Department of Public Health Sciences, Graduate School, Korea University, Seoul 02708, Korea; nana8931@naver.com (N.-N.J.); sohn1956@korea.ac.kr (J.-R.S.)
[4] Department of Cogno-Mechatronics Engineering, College of Nanoscience & Nanotechnology, Pusan National University, Busan 46241, Korea; nanohan@pusan.ac.kr
[5] Department of Nanofusion Technology, College of Nanoscience & Nanotechnology, Pusan National University, Busan 46241, Korea
* Correspondence: jsmoon@pusan.ac.kr (J.-S.M.); ojw@pusan.ac.kr (J.-W.O.)
† These authors contributed equally to this work.

Received: 3 September 2019; Accepted: 3 October 2019; Published: 11 October 2019

Abstract: Recently, new virus-based sensor systems that operate on M13 bacteriophage infrastructure have attracted considerable attention. These systems can detect a range of chemicals with excellent sensitivity and selectivity. Filaments consistent with M13 bacteriophages can be ordered by highly established forms of self-assembly. This allows M13 bacteriophages to build a homogeneous distribution and infiltrate the network structure of nanostructures under mild conditions. Phage display, involving the genetic engineering of M13 bacteriophages, is another strong feature of the M13 bacteriophage as a functional building block. The numerous genetic modification possibilities of M13 bacteriophages are clearly the key features, and far more applications are envisaged. This paper reviews the recent progress in the application of the M13 bacteriophage self-assembly structures through to sensor systems and discusses future M13 bacteriophage technology.

Keywords: virus; biosensor; M13 bacteriophage; color sensor; phage display

1. Introduction

1.1. M13 Bacteriophage

Distinguishing infinitesimal amounts of chemicals or microbials in a precise and simple manner is one of the most important techniques in academia and industry. Classical sensor technology is categorized as electromagnetic (gamma radiation, optics, microwave, radio wave, and Eddy current), mechanical (sound, MEMS, and fluid), magnetic, chemical (affinity, catalytic reactions, electrochemistry, and biochemistry), nuclear (nuclear magnetic resonance), and a combination of them (opto-acoustics and membrane technology) [1,2]. Generally, the key features of sensor technology are the low cost, small size, robustness, selectivity, and sensitivity [1,3].

Recently, the use of ecofriendly materials has been one of the main points in research and industrial manufacturing. The M13 bacteriophage, which can be categorized as a natural polymer, has been a major research focus owing to its target-specific response in chemical reactions with outstanding sensitivity [4]. The bacteriophage is a human-safe viral material that only infects certain criteria of bacteria, such as *Escherichia coli.* [5].

The filamentous type of M13 consists of approximately 2700 copies of the spirally arranged pVIII major coat protein on the body, with 5 to 7 copies of pIII, pVI, pIX, and pVII minor coat proteins at each end. M13 has a regular form, with a length and diameter of ~880 nm and ~6.6 nm, respectively. The genome structure and protein sequences of the M13 bacteriophage are fully understood. Therefore, M13 bacteriophages are very simple to genetically engineer to achieve the desired biochemical properties [6–9]. Figure 1 shows the basic protein structures and genetic engineering pathways of M13 bacteriophages. For example, genetically modifying the pVIII protein of M13 bacteriophage to possess the tryptophan–histidine–tryptophan (WHW) peptide sequence results in it showing specific binding affinity to nitrotoluene derivatives [4,10–14]. Consequently, the specific binding affinity to target materials, including chemical and biological materials, by simple genetic engineering provides great possibilities to use M13 bacteriophages as the core material of sensors.

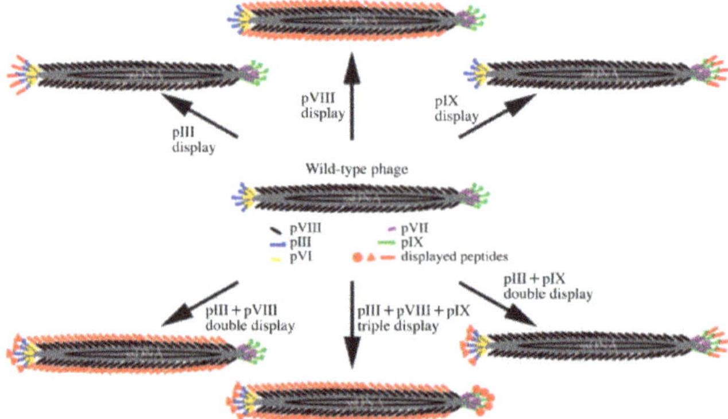

Figure 1. Basic structure of the M13 bacteriophage and the possible pathway of generic engineering. Reproduced from [9], with permission from WILEY-VCH, 2009.

M13 bacteriophages have a filamentous shape and different protein configurations at each end. One end has the pVI and pIII protein and the other has the pVII and pIX protein. By the well-defined, anisotropic structure and monodispersity caused by the natural character of M13, the material can be fabricated easily in hierarchical self-assembled structures using appropriate techniques [4,14]. The structure and behavior of M13 bacteriophages during fabrication are similar to those of liquid crystals. The bacteriophage can also be modified chemically and genetically to reveal a range of features, such as desired functionality and target-specific affiliation [11–13].

The target-specific reactivity character of the M13 bacteriophage makes it advantageous for sensor applications. The other strong point of the M13 bacteriophage is its ability to form higher-order dimensional structures in a specified order over self-assembly technology [13]. High-dimensional structures using natural building blocks, such as the M13 bacteriophage, were inspired by natural systems, such as the collagen bundle structure [4]. Although the imitation of a natural assembly structure on the micro scale using artificial building blocks, such as polymers or artificial nanomaterials, is progressing slowly, due to uncertainty and complexity, using natural building blocks such as M13 bacteriophage can be a solution. The filament structure and short dispersion of the M13 bacteriophage allow liquid crystal-like behavior in suspension under controlled conditions [15,16]. This liquid crystal behavior of the M13 bacteriophage can be managed mostly by the concentration of the M13 bacteriophage aqueous suspension. The crystal structure of self-assembled M13 bacteriophages shows nematic-, cholesteric-, and smectic-phase at low, medium, and high concentrations, respectively [16]. M13 is also attractive owing to its easy growth and handling characteristics. The use of natural substances as a building block makes a synthetic process possible at low cost under a non-toxic environment. The fabrication process associated with M13 bacteriophages using aqueous self-assembly technology allows easy processing without the need for organic solvents and additional processing.

By using the pulling technique, concentration of M13 bacteriophage aqueous solution and polling speed of substrate are main factors in the process [17,18].

An additional outstanding benefit of M13 bacteriophages is the easy genetic modification for the specific binding affinity to definite target substances. Each coat protein can be modified genetically using phage display technology. Random phage libraries containing more than 1.0×10^{11} random peptide sequences of M13 bacteriophages are screened for specific target materials [19,20]. The binding phage on the target material is then selected and the bacteria are infected and amplified. After repeating this process to isolate the target-specific binding phage, the target-specific peptide can be recognized by DNA analysis [21,22]. Recently, the technology has developed a M13 bacteriophage that binds specifically to several inorganic resources, e.g., GaAs, GaN, Ag, Pt, Au, Pd, Ge, Ti, SiO_2, quartz, $CaCO_3$, ZnS, CdS, Co, TiO_2, ZnO, CoPt, FePt, $BaTiO_3$, $CaMoO_4$, and hydroxyapatite [21–25]. The M13 bacteriophage can be made to have specific functionality by genetically engineering to carbon-based nanomaterials, such as C_{60}, graphene, and carbon nanotubes [24,25]. Moreover, these specific binding capacities of the M13 bacteriophage can be used as templates to produce various nanostructures, such as highly monodisperse nanostructures [26–28]. This review introduces the current advances of M13 bacteriophage-based biosensors.

1.2. M13 Bacteriophage-Based Biosensor

The M13 bacteriophage has been used to improve existing sensor systems. Förster resonance energy transfer (FRET) is based on long-term bipolar interactions between excited state donors and ground state recipients. The distance of each N-terminal to the end of the peptide on the surface of a M13 bacteriophage is approximately 3.2 nm (oa) and approximately 2.3 nm (ob), respectively. Wang et al. recently reported the fabrication of FRET-based lattice probes using the M13 bacteriophage [29,30]. Owing to the regular organization of the M13 bacteriophage on the substrate, a thin coating film of M13 bacteriophage can be used for surface plasmon resonance (SPR) measurements. Genetically engineered M13 bacteriophages, which possess the Arg-Gly-Asp (RGD) peptide sequence, have been used to detect the SPR signal of the cell proliferation rate and the morphology of cells [31]. The signal strength of surface-enhanced Raman spectroscopy (SERS) depends strongly on the distance between the target molecules and novel metal surface. Therefore, the controlled organization of metal nanoparticles and Raman active dye is the critical point. Cha et al. introduced an Au@Ag-core shell nanoparticle using an M13 bacteriophage as the template. In this approach, nanoparticles were functionalized by DNA-conjugated M13 bacteriophage, and showed a 75 times stronger Raman signal than DNA-functionalized nanoparticles without the M13 bacteriophage. This was mainly caused by the large number of functional moieties on pVIII protein of the M13 bacteriophage [32]. The M13 bacteriophage could also be used medically because of the strong specific reactivity to a particular reactant. Mao et al. fabricated M13–liposome–ZnPc (zinc phthalocyanine) for a more stable and upgraded cancer drug delivery system [33–35]. Simple genetic engineering could reveal target-specific affinity to the M13 bacteriophage. Oh et al. engineered the pVIII protein of M13 bacteriophage to have the AXXXWHWQXXDP (WHW) sequence and showed excellent binding affinity to trinitrotoluene (TNT), as shown in Figure 2a,b [36,37]. The modified M13 bacteriophage was then fabricated as a color film structure through a simple pulling technique. As shown in Figure 2c,d, the M13 bacteriophage-based color sensor could detect the gas phase of TNT down to the 300 p.p.b level and showed superior selectivity among TNT, Dinitrotoluene (DNT), and Nitrotoluene (MNT) [4].

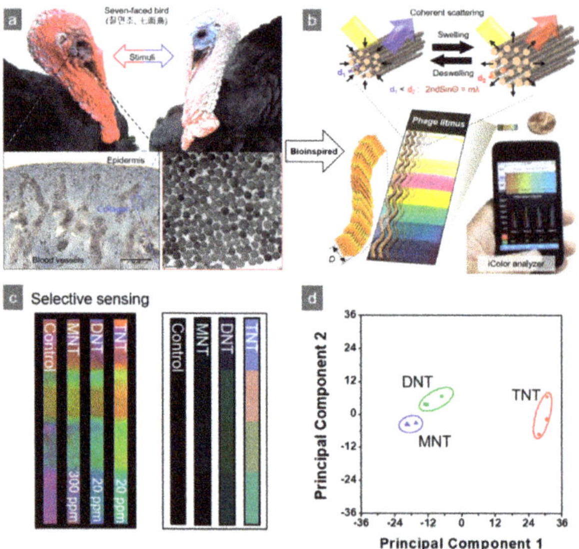

Figure 2. Nitrotoluene derivative detection using a M13 bacteriophage-based color sensor. (**a**) Bio-mimic structure of turkey collagen, (**b**) reversible color responding of a color sensor, (**c**) sensitivity of M13 bacteriophage-based color sensor upon exposure to nitrotoluene derivatives, and (**d**) principal component analysis (PCA) plot of the color changes. Reproduced from [4], with permission from Springer Nature, 2014.

2. M13 Bacteriophage-Based Protein and Microorganism Sensing

Indeed, using a natural M13 bacteriophage without any genetic engineering provides a great opportunity because of the abundant negative charge on the surface protein. The α-helical major coat protein (pVIII) has glutamate amino acid on the end of its sequence. Therefore, the dipole moment of the whole pVIII protein is directed from the outside (*N*-terminus) to the core (*C*-terminus). The charge distribution of the *C*-terminus (positive) and *N*-terminus (negative) induces a strong dipole to M13 bacteriophages possessing a natural negative charge [38]. This property allows the M13 bacteriophage to mix easily with positively charged materials, such as carbon nanofiber (CNF). Recently, Niedziòłka-Jönsson et al. reported a M13 bacteriophage and CNF complex structure for cysteine detection. The M13 bacteriophage and CNF were mixed in an aqueous solution, and it was assumed that the electrostatic interaction and π-π interaction between the M13 bacteriophage and CNF promoted an even distribution of CNF without aggregation, which occurred constantly when CNF was used alone in solution [39–41]. As a result, the addition of M13 bacteriophage to CNF increased the capacitive current with the growth of the faradaic current. An overall improvement of the electrochemical properties of the glassy carbon electrode (GCE)–CNF–M13 bacteriophage complex electrode applied to the electrocatalytic oxidation of the cysteine was observed and enhanced results were obtained, as shown in Figure 3 [39]. The fabrication in this study simply dropped CNF solution on the substrate (indium tin oxide (ITO) glass) then M13 bacteriophage dropped on the dried CNF. Although the process is not yet optimized, the study shows promising results. Using wild type M13 bacteriophage for this experiment was another strong point due to the simple preparation.

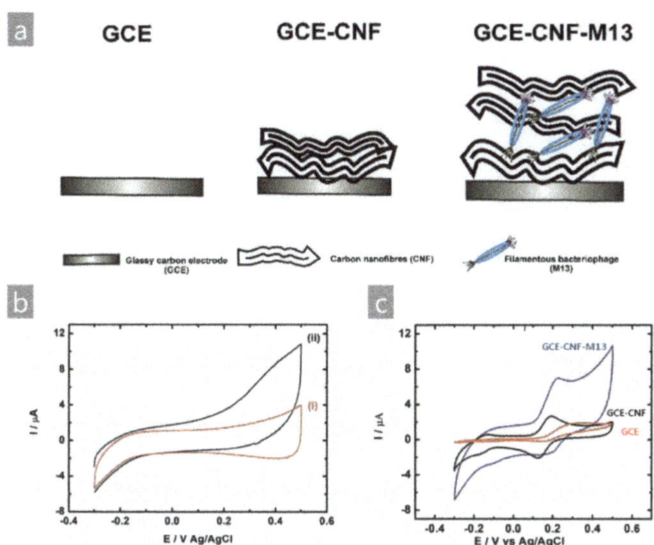

Figure 3. (**a**) Schematic diagram of the device fabrication of the CNF and M13 bacteriophage complex electrode, (**b**) cyclic voltammograms (3rd scan) of (i) glassy carbon electrode (GCE)–CNF and (ii) GCE–CNF–M13 bacteriophage treated in 1 mM cysteine solution in PBS, and (**c**) cyclic voltammograms (2nd scan) of the bare GCE, GCE–CNF, and GCE–CNF–M13 bacteriophage. Reproduced from [39], with permission from Elsevier, 2019.

As mentioned above, simple genetic engineering provides a specific interaction to a designated target material. Lladser et al. genetically engineered colorectal cancer (CRC)-specific M13 bacteriophages. The M13 bacteriophage was modified to have carcinoembryonic antigen (CEA)-specific moiety. CEA exists abundantly in colorectal cancer and is known to support the malignant features of colorectal cancer cells such as cell adhesion, cell-to-cell interaction, and signal transduction [42,43]. The CEA level clearly has metastatic potential, cancer progression, differentiation, and apoptosis of CRC cells [44–46]. In this study, CEA-specific (αCEA) M13 bacteriophages were applied to tumor cells for the tumor infiltration of neutrophils, macrophages, and the maturing of dendritic cells in tumor-draining lymph nodes [44]. For M13 bacteriophage genetic engineering, they transformed *E. coli* with a pSEX81 surface expression phagemid vector, which possesses CEA-specific single-chain fragment variable (scFv). The scFv was inserted at the NcoI and BamHI restriction sites upstream of the pIII protein sequence. *E. coli* were infected with a multiplicity of infection by a hyperphage that contains all genes of structural proteins of M13 bacteriophage except for the pIII protein [47]. The αCEA-M13 bacteriophage exhibited strong binding affinity to both CEA and CEA-expressing tumor cells (CT26) in vitro. The applied αCEA-M13 bacteriophage effectively suppressed a mouse cancer model compared to the phosphate-buffered saline (PBS)-control and wild type M13 bacteriophage experiment by successful intratumoral and systemic administration. Here, they confirmed that the tumor protection provided by αCEA-M13 bacteriophage occurred via CD8$^+$ T cells because the reduction of circulating CD8$^+$ T cells completely eliminated the antitumor protection. Figure 4 shows that the macrophage (F4/80$^+$) increased from approximately 24% to 54%; the neutrophils increased from approximately 6% to approximately 42%, and the tumor-infiltrating dendritic cells decreased from approximately 6% to approximately 2%. In other words, the application of αCEA-M13 bacteriophages to tumor cells resulted in an increase in the tumor infiltration of innate immune cells and maturing of dendritic cells at lymph nodes. Figure 5 shows decreased tumor volume and increased survival rate of mice by treating with αCEA-M13 bacteriophage. Their studies can be regarded as a successful suggestion of potential immunotherapy against CRC [44]. Antigen–antibody reaction is one of the most simple and widely used treatment methods in the medical field. However, it is still the most useful clinical method. Using highly concentrated functional moiety on the M13 bacteriophage, these types of treatments can be more effective as immunotherapy against CRC.

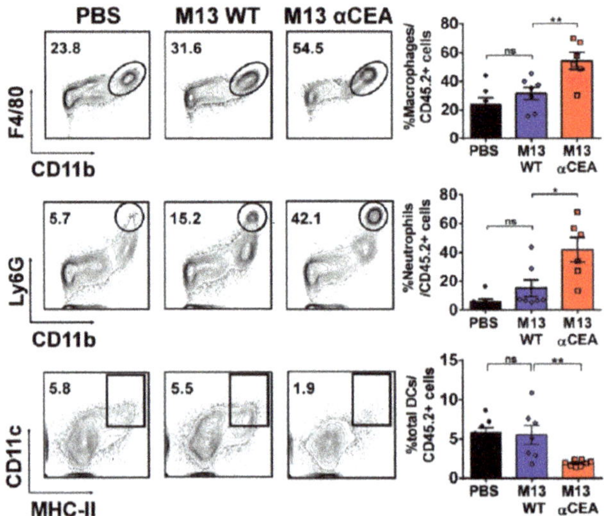

Figure 4. Treating the tumor microenvironment using PBS, wild type M13 bacteriophage, and αCEA-M13 bacteriophages for 24 h. Contour plots shows tumor infiltrating macrophages (F4/80$^+$CD11b$^+$), (**upper left**), neutrophils (CD11b$^+$Ly6G^{high}), (**middle left**), and dendritic cells (CD11c$^+$MHC-II$^+$), (**lower left**). The percentage of each plot (**right-hand panel**). Reproduced from [44], with permission from Springer Nature, 2017.

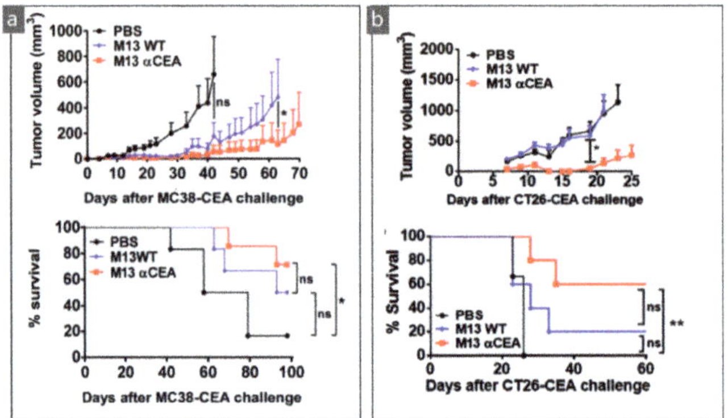

Figure 5. Growth of different CEA tumors ((**a**) MC38-CEA, (**b**) CT26-CEA) and survival of mice treated with M13 bacteriophage with or without αCEA. Reproduced from [44], with permission from Springer Nature, 2017.

Salmonella detection using a M13 bacteriophage is one of the largest fields of M13 bacteriophage research and could feasibly replace conventional methods, such as the polymerase chain reaction (PCR) and immunology-based assays [48–50]. Among the many types of pathogenic bacteria, *Salmonella* detection has received more attention because of its prevalence [51]. Recently, Thavarungkul et al. described a capacitive flow injection system for *Salmonella* spp. detection using a *Salmonella*-specific M13 bacteriophage in a working electrode [48]. They genetically modified the wild pVIII protein to have the NRPDSAQFWLHHGG sequence based on the phage display technique. They simply placed the *Salmonella*-specific M13 bacteriophage on a polytyramine (Pty)-coated gold electrode, and the amount of injected target *Salmonella* was determined by the capacitance of *Salmonella* over the total capacitance. Figure 6a shows the overall sensing procedure of this sensor device. The M13 bacteriophage-based electrode sensor was located in the running flow buffer. When the target *Salmonella* spp. was injected, the capacitance rolled up and down, and the new capacitance could be obtained. This capacitance was lower than that without a reaction to *Salmonella*. The difference in capacitance is denoted as ΔC and the total capacitance could be calculated as follows:

$$\frac{1}{C_{total}} = \frac{1}{C_{Pty}} + \frac{1}{C_{phage}} + \frac{1}{C_{Slamonella}} \tag{1}$$

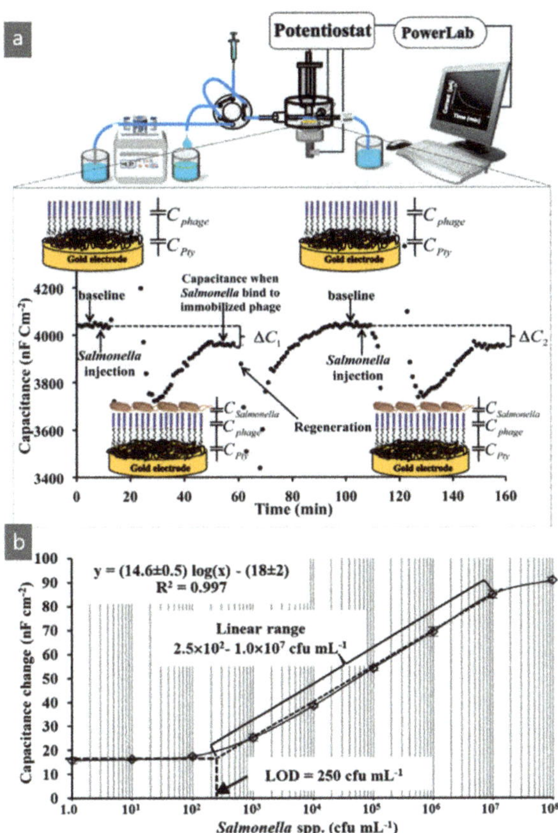

Figure 6. (a) Schematic procedure of the capacitive flow injection system for sensing *Salmonella* spp. using a M13 bacteriophage-based electrode sensor and (b) linear response of the sensor by number of *Salmonella* spp. Reproduced from [48], with permission from Elsevier, 2018.

The M13 bacteriophage-based electrode sensor was then washed with a recovery solution (5.0 mM of NaOH solution) to break the binding between the M13 bacteriophage and *Salmonella*. The M13 bacteriophage-based *Salmonella* sensor showed great selectivity to the *Salmonella* samples and exhibited an excellent recovery ratio, almost 100%. As a result, in Figure 5b, the sensor showed good sensing capability to detect *Salmonella* in chicken samples with linear range from 1.0×10^3 to

1.0×10^7 cfu/mL [48]. *Salmonella* detection is one of the main M13 bacteriophage biosensor research areas. Meanwhile, this research was performed using magnetoelastic, SPR, and microcantilever technique. The capacitive flow system was introduced very recently, and showed great results.

The target-specific binding affinity of the M13 bacteriophage could alter the macro-structure or morphology as well as the charge density of the M13 bacteriophage suspension. Therefore, the acoustic signal that passes through the M13 bacteriophage suspension can indicate the target-specific reaction both quantitatively and qualitatively [52–54]. Formin et al. reported an M13 bacteriophage-based biosensor for the detection of microbial cells with antibodies on the M13 bacteriophage [52,53]. They reacted initially with the specific antibody on the M13 bacteriophage to have a specific interaction with the target cell line (*Azospirllum basilense* strain Sp245). The cell suspension was located directly on the thin piezoelectric plate, which propagates the piezoactive acoustic wave by the input signal. Figure 7 presents a schematic diagram. After a specific interaction of M13 bacteriophages with the target microbial, they could measure significant insertion loss and phase change by a reaction. The parameters were also changed sensitively by different microbial numbers. Figure 8 shows increase conductivity of the solution caused a decrease in change of the phase and insertion loss of the output signal by increasing the number of cells. By increasing cell numbers, the interaction between M13 bacteriophage and cells occurs. Resulting clear signal changes were observed from 10^4 to 10^7 cells/mL [52]. Thus far, this is one of a few results using electroacoustic techniques to detect microbial target-specific interaction. Since this technique can be performed directly in suspension, it can be a useful supplement method for existing methods.

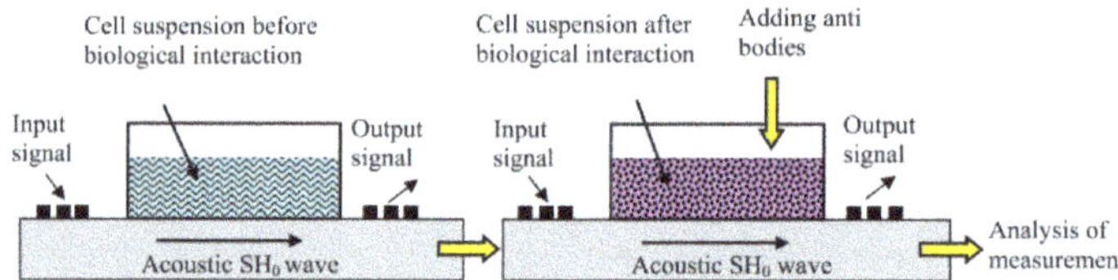

Figure 7. Schematic diagram of the electroacoustic measurements using an X–Y lithium niobate piezoelectric plate and target cell-specific M13 bacteriophage solution. Reproduced from [52], with permission from Elsevier, 2018.

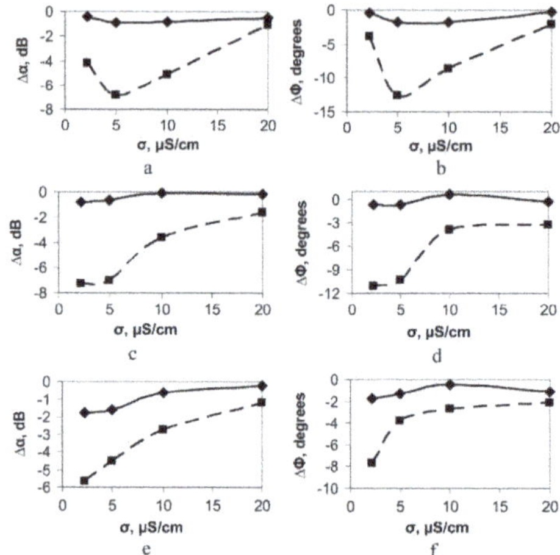

Figure 8. The change in the insertion loss ($\Delta\alpha$, left) and phase of the output signal ($\Delta\Phi$) of the conductivity of the buffer solution. (**a,b**) 10^4 cells/mL, (**c,d**) 10^6 cells/mL, and (**e,f**) 10^8 cells/mL. The solid and dotted lines correspond to the electrically open and shored channels, respectively. Reproduced from [52], with permission from Elsevier, 2018.

3. M13 Bacteriophage-Based Chemical Sensing

The M13 bacteriophage has a systemically regular shape because it is a natural substance. The highly certain shape and specific reactivity to the target material provided to the M13 bacteriophage have great potential to become a building template [6]. Haberer et al. discussed the gold nanopeapod structure covered with polypyrrole (PPy) located directly between platinum electrodes. In this particular experiment, an M13 bacteriophage was genetically engineered to have the VSGSSPDS protein sequence on its pVIII major coat protein. A gold-specific M13 bacteriophage was covered and formed a bridge between two pre-fabricated platinum electrodes, and then reacted with a gold solution. A commercial gold electroless deposition reagent was used to increase the size of the gold nanoparticle. Finally, electrodeposition of PPy was conducted to complete the gold PPy nanopeapod structure between the electrodes [55,56]. The product was applied as an NH_3 gas sensor due to nucleophilic NH_3 attack of the polymer backbone, decreasing the conjugation length, and irreversibly increasing the resistance. In other words, NH_3 molecules in this reaction act as an electron donor. Consequently, decreasing the electron hole concentration of the polymer results in a reversible de-doping process [57–59]. In this experiment, the irreversible reaction has competitive advantage over the reversible reaction, and partial recovery after gas sensing was achieved. This study showed an excellent detection limit of 0.005 ppm by NH_3 exposure to a M13 bacteriophage template gas sensor [55]. This technique implies resistance of one-dimensional nanowires which were fabricated using M13 bacteriophage as gas sensor. Due to the extremely thin layer of PPy and the catalytic effect of Au NPs, this technique showed great sensitivity. This structure highly depends on random connection between electrodes. Better performance can be expected from this structure by optimized structure such as nano pattern between two electrodes.

SERS technology based on the nanowire structure can be improved significantly by controlling the density of "hot spots", where Raman signal amplification occurs [60,61] For more effective signal amplification, the sensor surface needs to be highly functionalized, particularly around hot spots [62,63]. Jung et al. recently reported an M13 bacteriophage-covered silver nanowire SERS sensor to detect a certain pesticide, paraquat (PQ), which is a widely used herbicide. The M13 bacteriophage was functionalized genetically to have the WHW peptide sequence for the PQ specific binding affinity on its pVIII major coat protein. When the sensor was compared with the wild type M13 bacteriophage sensor and a sensor without M13 bacteriophage, the genetically functionalized sensor showed superior selectivity to the PQ sample [60]. They also tested the selectivity upon different pesticides of a functionalized M13 bacteriophage-SERS sensor using bipyridylium, which is a common herbicide. The WHW-type M13 bacteriophage-functionalized SERS sensor showed a better Raman signal when PQ was detected than bipyridylium at the same concentration. In Figure 9, using a hand-held Raman spectrometer, they could detect PQ concentrations as low as 0.1 $\mu g/cm^3$, which is far below the guidelines of most countries. Raman intensity changes before and after the washing step showed best results for WHW type M13 bacteriophage sensors. It also showed better efficiency when compared with wild type and nanowire without M13 bacteriophage. By measuring different pesticides (diquat (DQ) and difenzoquat (DIF)), PQ was the most suitable pesticide. [60,64]. This practical research showed detection results using a hand-held Raman spectrometer and agricultural product without pretreatment.

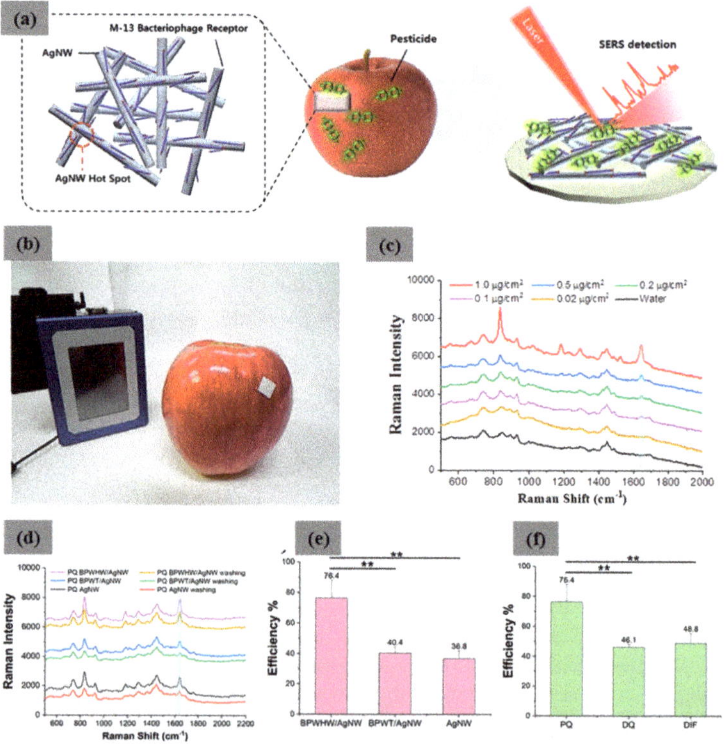

Figure 9. (**a**) Illustration of surface-enhanced Raman spectroscopy (SERS) by a functionalized M13 bacteriophage, (**b**) practical application of an M13 bacteriophage-based SERS sensor on an apple surface, (**c**) Raman spectra of paraquat (PQ) at different PQ concentrations on an apple surface, (**d**) Raman spectra of PQ using tryptophan–histidine–tryptophan (WHW), wild type of bacteriophage and without M13 bacteriophage before and after the washing step, (**e**) efficiency comparison of WHW, wild type and nanowire only, and (**f**) efficiency comparison of different pesticide (PQ, diquat (DQ), and difenzoquat (DIF)) using WHW M13 bacteriophage. Reproduced from [60], with permission from American Chemical Society, 2018.

4. M13 Bacteriophage-Based Color Sensor

The color sensor is one of the fields of M13 bacteriophage research, which has great potential in terms of sensitivity, selectivity, and portability [4,65,66]. Endocrine-disrupting chemicals were detected using an M13 bacteriophage-based color sensor. Genetically engineered M13 bacteriophages can be manufactured with a color sensor that has a structural color using a simple pulling technique, and applied to endocrine disrupting substance detection [65]. Although existing research focused on one specific substance, such as TNT [NC 13], this research focused on classifying various types of endocrine disruptors with similar structures. In this study, polychlorinated biphenyl (PCB) and phthalate compounds, traditional flame retardants, and plasticizers, were selected. These substances, which have been selected as regulated substances, are still included in pre-regulated polymer products, which act as environmental pollutants for use and disposal [65,67]. The M13 bacteriophage-based structural color sensor was modified with phage display technology. The protein sequence genetically engineered on the pVIII major coat protein is specific for covalent binding and is effective in the detection of aromatic compounds. In this study, four phthalate compounds and five PCB compounds were distinguished with high sensitivity, and the entire phthalate and PCB compound groups were distinguished with high accuracy (Figure 10) [65]. The detection limit of this color sensor was approximately 100 ppm, and the first two principal components of PCA analysis accounted for 54.89% of phthalate derivatives and 67.38% of PCB derivatives. The linear discriminant analysis (LDA) error between phthalate and PCB derivatives was 3.7% [65,68].

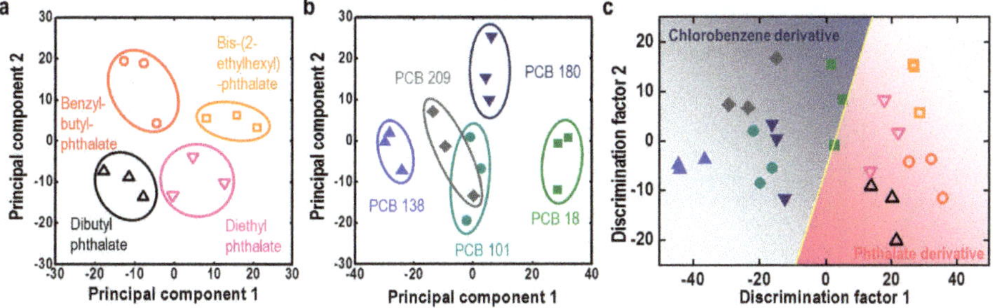

Figure 10. (a) PCA analysis of phthalate derivatives, (b) PCB derivatives, and (c) LDA analysis of phthalate and PCB derivatives using a M13 bacteriophage-based color sensor. Reproduced from [65], with permission from Elsevier, 2017.

Drug abuse causes serious health and environment problems [69–71]. Color sensor systems using M13 bacteriophages have shown promise for the detection of toxic substances [4,65]. Owing to the specific selectivity and sensitivity to target materials, M13 bacteriophage-based color sensors have the potential to detect various drugs [69]. Oh et al. discussed the structural color sensor to distinguish a range of antibiotics. The M13 bacteriophage-based structural color sensor was manufactured by inducing self-assembly using a simple pulling technique. Briefly, substrate (Au coated Si wafer) was deepen into M13 bacteriophage aqueous solution and pulled out slowly. By controlling the concentration of solution and pulling speed, a structural color band could be achieved through evaporation of M13 bacteriophage solution [4,65,69]. When the color sensor was exposed to an organic solvent, the bacteriophage bundle nanostructure constituting the microstructure of the sensor expanded, resulting in a structural color change. A genetically engineered M13 bacteriophage surface protein structure called WHWQ in this study could clearly distinguish three types of commercial antibiotics and three types of core ingredients of these antibiotics, which were confirmed by color and PCA analysis. Cefadroxil, amoxicillin, and rifampicin are some of the most popular antibiotics that a large number of pharmaceutical companies produce, and an absolute majority of patients consume. In Figure 11, PCA analysis produced clear discrimination between the three types of antibiotics as well as commercial and reagent grade. The first two discriminant factors accounted for 86.89% in this study [69].

An M13 bacteriophage-based color sensor also could be applied to microbial detection. Oh et al. reported a color sensor capable of detecting cancer cell types based on the M13 bacteriophage [72]. The M13 bacteriophage-based color sensor works in the form of a photonic nose, which functions similarly to other researched electronic noses, and shows structural color-based color changes by contact with the specific target materials. The target material in this study was a specific volatile compound that is unusually abundant in various cancer cells. The respiratory by-products of cancer cells, which are distinguished uniquely from normal cells, include hydrazine, xylene, ethylbenzene, ethanol, and toluene [72,73]. Therefore, they measured a range of organic reagents, populations of *E. coli*, and CO_2 concentration using the M13 bacteriophage-based color sensor. Here, the M13 bacteriophage-based color sensor showed significant sensitivity and selectivity for organic solvents, cell population, and CO_2 concentration. In this study, selectivity experiment has been done for hydrazine, xylene, ethylbenzene, ethanol, and toluene, volatile organic chemicals. These chemicals can be found in abnormally high concentration of in the lung cancer patient's respiration [72–75]. These organic compounds can act like an analytic marker for different cancer cells when released through respiration of cancer cells. Therefore, selective detection of these organic solvents was important to the selective detection of various cancer cells. Finally, each cancer cell's respiration by-product was analyzed using adjacent color sensors while culturing various types of cancer cells. As a result, they showed excellent discrimination through PCA and color analysis, with the first two discriminant factors accounting for 99.8% (Figure 12) [72].

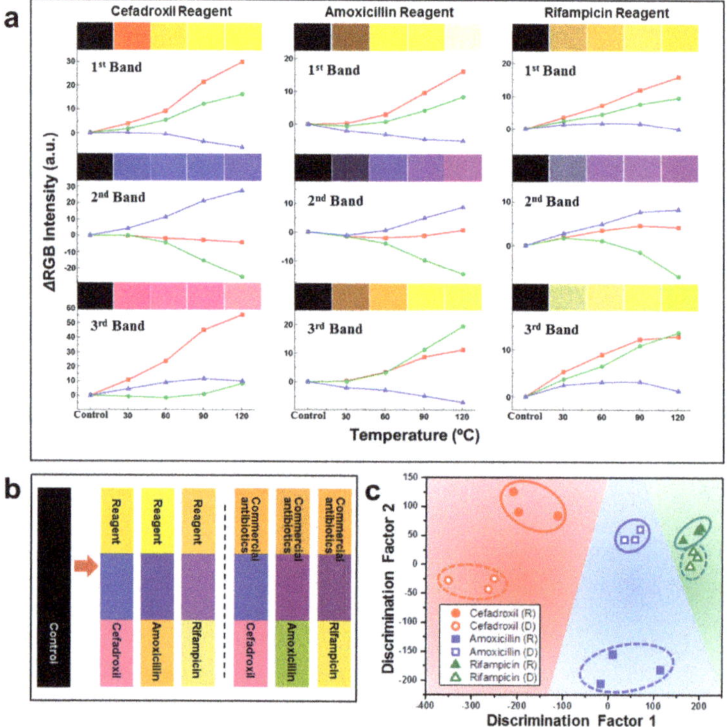

Figure 11. Measurement of various antibiotics using M13 bacteriophage-based color sensors. (**a**) Sensitivity measurement at different temperatures using fixed amount of samples, (**b**) color change translation of samples, and (**c**) PCA analysis of antibiotic sensing. Reproduced from [69], with permission from Wiley-VCH, 2016.

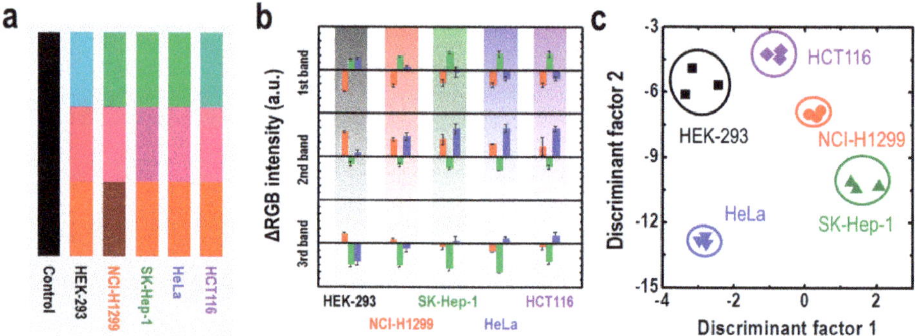

Figure 12. Cancer cell type identification using M13 bacteriophage-based color sensors. (**a**) Red, Green, and blue (RGB) color patterns of each cancer cell, (**b**) RGB intensity change, and (**c**) PCA analysis of samples. Reproduced from [72], with permission from The Royal Society of Chemistry, 2017.

5. Expansion of Analytic Method from Direct Sensing Using M13 Bacteriophage

A range of variations of WHW type M13 bacteriophages have common specific affinity to TNT molecules [4]. A π-π interaction was assumed among TNT molecules, histidine, and tryptophan based on experiment [4,6]. On the other hand, there is a lack of consideration as to whether WHW is really the best structure for TNT molecules or how this specific interaction occurs systemically and theoretically. Recently, Lee et al. reported the results of computational calculations of this interaction among TNT molecules, histidine, and tryptophan, using quantum mechanics (QM) calculations [76]. The theoretical candidates for comparison were WHW-, WAW-, WHA-, and AHW-simulated peptide sequences (A: Alanine). The specific binding affinity between the WHW and TNT molecules was confirmed experimentally using an SPR sensor system due to the significantly lower detection limit (500 fM) than any other sensor system based on M13 bacteriophages [4,77–80]. The dissociation constants for TNT molecules using the WHW-, WAW-, WHA-, and AHW-type M13 bacteriophages were 7.0×10^{-12}, 2.7×10^{-11}, 1.1×10^{-10}, and 3.3×10^{-8}, respectively. By QM calculations using Jaguar v8.4 software with the M06-2X/6-31G ** level of density functional theory, the binding energy of WHW-TNT was calculated to be 22.7 kcal mol^{-1}. Figure 13 shows the active binding mechanism for WHW and alanine substituted peptide sequences. For WAW, WHA, and AHW, the binding energy was 19.9, 16.0, and 18.8 kcal mol^{-1}, respectively. Finally, they confirmed the strong linear correlation between the experimental Gibbs free energy and calculated binding energy [76].

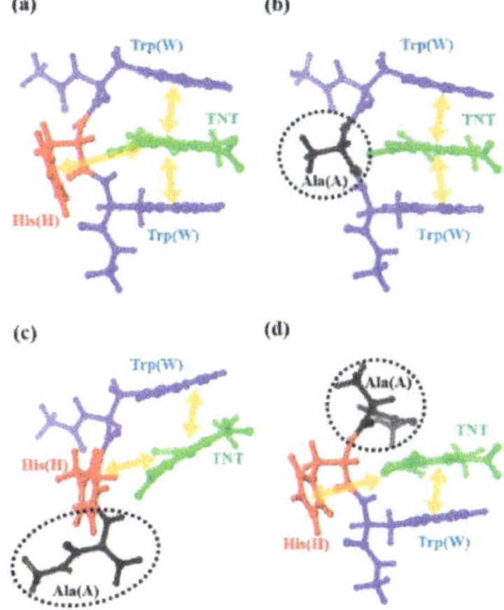

Figure 13. The stable constructions of TNT-specific peptides and their reactivity (red: histidine, blue: tryptophan, black: alanine, green: TNT, yellow, and arrow: π-π interaction). (a) WHW–TNT, (b) WAW–TNT, (c) WHA–TNT, and (d) AHW–TNT. Reproduced from [76], with permission from The Royal Society of Chemistry, 2019.

6. Conclusions

M13 bacteriophages have attracted considerable attention because of their simplicity of structure and functionality. The additional merit of M13 bacteriophages is the customized functionality, which can be provided easily through simple genetic engineering. In Table 1, large numbers of research groups have examined various fields using M13 bacteriophages. In particular, a biosensor using M13 bacteriophages has attracted serious interest because of the portability and intuitiveness. M13 bacteriophage has been utilized as an electrode support material for cysteine and *Salmonella* spp. detection [39,44,48]. Functionalized M13 bacteriophages have the potential to detect various cancer cells effectively [44,72]. Acoustic and SERS signals could be amplified by adding M13 bacteriophages to existing sensor systems [52,60]. In addition, M13 bacteriophage-based color sensors have been evaluated as a key tool to apply M13 bacteriophages to sensor systems [65,69,72]. Application of the M13 bacteriophage is still in its infancy. There remain many more protein sequences to be found, and low dimensional fabrication control should be studied. Mass production and reproducibility of nanostructures also need to be studied and overcome. Active theoretical studies on the function of genetically engineered M13 bacteriophages for specific target materials will open more systematic approaches as well as novel applications of M13 bacteriophages [12,76].

Table 1. Biosensor applications of M13 bacteriophages.

M13 Bacteriophage Immobilizations	Detection Technique	Analytes	Real Sample	Ref.
M13/Au coated Si wafer	Color analysis	Nitrotoluene	TNT, DNT, MNT	[4]
M13/donor and acceptor dye	FRET	Intracellular pH	RAW264.7 macrophage	[30]
M13/Au coated glass	SPR	Cell proliferation signal	NIH3T3 mouse fibroblast	[31]
M13/Au@Ag NPs/Raman dye/DNA	SERS	Antibody concentration	Sterptavidin/anti-goat IgG	[32]
M13/zinc phthalocyanine/methyl viologen	Fluorescence image	Breast cancer cells	SKBR-3 cell line	[35]
M13/CNF/GCE	Electrochemical	Cysteine	L-Cysteine solution in PBS	[39]
M13/scFv	Optical/surgical	CEA tumor cells	MC38- and CT26-CEA	[44]
M13/Pty/Au electrode	Capacitive	Salmonella	Chicken	[48]
M13/antibody	Electroacoustic	Cell number	Sp245 cell line	[52]
M13/Ag nanowire	SERS	Pesticides	PQ, DQ, and DIF	[60]
M13/Au coated Si wafer	Color analysis	Endocrine disrupting chemicals	Phthalate and PCB derivatives	[65]
M13/Au coated Si wafer	Color analysis	Antibiotics	Commercial and reagent antibiotics	[69]
M13/Au coated Si wafer	Color analysis	Cancer cells	HEK-293, NCI-H1299, SK-Hep-1, HeLa, HCT116 cancer cell lines	[72]

Author Contributions: J.-S.M., E.J.C., N.-N.J., and J.-W.O. developed the idea and wrote of the review article. J.-R.S. and D.-W.H. revised and improved the manuscript. J.-S.M. and J.-W.O. supervised the manuscript. All authors have given approval to the final version of the manuscript.

Funding: This research was supported by a grant of the Korea Health Technology R&D Project through the Korea Health Industry Development Institute (KHIDI), funded by the Ministry of Health and Welfare, Korea (HI17C1662). This work was also supported by the National Research Foundation of Korea (NRF) Grant funded by the Korean Government (MOE) (No. NRF-2016R1A6A3A11931113 and NRF-2017M3A9E4048170).

Conflicts of Interest: The authors declare no conflict of interest.

References

1. Andersen, P.D.; Jørgensen, B.H.; Lading, L.; Rasmussen, B. Sensor foresight—Technology and market. *Technovation* **2004**, *24*, 311–320. [CrossRef]

2. Hollingum, J. Foresight launches three sensor programmes. *Sens. Rev.* **1999**, *19*, 116–120. [CrossRef]

3. Sensors: The next wave of infotech innovation. *Ten-Year Forecast* Institute for the future. **1997**, 115–122.

4. Oh, J.-W.; Chung, W.-J.; Heo, K.; Jin, H.-E.; Lee, B.Y.; Wang, E.; Zueger, C.; Wong, W.; Meyer, J.; Kim, C.; et al. Biomimetic virus-based colorimetric sensors. *Nat. Commun.* **2014**, *5*, 3043. [CrossRef]

5. Martin-Herranz, A.; Ahmad, A.; Evans, H.M.; Ewert, K.; Schulze, U.; Safinya, C.R. Surface functionalized cationic lipid-DNA complexes for gene delivery: PEGylated lamellar complexes exhibit distinct DNA-DNA Interaction Regimes. *Biophys. J.* **2004**, *86*, 1160–1168. [CrossRef]

6. Moon, J.S.; Kim, W.G.; Kim, C.; Park, G.-T.; Heo, J.; Yoo, S.Y.; Oh, J.-W. M13 bacteriophage-based self-assembly Structures and their functional capabili-ties. *Mini Rev. Org. Chem.* **2015**, *12*, 271–281. [CrossRef]

7. Chung, W.J.; Lee, D.Y.; Yoo, S.Y. Chemical modulation of M13 bacteriophage and its functional opportunities for nanomedicine. *Int. J. Nanomed.* **2014**, *9*, 5825–5836.

8. Lee, Y.J.; Yi, H.; Kim, W.-J.; Kang, K.; Yun, D.S.; Strano, M.S.; Ceder, G.; Belcher, A.M. Fabricating genetically engineered high-power lithium-ion batteries using multiple virus genes. *Science* **2009**, *324*, 1051–1055.

9. Lee, Y.; Kim, J.; Yun, D.S.; Nam, Y.S.; Yang, S.-H.; Belcher, A.M. Virus-templated Au and Au-Pt core-shell nanowires and their electrocatalytic activities for fuel cell applications. *Energy Sci.* **2012**, *5*, 8328–8334. [CrossRef]

10. Mao, C.; Liu, A.; Cao, B. Virus-based chemical and biological sensing. *Angew* **2009**, *48*, 6790–6810. [CrossRef]

11. Sundar, V.C.; Yablon, A.D.; Grazul, J.L.; Ilan, M.; Aizenberg, J. Fibre-optical features of a glass sponge. *Nature* **2003**, *424*, 899–900. [CrossRef]

12. Espinosa, H.D.; Luster, A.L.; Latourte, F.J.; Loh, O.Y.; Gregoire, D.; Zavattieri, P.D. Tablet-level origin of toughening in abalone shells and translation to synthetic composite materials. *Nat. Commun.* **2011**, *2*, 173. [CrossRef] [PubMed]

13. Aizenberg, J.; Weaver, J.C.; Thanawala, M.S.; Sundar, V.C.; Morse, D.E.; Fratzl, P. Skeleton of euplectella sp.: Structural hierarchy from the nanoscale to the macroscale. *Science* **2005**, *309*, 275–278. [CrossRef] [PubMed]

14. Chung, W.-J.; Oh, J.-W.; Kwak, K.; Lee, B.Y.; Meyer, J.; Wang, E.; Hexemer, A.; Lee, S.-W. Biomimetic self-templating supramolecular structure. *Naure* **2011**, *478*, 364–368. [CrossRef] [PubMed]

15. Dogic, Z.; Fraden, S. Ordered phases of filamentous viruses. *Curr. Opin. Colloid Interface Sci.* **2006**, *11*, 47–55. [CrossRef]

16. Yang, S.H.; Chung, W.J.; Mcfarland, S.; Lee, S.W. Assembly of Bacteriophage into Functional Materials. *Chem. Rec.* **2013**, *13*, 43–59. [CrossRef]

17. Gimenez, S.; Lana-Villarreal, T.; Gómez, R.; Agouram, S.; Muñoz-Sanjosé, V.; Mora-Seró, I. Determination of limiting factors of photovoltaic efficiency in quantum dot sensitized solar cells: Correlation between cell performance and structural properties. *J. Appl. Phys.* **2010**, *108*, 064310. [CrossRef]

18. Karpan, V.M.; Khomyakov, P.A.; Starikov, A.A.; Giovannetti, G.; Zwierzycki, M.; Talanana, M.; Brocks, G.; van den Brink, J.; Kelly, P.J. Theoretical prediction of perfect spin filtering at interfaces between close-packed surfaces of Ni or Co and graphite or graphene. *Phys. Rev. B* **2008**, *78*, 195419. [CrossRef]

19. Smith, G.P.; Petrenko, V.A. Phage display. *Chem. Rev.* **1997**, *97*, 391–410. [CrossRef]

20. Smith, G.P. Filamentous fusion phage: Novel expression vectors that display cloned antigens on the virion surface. *Science* **1985**, *228*, 1315–1317. [CrossRef]

21. Cao, B.; Mao, C. *Phage Nanobiotechnology*; Petrenko, V., Smith, G.P., Eds.; RSC publishing: Cambridge, UK, 2011.

22. Brissette, R.; Goldstein, N.I. *Methods in Molecular Biology*; Fisher, P., Ed.; Humana: Totowa, NJ, USA, 2007.

23. Flynn, C.E.; Lee, S.W.; Peelle, B.R.; Belcher, A.M. Viruses as vehicles for growth, organization and assembly of materials. *Acta Mater.* **2003**, *51*, 5867–5880. [CrossRef]

24. Cui, Y.; Kim, S.N.; Jones, S.E.; Wissler, L.L.; Naik, R.R.; McAlpine, M.C. Chemical functionalization of graphene enabled by phage displayed peptides. *Nano Lett.* **2010**, *10*, 4559–4565. [CrossRef] [PubMed]

25. Kim, S.N.; Kuang, Z.; Slocik, J.M.; Jones, S.E.; Cui, Y.; Farmer, B.L.; McAlpine, M.C.; Naik, R.R. Preferential binding of peptides to graphene edges and planes. *J. Am. Chem. Soc.* **2011**, *133*, 14480–14483. [CrossRef] [PubMed]

26. Mao, C.; Solis, D.J.; Reiss, B.D.; Kottmann, S.T.; Sweeney, R.Y.; Hayhurst, A.; Georgiou, G.; Iverson, B.; Belcher, A.M. Virus-based toolkit for the directed synthesis of magnetic and semiconducting nanowires. *Science* **2004**, *303*, 213–217. [CrossRef] [PubMed]

27. Mio, C.; Flynn, C.E.; Hayhurst, A.; Sweeney, R.; Qi, J.; Georgiou, G.; Iverson, B.; Belcher, A.M. Viral assembly of oriented quantum dot nanowires. *Proc. Natl. Acad. Sci. USA* **2003**, *100*, 6946–6951. [CrossRef] [PubMed]

28. Huang, Y.; Chiang, C.-Y.; Lee, S.K.; Gao, Y.; Hu. E., L.; De Yoreo, J.; Belcher, A.M. Programmable assembly of nanoarchitectures using genetically engineered viruses. *Nano Lett.* **2005**, *5*, 1429–1434. [CrossRef] [PubMed]

29. Deo, S.; Godwin, H.A. A selective, ratiometric fluorescent sensor for $Pb2^+$. *J. Am. Chem. Soc.* **2000**, *122*, 174–175. [CrossRef]

30. Chen, L.; Wu, Y.; Lin, Y.; Wang, Q. Virus-templated FRET platform for the rational design of ratiometric fluorescent nanosensors. *Chem. Commun.* **2015**, *51*, 10190–10193. [CrossRef]

31. Yoo, S.Y.; Oh, J.-W.; Lee, S.W. Phage-chips for novel optically readable tissue engineering assays. *Langmuir* **2012**, *28*, 2166–2172. [CrossRef]

32. Lee, J.H.; Xu, P.F.; Domaille, D.W.; Choi, C.; Jin, S.; Cha, J.N. M13 Bacteriophage as materials for amplified surface enhanced Raman scattering protein sensing. *Adv. Funct. Mater.* **2014**, *24*, 2079–2084. [CrossRef]

33. Kawakami, K.; Nishihara, Y.; Hirano, K. Effect of hydrophilic polymers on physical stability of liposome dispersions. *J. Phys. Chem. B* **2001**, *105*, 2374–2385. [CrossRef]

34. Ruysschaert, T.; Germain, M.; Gomes, J.F.; Fournier, D.; Sukhorukov, G.B.; Meier, W.; Winterhalter, M. Liposome-based nanocapsules. *IEEE Trans. Nanobiosci.* **2004**, *3*, 49–55. [CrossRef]

35. Ngweniform, P.; Abbineni, G.; Cao, B.; Mao, C. Self-assembly of drug-loaded liposomes on genetically engineered target-recognizing M13 phage: A novel nanocarrier for targeted drug delivery. *Small* **2009**, *5*, 1963–1969. [CrossRef] [PubMed]

36. Jaworski, J.W.; Raorane, D.; Huh, J.H.; Majumdar, A.; Lee, S.-W. Evolutionary screening of biomimetic coatings for selective detection of explosives. *Langmuir* **2008**, *24*, 4938–4943. [CrossRef] [PubMed]

37. Jin, H.; Won, N.; Ahn, B.; Kwang, J.; Heo, K.; Oh, J.-W.; Sun, Y.; Cho, S.G.; Lee, S.-W.; Kim, S. Quantum dot-engineered M13 virus layer-by-layer composite films for highly selective and sensitive turn-on TNT sensors. *Chem. Commun.* **2013**, *49*, 6045–6047. [CrossRef]

38. Lee, B.Y.; Zhang, J.; Zueger, C.; Chung, W.-J.; Yoo, S.Y.; Wang, E.; Meyer, J.; Ramesh, R.; Lee, S.-W. Virus-based piezoelectric energy generation. *Nat. Nanotechnol.* **2012**, *7*, 351–356. [CrossRef]

39. Szot-Karpińska, K.; Leśniewski, A.; Jönsson-Niedziółka, M.; Marken, F.; Niedziółka-Jönsson, J. Electrodes modified with bacteriophages and carbon nanofibres for cysteine detection. *Sens. Actuat B Chem.* **2019**, *287*, 78–85. [CrossRef]

40. Szot, K.; Lesniewski, A.; Niedziolka, J.; Jönsson, M.; Rizzi, C.; Gaillon, L.; Marken, F.; Rogalski, J.; Opallo, M. Sol–gel processed ionic liquid—Hydrophilic carbon nanoparticles multilayer film electrode prepared by layer-by-layer method. *J. Electroanal. Chem.* **2008**, *623*, 170–176. [CrossRef]

41. Macdonald, S.; Szot, K.; Niedziolka, J.; Marken, F.; Opallo, M. Introducing hydrophilic carbon nanoparticles into hydrophilic sol-gel film electrodes. *J. Solid State Electr.* **2008**, *12*, 287–293. [CrossRef]

42. Bajenova, O.; Chaika, N.; Tolkunova, E.; Davydov-Sinitsyn, A.; Gapon, S.; Thomas, P.; O'Brien, S. Carcinoembryonic antigen promotes colorectal cancer progression by targeting adherens junction complexes. *Exp. Cell Res.* **2014**, *324*, 115–123. [CrossRef]

43. Bajenova, O.; Gorbunova, A.; Evsyukov, I.; Rayko, M.; Gapon, S.; Bozhokina, E.; Shishkin, A.; O'Brien, S.J. The genome-wide analysis of carcinoembryonic antigen signaling by colorectal cancer cells using RNA sequencing. *PLoS ONE* **2016**, *11*, e0161256. [CrossRef]

44. Murgas, P.; Bustamante, N.; Araya, N.; Cruz-Gómez, S.; Durán, E.; Gaete, D.; Oyarce, C.; López, E.; Herrada, A.A.; Ferreira, N.; et al. A filamentous bacteriophage targeted to carcinoembryonic antigen induces tumor regression in mouse models of colorectal cancer. *Cancer Immunol. Immun.* **2018**, *67*, 183–193. [CrossRef] [PubMed]

45. Ordoñez, C.; Screaton, R.A.; Ilantzis, C.; Stanners, C.P. Human carcinoembryonic antigen functions as a general inhibitor of anoikis. *Can. Res.* **2000**, *60*, 3419–3424.

46. Taheri, M.; Saragovi, H.U.; Stanners, C.P. The adhesion and differentiation-inhibitory activities of the immunoglobulin super-family member, carcinoembryonic antigen, can be independently blocked. *J. Biol. Chem.* **2003**, *278*, 14632–14639. [CrossRef] [PubMed]

47. Rondot, S.; Koch, J.; Breitling, F.; Dubel, S. A helper phage to improve single-chain antibody presentation in phage display. *Nat. Biotechnol.* **2001**, *19*, 75–78. [CrossRef]

48. Niyomdecha, S.; Limbut, W.; Numnuam, A.; Kanatharana, P.; Charlermroj, R.; Karoonuthaisiri, N.; Thavarungkul, P. Phage-based capacitive biosensor for Salmonella detection. *Talanta* **2018**, *188*, 658–664. [CrossRef]

49. Wolffs, P.F.G.; Glencross, K.; Thibaudeau, R.; Griffths, M.W. Direct quantitation and detection of salmonellae in biological samples without enrichment, using two-step filtration and real-time PCR. *Appl. Environ. Microbiol.* **2006**, *72*, 3896–3900. [CrossRef]

50. Bennett, A.R.; Greenwood, D.; Tennant, C.; Banks, J.G.; Betts, R.P. Rapid and definitive detection of Salmonella in foods by PCR. *Lett. Appl. Microbiol.* **1998**, *26*, 437–441. [CrossRef]

51. Herikstad, H.; Motarjemi, Y.; Tauxe, R.V. Salmonella surveillance: A global survey of public health serotyping. *Epidemiol. Infect.* **2002**, *129*, 1–8. [CrossRef]

52. Guliy, O.; Zaitsev, B.D.; Borodina, I.A.; Shikhabudinov, A.M.; Teplykh, A.A.; Staroverov, S.A.; Fomin, A.S. The biological acoustic sensor to record the interactions of the microbial cells with the phage antibodies in conducting suspensions. *Talanta* **2018**, *178*, 569–576. [CrossRef]

53. Kanoatov, M.; Krylov, S.N. Analysis of DNA in phosphate buffered saline using kinetic capillary electrophoresis. *Anal. Chem.* **2016**, *88*, 7421–7428. [CrossRef]

54. Smith, G.P.; Scott, J.K. Libraries of peptides and proteins displayed on filamentous phage. *Methods Enzymol.* **1993**, *217*, 228–257. [PubMed]

55. Yan, Y.; Zhang, M.; Moon, C.H.; Su, H.C.; Myung, N.V.; Haberer, E.D. Viral-templated gold/polypyrrole nanopeapods for an ammonia gas sensor. *Nanotechnology* **2016**, *27*, 325502. [CrossRef] [PubMed]

56. Oh, D.; Dang, X.; Yi, H.; Allen, M.A.; Xu, K.; Lee, Y.J.; Belcher, A.M. Graphene sheets stabilized on genetically engineered M13 viral templates as conducting frameworks for hybrid energy-storage materials. *Small* **2012**, *8*, 1006–1011. [CrossRef] [PubMed]

57. Gustafsson, G.; Lundström, I. The effect of ammonia on the physical properties of polypyrrole. *Synth. Met.* **1987**, *21*, 203–208. [CrossRef]

58. Gustafsson, G.; Lundström, B.; Liedberg, I.; Wu, C.R.; Inganäs, O.; Wennerström, O. The interaction between ammonia and poly(pyrrole). *Synth. Met.* **1989**, *31*, 163–179.

59. Choi, J.; Hormes, J.; Kahol, P.K. Study of ammonia-gas-induced irreversibility in polypyrrole films. *Appl. Phys. Lett.* **2003**, *83*, 2288–2290. [CrossRef]

60. Koh, E.H.; Mun, C.; Kim, C.; Park, S.G.; Choi, E.J.; Kim, S.H.; Dang, J.; Choo, J.; Oh, J.-W.; Kim, D.H.; et al. M13 bacteriophage/silver nanowire surface-enhanced Raman scattering sensor for sensitive and selective pesticide detection. *ACS Appl. Mater. Interfaces* **2018**, *10*, 10388–10397. [CrossRef]

61. Park, S.-G.; Mun, C.W.; Lee, M.K.; Jeon, T.Y.; Shim, H.-S.; Lee, Y.-J.; Kwon, J.-D.; Kim, C.S.; Kim, D.-H. 3D Hybrid Plasmonic Nanomaterials for Highly Efficient Optical Absorbers and Sensors. *Adv. Mater.* **2015**, *27*, 4290–4295. [CrossRef]

62. Nagy-Simon, T.; Tatar, A.-S.; Craciun, A.-M.; Vulpoi, A.; Jurj, M.-A.; Florea, A.; Tomuleasa, C.; Berindan-Neagoe, I.; Astilean, S.; Boca, S. Antibody Conjugated, Raman Tagged Hollow Gold–Silver Nanospheres for Specific Targeting and Multimodal Dark-Field/SERS/Two Photon-FLIM Imaging of CD19(+) B Lymphoblasts. *ACS Appl. Mater. Interfaces* **2017**, *9*, 21155–21168. [CrossRef]

63. Zhao, J.; Zhang, K.; Li, Y.; Ji, J.; Liu, B. High-Resolution and universal visualization of latent fingerprints based on aptamer-functionalized core–shell nanoparticles with embedded SERS reporters. *ACS Appl. Mater. Interfaces* **2016**, *8*, 14389–14395. [CrossRef]

64. Fang, H.; Zhang, X.; Zhang, S.J.; Liu, L.; Zhao, Y.M.; Xu, H.J. Ultrasensitive and quantitative detection of paraquat on fruits skins via surface-enhanced Raman spectroscopy. *Sens. Actuators B* **2015**, *213*, 452–456. [CrossRef]

65. Moon, J.-S.; Park, M.; Kim, W.-G.; Kim, C.; Hwang, J.; Seol, D.; Kim, C.-S.; Sohn, J.-R.; Chung, H.; Oh, J.-W. M-13 bacteriophage based structural color sensor for detecting antibiotics. *Sens. Actuators B* **2017**, *240*, 757–762. [CrossRef]

66. Slavik, R.; Homola, J. Ultrahigh resolution long range surface plasmon-based sensor. *Sens. Actuators B* **2007**, *123*, 10–12. [CrossRef]

67. Ferber, D. WHO advises kicking the livestock antibiotic habit. *Science* **2003**, *301*, 1027. [CrossRef] [PubMed]

68. Nam, K.T.; Kim, D.-W.; Yoo, P.J.; Chiang, C.-Y.; Meethong, N.; Hammond, P.T.; Chiang, Y.-M.; Belcher, A.M. Virus-enabled synthesis and assembly of nanowires for lithium ion battery electrodes. *Science* **2006**, *312*, 885–888. [CrossRef] [PubMed]

69. Moon, J.S.; Lee, Y.; Shin, D.M.; Kim, C.; Kim, W.G.; Park, M.; Han, J.; Song, H.; Kim, K.; Oh, J.-W. Identification of endocrine disrupting chemicals using a virus-based colorimetric sensor. *Chem. Asian J.* **2016**, *11*, 3097–3101. [CrossRef]

70. Sadik, O.A.; Witt, D.M. Peer reviewed: Monitoring endocrine-disrupting chemicals. *Environ. Sci. Technol.* **1999**, *33*, 368A–374A. [CrossRef]

71. Yoon, Y.; Westerhoff, P.; Snyder, S.A.; Esparza, M. HPLC-fluorescence detection and adsorption of bisphenol A, 17beta-estradiol, and 17alpha-ethynyl estradiol on powdered activated carbon. *Water Res.* **2003**, *37*, 3530–3537. [CrossRef]

72. Moon, J.-S.; Kim, W.-G.; Shin, D.-M.; Lee, S.Y.; Kim, C.; Lee, Y.; Han, J.; Kim, K.; Yoo, S.Y.; Oh, J.-W. Bioinspired M-13 bacteriophage-based photonic nose for differential cell recognition. *Chem. Sci.* **2017**, *8*, 921–927. [CrossRef]

73. Peng, G.; Tisch, U.; Adams, O.; Hakim, M.; Shehada, N.; Broza, Y.Y.; Billan, S.; Bortnyak, R.R.; Kuten, A.; Haick, H. Diagnosing lung cancer in exhaled breath using gold nanoparticles. *Nat. Nanotechnol.* **2009**, *4*, 669–673. [CrossRef]

74. Yun, J.M.; Kim, K.N.; Kim, J.Y.; Shin, D.O.; Lee, W.J.; Lee, S.H.; Lieberman, M.; Kim, S.O. DNA origami nanopatterning on chemically modified graphene. *Angew* **2011**, *51*, 912–915. [CrossRef] [PubMed]
75. Yun, J.M.; Ganesan, R.; Choi, J.-H.; Kim, J.-B. Local pH-Responsive Diazoketo-Functionalized Photoresist for Multicomponent Protein Patterning. *ACS Appl. Mater. Interfaces* **2013**, *5*, 10253–10259. [CrossRef] [PubMed]
76. Kim, W.G.; Zueger, C.; Kim, C.; Wong, W.; Devaraj, V.; Yoo, H.W.; Hwang, S.; Oh, J.-W.; Lee, S.W. Experimental and numerical evaluation of a genetically engineered M13 bacteriophage with high sensitivity and selectivity for 2,4,6-trinitrotoluene. *Org. Biomol. Chem.* **2019**, *17*, 5666–5670. [CrossRef] [PubMed]
77. Blaik, R.A.; Lan, E.; Huang, Y.; Dunn, B. gold-coated m13 bacteriophage as a template for glucose oxidase biofuel cells with direct electron transfer. *ACS Nano* **2015**, *10*, 324–332. [CrossRef] [PubMed]
78. Aguilar, A.D.; Forzani, E.S.; Leright, M.; Tsow, F.; Cagan, A.; Iglesias, R.A.; Nagahara, L.A.; Amlani, I. A hybrid nanosensor for tnt vapor detection. *Nano Lett.* **2010**, *10*, 380–384. [CrossRef]
79. Rose, A.; Zhu, Z.; Madigan, C.F.; Swager, T.M.; Bulović, V. Sensitivity gains in chemosensing by lasing action in organic polymers. *Nature* **2005**, *434*, 876–879. [CrossRef]
80. Riskin, M.; Tel-Vered, R.; Lioubashevski, O.; Willner, I. Ultrasensitive surface plasmon resonance detection of trinitrotoluene by a bis-aniline-cross-linked au nanoparticles composite. *J. Am. Chem. Soc.* **2009**, *131*, 7368–7378. [CrossRef]

 *nanomaterials*

Review

Recent Developments and Prospects of M13-Bacteriophage Based Piezoelectric Energy Harvesting Devices

In Woo Park, Kyung Won Kim, Yunhwa Hong, Hyun Ji Yoon, Yonghun Lee, Dham Gwak and Kwang Heo *

Department of Nanotechnology and Advanced Materials Engineering, Sejong University, Seoul 05006, Korea; lovepiw@naver.com (I.W.P.); saheon7@gmail.com (K.W.K.); monkey1434@gmail.com (Y.H.); hyunji9193@naver.com (H.J.Y.); alwayscoast@gmail.com (Y.L.); whalephant13@gmail.com (D.G.)
* Correspondence: kheo@sejong.ac.kr

Received: 28 November 2019; Accepted: 30 December 2019; Published: 2 January 2020

Abstract: Recently, biocompatible energy harvesting devices have received a great deal of attention for biomedical applications. Among various biomaterials, viruses are expected to be very promising biomaterials for the fabrication of functional devices due to their unique characteristics. While other natural biomaterials have limitations in mass-production, low piezoelectric properties, and surface modification, M13 bacteriophages (phages), which is one type of virus, are likely to overcome these issues with their mass-amplification, self-assembled structure, and genetic modification. Based on these advantages, many researchers have started to develop virus-based energy harvesting devices exhibiting superior properties to previous biomaterial-based devices. To enhance the power of these devices, researchers have tried to modify the surface properties of M13 phages, form biomimetic hierarchical structures, control the dipole alignments, and more. These methods for fabricating virus-based energy harvesting devices can form a powerful strategy to develop high-performance biocompatible energy devices for a wide range of practical applications in the future. In this review, we discuss all these issues in detail.

Keywords: virus; M13 bacteriophage; energy generator; piezoelectric; self-assembly; genetic engineering

---

1. Introduction

Recently, energy harvesting from biomechanical movement has attracted a great deal of interest in wearable, sustainable, and biomedical technologies [1–3]. Especially, piezoelectric devices have been actively studied for bio-implantable application, because they can easily generate energy by using simple motions and vibrations without any other external source [4–6]. General piezoelectric materials have the ability to generate electrical charges from applied mechanical stress. Although the energy output from piezoelectric generators may not be as large as that from other alternative energy sources, these also have their own advantages, such as simple device structures and various material groups.

After Jacques and Pierre Curie discovered piezoelectricity from quartz in 1880, various piezoelectric materials were developed from ceramics to natural biomaterials. Among them, ceramic materials such as lithium niobate ($LiNbO_3$) [7], potassium niobate ($KNbO_3$) [7], lithium tantalate ($LiTaO_3$) [8], barium titanate ($BaTiO_3$) [9], lead zirconate titanate ($Pb[Zr_xTi_{1-x}]O_3$) [10], and etc., were most investigated due to their superior piezoelectric properties. Lead zirconate titanate (PZT) has become the most common piezoelectric material in practical application today. However, as the issue of toxicity in lead-containing devices is starting to appear, extensive study has been conducted to replace it with lead-free piezoelectric materials [11]. For this purpose, several ceramics (bismuth ferrite ($BiFeO_3$), sodium niobate ($NaNbO_3$), barium titanate ($BaTiO_3$), bismuth titanate ($Bi_4Ti_3O_{12}$), quartz, etc.) [12–14] and organic materials (polyvinylidene fluoride (PVDF), polyvinylidene chloride (PVDC), etc.) [15,16] were found. Particularly, single-crystal zinc oxide (ZnO) nanostructures with wurtzite structure exhibit larger piezoelectric constants than those of bulk ZnO. However, the poor biocompatibility and brittle characteristics of these piezoelectric materials limit their applications in wearable and biomedical applications [17].

Otherwise, biomaterials have been regarded as promising alternative materials due to their good biocompatibility, non-toxicity, and environmental friendliness. After the first discovery of the piezoelectric effect of bone, the piezoelectric properties of diverse natural biomaterials such as wood, bone, hair, dentin, tendon, and collage were investigated [18–23]. Since these natural materials exhibit very weak piezoelectricity and are difficult to mass-produce, there is a limit to use for practical applications. Recently, there has been a growing interest in the biopiezoelectric materials to overcome these limitations. Polysaccharide materials [24–27], viruses [28–31], and self-assembled biomaterials [32,33] have been identified as good candidates, because their piezoelectric constants are higher than those of previous natural biomaterials, and they are possible to mass-produce. Especially, it has been found their physical and chemical properties can be modulated by their morphology, surface charges, and phases, and the piezoelectric response is directly related to these. Furthermore, the discovery of piezoelectricity in bone [34–36], which has been the most frequently studied tissue, aroused great interest because it seemed to provide an important key to understanding bone physiology. Researchers hypothesized that bone's piezoelectric signal by physical stimulation could regulate bone growth, repair, wound healing, and tissue regeneration [37–39]. In addition, piezoelectric biomaterials also have several advantages for use in sensors [40], energy storage [41], energy harvesting, and other areas [42]. Despite these advantages of biopiezoelectric materials and their potential applications, a comprehensive review of virus-based piezoelectric energy harvesting devices have not been reported.

In this short review, we provide an overview of M13 bacteriophages (phages) as superior biopiezoelectric materials for piezoelectric energy harvesting applications. In addition, we discuss in detail the piezoelectric properties of M13 phages and the fabrication of M13 phage-based piezoelectric energy harvesting devices. It is expected that this review will inspire the design of novel biomaterials and the development of functional devices for energy harvesting, sensing, biomedical applications, and other applications.

2. Biological Building Block for Piezoelectric Energy Harvesting Devices: M13 Bacteriophages

Due to their unique structural, biological, and physical properties, the M13 phage is the most attractive candidate in biomaterials for mimicking natural structures and developing novel piezoelectric energy harvesting devices. Especially, from the engineering point of view, M13 phages have several advantages, such as (1) structural similarity with collagens; (2) mass-producibility by bacteria infection and mass-amplification; (3) surface tunability through genetic engineering; (4) possibility of forming a highly-ordered structure via self-assembly; and (5) superior piezoelectric properties.

The M13 phage is a filamentous bacteriophage composed of circular single-stranded deoxyribonucleic acid (ssDNA) and capsid proteins. ssDNA is encapsulated in approximately 2700 copies of the helically arranged major coat protein pVIII, and five to seven copies of two different minor coat proteins (pIX, pVI, pIII, pVII) on the ends (Figure 1a). The diameter and length of M13 phages are about 6.6 nm and 880 nm, respectively [43]. Since the structural characteristic of phages is very similar to the structure of the human collagen, the M13 phages are quite capable of mimicking nature's hierarchical structures based on collagen [44]. These M13 phages, which are perfectly identical copies, can be mass-produced using the living characteristic of viruses. M13 phages are any group of viruses which carry out a lysogenic infection in which the phage inserts its genome into the bacterial genome. The minor coat protein pIII attaches to the receptor of the host bacteria and infects the bacteria. A huge amount of phages can be produced in an infected bacteria using the metabolic reactions of the host cell (Figure 1b). The infected cells are not involved in the cell lysis, but a decrease in the rate of cell growth [45].

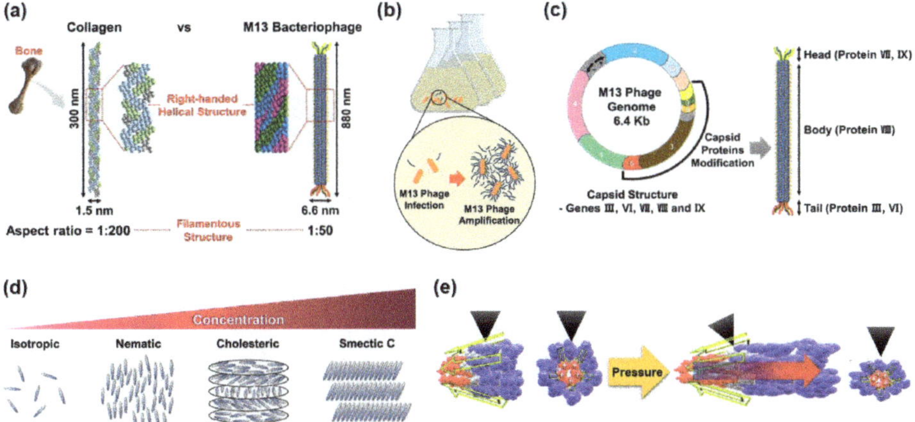

Figure 1. Schematic diagram showing the main characteristics of M13 bacteriophages (phages). (**a**) Structural similarity between natural collagen and M13 phages. Both biomaterials are filamentous structures with high aspect ratio and have right-handed helical structures. (**b**) Infection and mass-amplification of M13 phages. A huge amount of phages can be produced in infected bacteria by using the metabolic reactions of the host cell. (**c**) M13mp phage vector and surface modification through genetic engineering. We can design the molecular structures of the phage's outer surfaces in accordance with the desired properties and display the related peptide motif on the coat proteins of M13 phages. (**d**) Liquid crystalline phase transition of M13 phages. M13 phages exhibit a lyotropic liquid crystalline phase transition due to their helical structure, nanofibrous shape, monodispersity, and functional motifs. (**e**) Piezoelectric properties of M13 phages, which enable us to make piezoelectric energy harvesting devices.

Recent advances in genetic engineering make it possible to modulate the peptide sequence of phage proteins as desired. By using the recombinant DNA technique and M13mp phage vectors, we can design the molecular structures of surfaces according to the required properties and easily display the related peptide motif on the coat proteins of M13 phages (Figure 1c). This ability of M13 phages is a unique feature that distinguishes them from other nano and biomaterials [45].

In addition, M13 phages exhibit a lyotropic liquid crystalline phase due to their helical structure, nanofibrous shape, monodispersity, and expressed functional motifs (Figure 1d). According to the concentration of the phage suspension, the resulting structures of M13 phage films change from an isotropic phase to a cholesteric phase in a controlled manner [43]. Owing to these characteristics, we can prepare highly-ordered crystalline structures in a large area, which allows us to fabricate functional devices.

Lastly, recent studies have shown that M13 phages have excellent piezoelectric properties, which are larger than other natural biomaterials [28]. This makes it possible to fabricate high-performance piezoelectric energy harvesting devices (Figure 1e).

3. Introduction to Piezoelectric Effect

Piezoelectricity is a phenomenon of coupling between the electrical and mechanical states of a material by crystal deformation. When piezoelectric materials are mechanically stressed and deformed, the positive and negative charge centers shift in the materials, which then results in an external electrical field and a current flow. The opposite can also happen. When an electrical field is applied to the materials, the piezoelectric materials are stretched or compressed (Figure 2a).

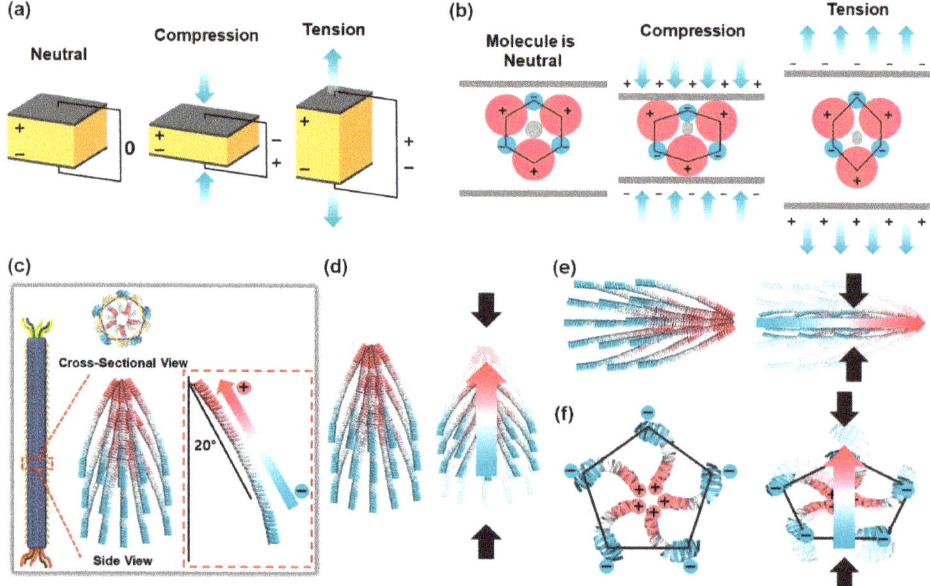

Figure 2. Schematic diagram showing the working mechanism of the piezoelectric effect. (**a**) Diagrammatic demonstration of the piezoelectric effect. (**b**) Schematic illustration showing the working mechanism of the piezoelectric effect in crystal structures. In equilibrium, the charges of the unit cell are neutral (no net dipole moment). When mechanical stresses are applied, a net dipole moment and electrical polarization arise in piezoelectric materials. (**c**) Schematic diagram showing structural characteristics of M13 bacteriophages. M13 phages have five-fold rotational symmetry, two-fold screw symmetry, and no inversion center. Schematic illustrations showing the working mechanism of the piezoelectric effect based on M13 phages when the stress is applied along the phage long axis (**d**) and phage short axis (**e**,**f**).

This direct piezoelectric effect was first discovered in 1880 by Paul-Jacques Curie, Pierre, and Marie Curie. They combined the knowledge of pyroelectricity with their understanding of crystal structures and behavior, and demonstrated the first piezoelectric effect by using crystals of quartz and Rochelle salt. Since then, many researchers have discovered and reported the piezoelectric properties of organic [15,16], inorganic [7–14,17], and biomaterials [18–33].

These piezoelectric characteristics originate from the deformation of the crystal structure and charge rearrangement within the material. In the equilibrium state, the arrangement of charges within the material lattice is neutral. However, there will be a charge redistribution within the unit cell and this induces net charges on the faces of the unit cell under the mechanical stress, which results in a net dipole moment. The sum of these dipole moments from all the unit cells leads to charge separation and generates electrical polarization in piezoelectric materials (Figure 2b). The most important thing in this system is that the materials must not have a center of symmetry, because the sum of net dipole moments is zero if the materials have a symmetry center.

Fortunately, M13 phages have structural properties that can make them suitable for piezoelectric properties. M13 phages are composed of ssDNA and capsid proteins, and each capsid protein has a dipole moment because these capsid proteins are made up of three parts—a positively charged area, a neutral charged area, and a negatively charged area. Especially, 2700 pVIII major coat proteins, which are the body-coat proteins, are assembled on the ssDNA with a 20° tilt angle with respect to the phage long axis and are arranged in right-handed helical structures. These major coat proteins form a pentagonal structure, which means that M13 phages have five-fold rotational symmetry, two-fold screw symmetry, and no inversion center (Figure 2c) [46,47]. Therefore, M13 phages can have piezoelectric properties. Interestingly, these complex structural characteristics of M13 phages allow us to use various piezoelectric properties. When stress is applied along the phage long axis, net dipole moments and electrical polarization are generated along the direction of applied stress (Figure 2d). On the other hand, when the stress is applied along the body (phage short axis), net dipole moments and electrical polarization are generated in two different directions (Figure 2e,f). Due to these characteristics, various types of piezoelectric energy harvesting devices can be developed.

4. Surface Modification of M13 Bacteriophages through Genetic Engineering

One of the great advantages of M13 phages as functional materials is the possibility of surface modification through genetic engineering. The most frequently used and well-established method to modify the genes of phages is recombinant DNA technology, which involves the insertion of foreign genes into the bacterial plasmids. Especially, M13mp phage vectors are usually used for engineering M13 phages. By incorporating foreign DNA, converting certain DNA into foreign DNA, and deleting specific DNA with enzymes, a high density of functional peptides and proteins can be simultaneously displayed on the M13 phage's coat proteins [45]. This technique enables us to design the surface molecular structures of M13 phages according to their purpose. For example, J.-W. Oh et al. developed the highly trinitrotoluene (TNT)-selective sensors based on phage colorimetric structures by expressing the AXXXWHWQXXDP (WHW) peptide sequence (which shows excellent binding affinity to TNT molecules) on pVIII major coat proteins [48]. J. Wang et al. reported that RGD phages induce osteogenesis and angiogenesis by activating the endothelialization and osteogenic differentiation of mesenchymal stem cells. In this work, RGD and RGD/PHSRN (combination of RGD and PHSRN) peptides, which interact with integrin, have a key role in adhesion with fibronectin [49,50].

These genetic modifications are also very useful in fabricating energy harvesting devices. Most energy harvesting devices (e.g., piezoelectric and triboelectric devices) have a direct correlation to surface charges and dipole moments. Therefore, the number of charges on the outer surfaces of M13 phages should be increased to improve the power of energy generators. For this purpose, B. Y. Lee et al. expressed AEGDP (1E), AEEGDP (2E), AEEEGDP (3E), and AEEEEDP (4E) peptide sequences on the outer surfaces of the pVIII major coat protein of M13 phages (Figure 3a) [28]. In the case of vertically aligned phages, the HHHHHH peptide sequence was expressed at the N-terminus of the pIII minor coat protein with a spacer GGGS as a specific binder with Ni-NTA surface (Figure 3b). The YEEE peptide was also expressed between the first and fourth residues at the N-terminal of the pVIII major coat protein for enhancing mechanical stability by Y-Y cross-linkage between phages through UV illumination [29].

Likewise, we can improve the physical and chemical properties and extend the range of applications through genetic molecular design. In the future, we expect that chemical modification (e.g., bioconjugate techniques and cross-linking) as well as genetic modification will be used for the further improvement of physical properties.

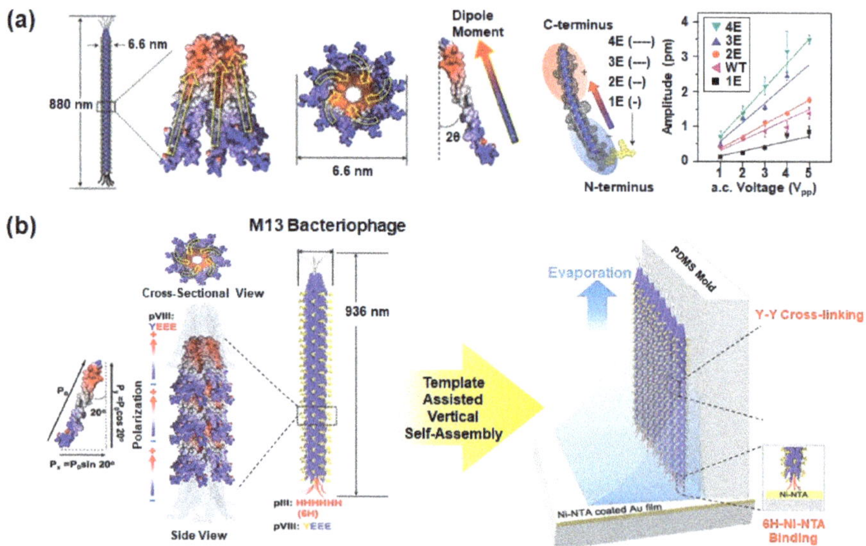

Figure 3. Surface modification of M13 bacteriophages through genetic engineering. (**a**) AEGDP (1E), AEEGDP (2E), AEEEGDP (3E), and AEEEEDP (4E) peptide sequences were expressed on the outer surfaces of the pVIII major coat protein of phages to enhance the piezoelectric properties. The M13 phage is 880 nm in length and 6.6 nm in diameter, which is covered by 2700 pVIII coat proteins and has five copies each of pIII and pIX proteins. The dipole moment is directed from the N-terminus (blue) to the C-terminus (red). Reproduced with permission from [28]. Copyright Nature Research, 2012. (**b**) HHHHHH peptide sequence expressed at the end of the N-terminus of the pIII minor coat protein for specific binding of phages on the Ni-NTA surface. The YEEE peptide motif is expressed between the first and fourth residues at the N-terminal of the pVIII major coat protein to enhance the mechanical stability by Y-Y cross-linkage. Reproduced with permission from [29]. Copyright American Chemical Society, 2019.

5. Piezoelectric Properties of M13 Bacteriophages

In general, the M13 phage is covered with 2700 pVIII coat major proteins, and the individual coat protein of phages is roughly divided into three sections: a positively charged region (C-terminus), a neutral region, and a negatively charged region (N-terminus). Owing to this adequate arrangement of charges, each coat protein has a dipole moment which is directed from the N-terminus to the C-terminus. Furthermore, the positively charged region of coat proteins is bound to central single-stranded DNA with a 20° tilt angle with respect to the phage long axis and α-helical structure when they are released from the host cell. The resulting structures of assembled pVIII coat proteins have five-fold rotational symmetry, two-fold screw symmetry, and no inversion center. Based on this fundamental study of phage structures, we can easily predict that M13 phages can present strong piezoelectric properties due to their permanent axial polarization caused by the net dipole moment in the pVIII proteins [28]. In 2012, B.Y. Lee et al. successfully observed the piezoelectric properties of M13 phages by using piezoresponse force microscopy (PFM) (Figure 4a). For this study, they prepared the phage monolayer sample by vertically pulling an octadecanethiol (ODT)/cysteamine patterned substrate from the phage suspension at a constant speed. Then, the piezoelectric properties of wild-type, 1E, 2E, 3E and 4E phages were measured by PFM. Vertical PFM measurements revealed that the effective piezoelectric coefficients (d_{eff}) of the wild-type phage was 0.30 ± 0.03 pm V^{-1}. The coefficients of 1E, 2E, 3E, and 4E phages were 0.14 ± 0.03 pm V^{-1}, 0.35 ± 0.03 pm V^{-1}, 0.55 ± 0.03 pm V^{-1}, and 0.70 ± 0.05 pm V^{-1}, respectively. This indicates that the coefficients improve as the surface charges of phages increase [28].

To further enhance the piezoelectric properties, they fabricated a multilayer phage film with 100 nm thickness. The multilayer film exhibited an increased effective piezoelectric coefficient (3.9 ± 0.05 pm V^{-1}). Although this value is lower than the d_{33} values of periodically-poled lithium niobate (PPLN) (13.2 pm V^{-1}), it is higher than collagen (1.1 pm V^{-1}) and other natural biomaterials. In addition, the effective piezoelectric coefficient of M13 phage films is further enhanced by fabricating vertically aligned phage nanostructures. These vertically assembled phages exhibited unidirectionally oriented piezoelectric polarization with an effective vertical piezoelectric coefficient of 13.2 pm V^{-1} (Figure 4b). Therefore, M13 phages are the best natural biomaterials for developing piezoelectric energy generators based on biomaterials [28].

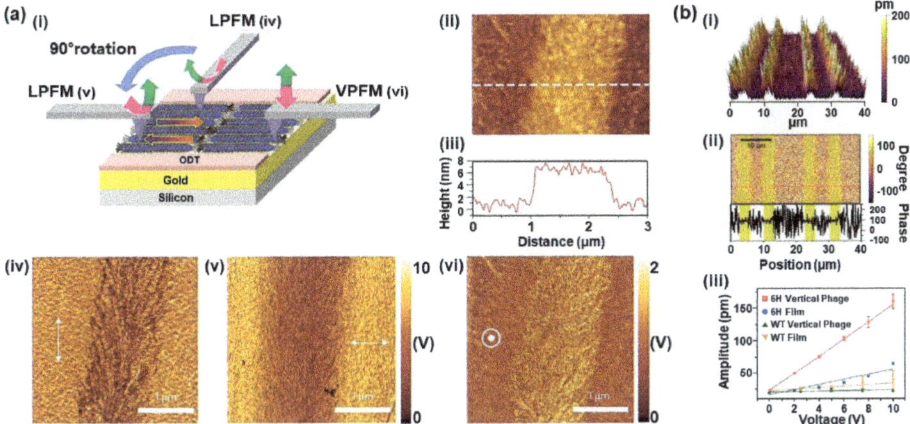

Figure 4. Piezoelectric properties of M13 bacteriophages. (**a**) Schematic of piezoresponse force microscopy (PFM) measurement (i); AFM topography (ii); height profile (iii); lateral PFM image along the phage long axis direction (iv); lateral PFM image obtained after changing the scanning direction by 90° (v); and vertical PFM image (vi) of the phage monolayer film. The resulting effective piezoelectric coefficients of 1E, 2E, 3E, and 4E phages were 0.14 ± 0.03 pm V^{-1}, 0.35 ± 0.03 pm V^{-1}, 0.55 ± 0.03 pm V^{-1}, and 0.70 ± 0.05 pm V^{-1}, respectively. Reproduced with permission from [28]. Copyright Nature Research, 2012. (**b**) PFM image (i), PFM phase image (ii) of vertically aligned phages which exhibits unidirectional polarization in the out-of-plane direction, and comparison of out-of-plane PFM amplitude versus applied voltage along the aligned direction (iii). The resulting effective piezoelectric coefficients (d_{eff}) of 6H vertical phage, 6H film, WT vertical phage, and WT film were ~13.2 pm V^{-1}, ~3.96 pm V^{-1}, ~0.35 pm V^{-1} and 1.22 pm V^{-1}, respectively. Reproduced with permission from [29]. Copyright American Chemical Society, 2019.

6. Developments and Applications of M13 Bacteriophage Based Piezoelectric Energy Harvesting Devices

Due to their excellent piezoelectric properties, the group of Prof. Lee at UC Berkeley first fabricated M13 phage-based piezoelectric energy generators in 2012 [28]. They prepared well-ordered self-assembled multilayer films based on M13 phages onto gold-coated flexible substrates by using a drop and evaporation method (Figure 5a). During the evaporation process, M13 phages were self-assembled and formed long-range ordered smectic-phase liquid-crystalline films by their chiral and monodisperse characteristics. When they overlaid a counter gold-coated flexible substrate on the film and embedded the device between two 2.5-mm-thick polydimethylsiloxane (PDMS) matrices, they could fabricate phage-based energy generators (Figure 5b). The generating power of the device can be modulated by the surface modification of M13 phages through genetic engineering, and the device produced a current of 6 nA and a voltage of 400 mV when they use 4E phages (Figure 5c). This was a sufficient energy output to turn on a liquid-crystal display [28].

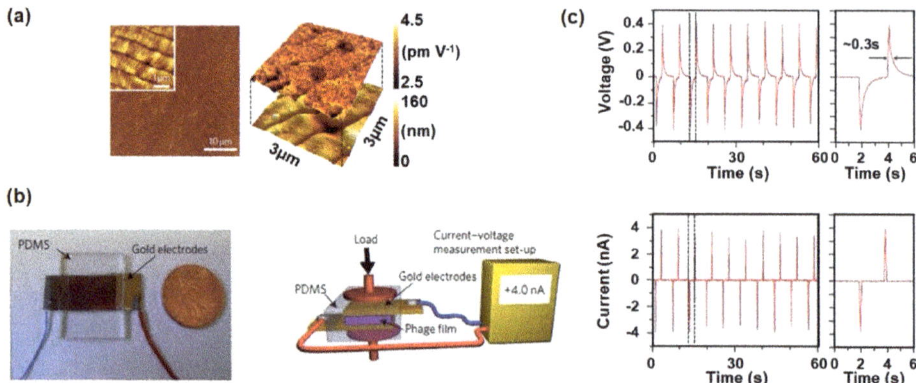

Figure 5. Characterization of an M13-bacteriophage-based piezoelectric energy generator (drop-casted film). (**a**) AFM topography image of phage film prepared by drop-casting, showing the long-range ordered smectic-phase liquid-crystalline film structure. Reproduced with permission from [28]. Copyright Nature Research, 2012. (**b**) Photograph of an M13 phage-based piezoelectric energy generator and schematic of piezoelectric energy generation measurement set-up. Reproduced with permission from [28]. Copyright Nature Research, 2012. (**c**) Open-circuit voltage and short-circuit current signal from the M13 phage-based piezoelectric energy generator. The device produced a current of 6 nA and a voltage of 400 mV when they use 4E phages. Reproduced with permission from [28]. Copyright Nature Research, 2012.

One of effective strategy for enhancing the power of phage-based energy generators is modulating the film morphology. Recently, K. Heo et al. reported a novel biomimetic assembly method for fabricating phage-based hierarchical structures with diverse surface morphologies by mimicking nature's self-assembly system [30]. They modulated the meniscus by controlling the thermodynamic and kinetic parameters (i.e., phage concentration, ionic concentration, phage surface charge, and pulling speed) and created a hierarchically organized phage film with diverse morphology in a controlled manner as the meniscus can serve as a transient scaffold to guide phage self-assembly (Figure 6a). In this process, the shape of the meniscus can be systematically modulated due to a combination of multiple factors, such as fingering instability, Rayleigh instability, and elastocapillary instability. All of the resulting phage structures were long-range-ordered chiral phage films showing multiple levels of hierarchical organization from nano- to macro-scale (single phage–phage filaments–fiber bundles–mesoscale periodic structure–macroscale band) (Figure 6b) [30]. When they fabricated piezoelectric energy generators based on these hierarchically organized phage films, the device power was improved compared to previous drop-casted phage films. The continuous and line film patterns exhibited 6.3 and 56 nA short-circuit current and 0.36 and 0.75 V open-circuit voltage, respectively (Figure 6c). The 2D-dot patterns showed the highest piezoelectric performance, which exhibited peak values of 94 nA current and 0.95 V voltage. They claimed that these enhanced piezoelectric properties of 2D-dot phage patterns were mainly due to the enhanced crystallinity of the phage nanofilaments that were periodically organized in an active array, in contrast to other films. The enhanced power was available to display the words "VIRUS" and "LEE LAB" on a liquid-crystal display.

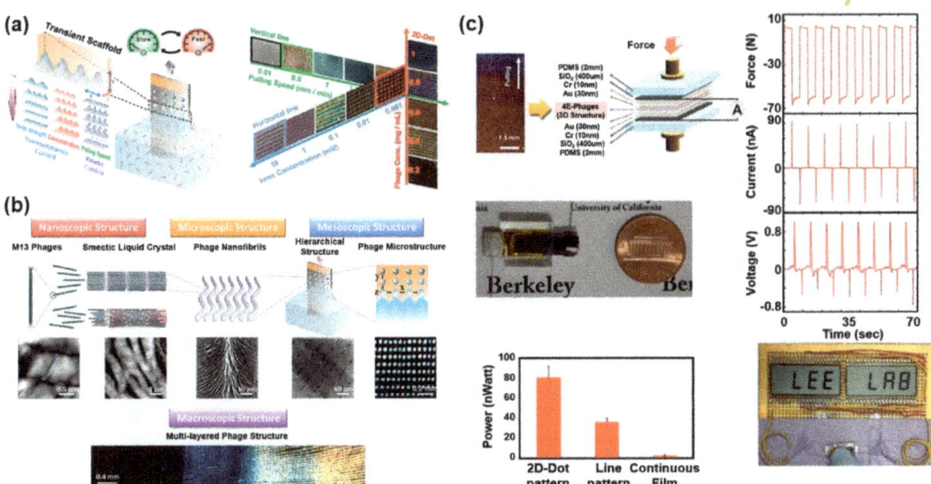

Figure 6. Characterization of an M13 bacteriophage-based piezoelectric energy generator. (biomimetic hierarchical structures). (**a**) Schematic illustration of the biomimetic hierarchical assembly process and structural transition diagram of phage films depending on phage concentration, ionic concentration, and pulling speed. Reproduced with permission from [30]. Copyright Elsevier, 2019. (**b**) Hierarchical structures of the phage-based films from nanoscale to macroscale. (single phage–phage filaments–fiber bundles–mesoscale periodic structure–macroscale band). (**c**) Piezoelectric characterization depending on phage structures. The 2D-dot patterns showed the highest piezoelectric performance, exhibiting peak values of 94 nA current and 0.95 V voltage. Reproduced with permission from [30]. Copyright Elsevier, 2019.

Another strategy for improving the piezoelectric power is to change the direction of the mechanical force applied to the phage. As mentioned in Section 3, if we consider the direction of the dipole moment in the individual M13 phages, it is predicted that the accumulated charges are maximized when the mechanical force is applied in the vertical direction rather than the lateral direction (Figure 2d). However, we could not carry out the related research because of the absence of an effective process to vertically align the phages. Recently, D.-M. Shin et al. developed a robust and facile method to prepare vertically aligned the phages for the first time (Figure 7a) [31]. They extruded phage suspension into a porous anodic aluminium oxide (AAO) template at precisely controlled speeds and repeated this process until all holes of the porous template were completely filled with phages, resulting in the formation of phage nanopillars (PNPs). During this process, negatively charged 4E phages were randomly adsorbed on the positively charged inner surface of the porous template and spontaneously accumulated inside the pores due to their liquid-crystalline characteristics. Afterwards, they deposited bottom and top Au electrodes on the AAO template including the PNPs and encapsulated the whole device using PDMS to improve their stability. The 4E PNP-based energy generator produced a 232 mV open-circuit voltage and 1.1 nA short-circuit current (Figure 7b) [31]. The relatively low power of this device compared to what was expected is presumed to be a result of the difficulty of forming well-ordered liquid crystalline phage structures, and the direction of the dipole moments of individual phages is therefore randomly oriented.

To overcome the limitation of the dipole alignment at the vertically oriented phage structures, J.H. Lee et al. developed a powerful method by combining the self-assembly of phages in a micro-fluidic channel and the surface modification of phages through genetic engineering (Figure 8a) [29]. To align the phages in the vertical direction, they tried to use a PDMS mold with micro-channels. When they dropped the phage suspension on the substrate and put on the PDMS mold, phage suspensions were infiltrated inside the micro-channels. The phages were vertically assembled on the wall of the

PDMS micro-channels and cross-linked with each other as the solvent was evaporated with UV light exposure. The morphology and filling density of vertical phage structure could be controlled by the initial phage concentration. Furthermore, they controlled the direction of the dipole moment of the phages by changing the peptide sequence of minor coat proteins at the same time. Because they inserted hexa-histidine (6H) at the N-terminal of the minor coat protein (pIII) of the phages through genetic modification, the pIII proteins of all phages were strongly specifically bound with the nickel-nitrilotriacetic acid (Ni-NTA) modified substrate, which polarized the dipole moment of the M13 phages. Finally, they fabricated piezoelectric energy harvesters using the resulting vertically aligned and unipolarized phages, and the peak voltage reached 2.8 V, with a current of 120 nA (Figure 8b). This is the largest power among phage-based energy generators. Five integrated energy generators demonstrated the operation of a liquid-crystal display reading "UC Berkeley" [29].

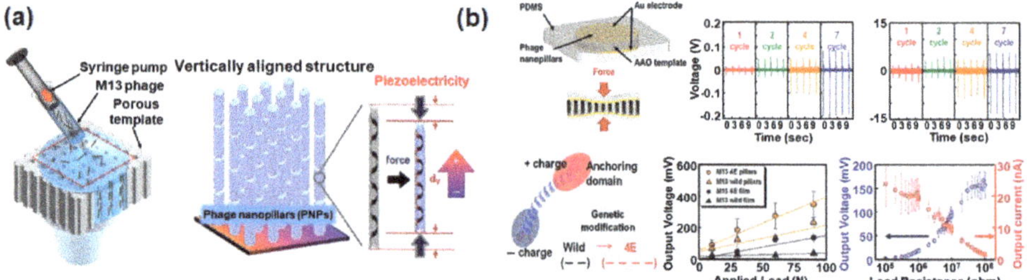

Figure 7. Characterization of an M13 bacteriophage-based piezoelectric energy generator (vertically aligned phage nanopillars). (**a**) Schematic showing the fabrication process and the resulting structures of vertically aligned M13 bacteriophage nanopillars. Reproduced with permission from [31]. Copyright The Royal Society of Chemistry, 2015. (**b**) Piezoelectric characterization of the phage nanopillar (PNP)-based energy generators. The 4E PNP-based energy generator produced 232 mV open–circuit voltage and 11.1 nA short-circuit current. Reproduced with permission from [31]. Copyright The Royal Society of Chemistry, 2015. AAO: anodic aluminium oxide.

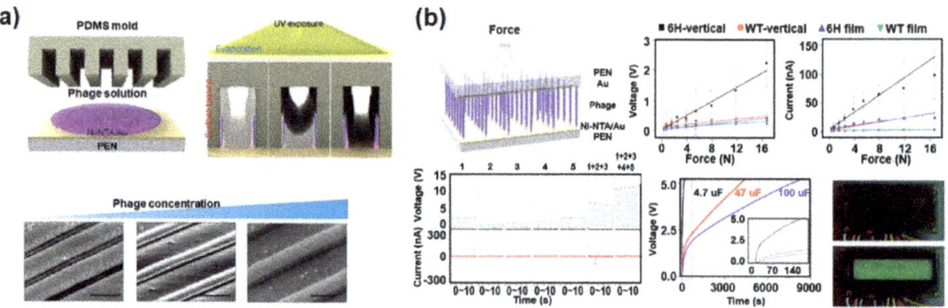

Figure 8. Characterization of an M13 bacteriophage-based piezoelectric energy generator (vertically aligned and unipolarized phage structures). (**a**) Schematic diagram of the fabrication process and scanning electron microscope (SEM) images of vertically aligned phages. Reproduced with permission from [29]. Copyright American Chemical Society, 2019. (**b**) Characterization of vertically aligned phage-based piezoelectric energy harvesters. This device exhibited a peak voltage of 2.8 V and a current of 120 nA. This is the largest power among phage-based energy generators. Reproduced with permission from [29]. Copyright American Chemical Society, 2019.

Although the power of phage-based energy generators is improving with diverse strategies, the power of the devices is still too low. For practical applications of these devices, many ways for enhancing electrical properties and developing mass-production methods should be contrived.

Because these methods are to prepare the M13 phage film by self-assembly, the performance of the devices may be reduced when they are used for a long time. Fortunately, laterally assembled structures are very stable and robust, allowing the devices to run reliably for a long time [28,30]. However, vertically aligned structures are likely to be vulnerable to long-term use. This can be solved by using chemically cross-linked M13 phages [29] and filling rubbery buffer materials in the empty space.

As this technique is at an early stage of research, studies of the devices' ideal operating conditions and toxicity issues have not been adequately carried out. However, by inferring from previous research based on M13 phages, the characteristics of these device can be predicted. Because the M13 phage is a biological material, optimized conditions for operating these devices will be room temperature (30 °C–70 °C) and low humidity [28]. However, these conditions can be modulated by surface modification. Through genetic engineering, we can modify the surface peptide motif to increase the hydrophobicity and cross-link phages to each other. Further, the M13 phage is known to be benign to humans because its host is *Escherichia coli* bacteria, not human cells [51–53]. Removing the infection motif in the pIII protein through genetic modification is also expected to be a good way to block the toxicity issues. Nevertheless, the study of M13 phages' toxicity should be conducted in the near future.

Since these technologies are still in their early stages of research, it is too early to discuss mass-production for practical applications. Most of the techniques discussed here are not suitable for mass production, because they use new process methods rather than conventional fabrication techniques. However, because these novel fabrication processes are very simple and facile, there is a strong possibility of mass-production and scale-up in the future. Although one of the main issues for scale-up is mass-production of the M13 phages, we can solve this problem using huge fermenters in the factories, like with biosimilar drug and alcohol manufacturing. Although the manufacturing cost of these devices is more expensive than existing devices, the M13 phage-based devices have several strong advantages which are very important in the biomedical fields. The M13 phage has very high piezoelectric coefficient compared to other biomaterials and their surfaces can be easily modified by genetic engineering. Further, it is also possible to mass-produce them.

7. Conclusions and Future Perspective

Even though the piezoelectric properties of biomaterials are lower than other inorganic materials, it is very important to design novel piezoelectric biomaterials and develop functional devices because of their specific applications in biomedical field. In particular, M13 bacteriophages are very attractive materials due to their unique features which distinguish them from other materials, such as their similar structures with collagens, mass-amplification, genetic modification, liquid-crystalline phase transition, and excellent piezoelectric properties. Recently, taking advantage of these characteristics, many researchers have made a great deal of efforts to fabricate M13 phage-based piezoelectric energy harvesting devices. Among these devices, vertically aligned phage films exhibited the highest performance—a peak voltage of 2.8 V and a peak current of 120 nA [29].

However, it is still a challenge to develop high-performance piezoelectric energy generators based on M13 phages owing to the limitations of surface modification, structural, and dipole alignment control. Thus, the novel design of phage structures through genetic and chemical modification may improve the performance of devices. Further, fabricating triboelectric devices based on M13 phages will also be an effective way to enhance the power of devices.

Another strategy for enhancing the power of devices is to develop composite structures composed of organic and inorganic biomaterials. Recently, novel methods for coating inorganic materials on biomaterial surfaces are attracting the attention of many researchers because of their various applications in biomedical field. For example, some researchers have reported effective methods to coat the inorganic materials on M13 bacteriophage surfaces via biomineralization [54–56], while other researchers developed the strategies to cast metals on the surface of biological materials by using protein cage systems and self-assembly [57–59]. These methods are expected to be used to produce precursors for energy-harvesting devices and maximize the power of devices.

High-performance energy harvesting devices based on biomaterials can be used in various fields, such as chemical/bio-sensors, artificial skin, bioimplantable energy devices, flexible electronics, soft robotics, and more (Figure 9). Especially, because there are many reports indicating that the surface charges and electrical signal can affect tissue regeneration, these piezoelectric biomaterials are also expected to be utilized for the development of biodegradable scaffolds for tissue engineering in the future.

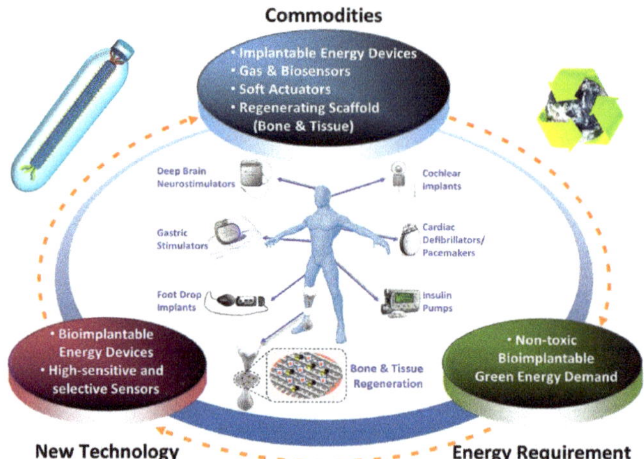

Figure 9. Schematic diagram showing the practical applications of M13 bacteriophage-based piezoelectric energy harvesting devices in the future.

Author Contributions: K.H., I.W.P., and K.W.K. developed the idea of the review article and wrote the paper. Y.H., H.J.Y., Y.L., and D.G. revised and improved the manuscript. K.H. supervised the manuscript. All authors have read and agreed to the published version of the manuscript.

Funding: This project was supported by Basic Science Research Program through the National Research Foundation of Korea (NRF) funded by the Ministry of Education (NRF–2019R1F1A1063020). K.H. acknowledges the Korea Institute of Energy Technology Evaluation and Planning (KETEP) and the Ministry of Trade, Industry &Energy (MOTIE) of the Republic of Korea (20184030202260).

Conflicts of Interest: The authors declare no conflict of interest.

References

1. Zheng, Q.; Zou, Y.; Zhang, Y.L.; Liu, Z.; Shi, B.J.; Wang, X.X.; Jin, Y.M.; Ouyang, H.; Li, Z.; Wang, Z.L. Biodegradable triboelectric nanogenerator as a life-time designed implantable power source. *Sci. Adv.* **2016**, *2*, e1501478. [CrossRef] [PubMed]

2. Parida, K.; Xiong, J.; Zhou, X.; Lee, P.S. Progress on triboelectric nanogenerator with stretchability, self-healability and bio-compatibility. *Nano Energy* **2019**, *59*, 237. [CrossRef]

3. Ali, F.; Raza, W.; Li, X.; Gul, H.; Kim, K.-H. Piezoelectric energy harvesters for biomedical applications. *Nano Energy* **2019**, *57*, 879. [CrossRef]

4. Zhou, Q.F.; Lau, S.T.; Wu, D.W.; Shung, K. Piezoelectric films for high frequency ultrasonic transducers in biomedical applications. *Prog. Mater. Sci.* **2011**, *56*, 139. [CrossRef] [PubMed]

5. Hwang, G.T.; Byun, M.; Jeong, C.K.; Lee, K.J. Flexible Piezoelectric Thin-Film Energy Harvesters and Nanosensors for Biomedical Applications. *Adv. Healthc. Mater.* **2015**, *4*, 646. [CrossRef] [PubMed]

6. Zheng, Q.; Shi, B.J.; Li, Z.; Wang, Z.L. Recent Progress on Piezoelectric and Triboelectric Energy Harvesters in Biomedical Systems. *Adv. Sci.* **2017**, *4*, 1700029. [CrossRef] [PubMed]

7. Guo, Y.P.; Kakimoto, K.; Ohsato, H. (Na$_{0.5}$K$_{0.5}$)NbO$_3$-LiTaO$_3$ lead-free piezoelectric ceramics. *Mater. Lett.* **2005**, *59*, 241. [CrossRef]

8.   Smith, R.T.; Welsh, F.S. Temperature Dependence of Elastic, Piezoelectric, and Dielectric Constants of Lithium Tantalate and Lithium Niobate. *J. Appl. Phys.* **1971**, *42*, 2219. [CrossRef]

9.   Setter, N.; Damjanovic, D.; Eng, L.; Fox, G.; Gevorgian, S.; Hong, S.; Kingon, A.; Kohlstedt, H.; Park, N.Y.; Stephenson, G.B.; et al. Ferroelectric thin films: Review of materials, properties, and applications. *J. Appl. Phys.* **2006**, *100*, 051606. [CrossRef]

10. Haertling, G.H. Ferroelectric ceramics: History and technology. *J. Am. Ceram. Soc.* **1999**, *82*, 797. [CrossRef]

11. Saito, Y.; Takao, H.; Tani, T.; Nonoyama, T.; Takatori, K.; Homma, T.; Nagaya, T.; Nakamura, M. Lead-free piezoceramics. *Nature* **2004**, *432*, 84. [CrossRef] [PubMed]

12. Egerton, L.; Dillon, D.M. Piezoelectric and Dielectric Properties of Ceramics in the System Potassium-Sodium Niobate. *J. Am. Ceram. Soc.* **1959**, *42*, 482. [CrossRef]

13. Leontsev, S.O.; Eitel, R.E. Dielectric and Piezoelectric Properties in Mn-Modified $(1-x)BiFeO_3-xBaTiO_3$ Ceramics. *J. Am. Ceram. Soc.* **2009**, *92*, 2957. [CrossRef]

14. Rodel, J.; Jo, W.; Seifert, K.T.P.; Anton, E.M.; Granzow, T.; Damjanovic, D. Perspective on the Development of Lead-free Piezoceramics. *J. Am. Ceram. Soc.* **2009**, *92*, 1153. [CrossRef]

15. Nunes, J.S.; Wu, A.; Gomes, J.; Sencadas, V.; Vilarinho, P.M.; Lanceros-Mendez, S. Relationship between the microstructure and the microscopic piezoelectric response of the alpha- and beta-phases of poly(vinylidene fluoride). *Appl. Phys. A Mater. Sci. Process.* **2009**, *95*, 875.

16. Guo, H.F.; Li, Z.S.; Dong, S.W.; Chen, W.J.; Deng, L.; Wang, Y.F.; Ying, D.J. Piezoelectric PU/PVDF electrospun scaffolds for wound healing applications. *Colloids Surf. B* **2012**, *96*, 29. [CrossRef]

17. Wang, Z.L. Towards Self-Powered Nanosystems: From Nanogenerators to Nanopiezotronics. *Adv. Funct. Mater.* **2008**, *18*, 3553. [CrossRef]

18. Fukada, E. Piezoelectric Properties of Biological Polymers. *Q. Rev. Biophys.* **1983**, *16*, 59. [CrossRef]

19. Minary-Jolandan, M.; Yu, M.F. Nanoscale characterization of isolated individual type I collagen fibrils: Polarization and piezoelectricity. *Nanotechnology* **2009**, *20*, 085706. [CrossRef]

20. Vivekananthan, V.; Alluri, N.R.; Purusothaman, Y.; Chandrasekhar, A.; Selvarajan, S.; Kim, S.J. Biocompatible Collagen Nanofibrils: An Approach for Sustainable Energy Harvesting and Battery-Free Humidity Sensor Applications. *ACS Appl. Mater. Interfaces* **2018**, *10*, 18650. [CrossRef]

21. Williams, W.S.; Breger, L. Analysis of Stress Distribution and Piezoelectric Response in Cantilever Bending of Bone and Tendon. *Ann. N. Y. Acad. Sci.* **1974**, *238*, 121. [CrossRef] [PubMed]

22. Liboff, A.R.; Shamos, M.H. Piezoelectric Effect in Dentin. *J. Dent. Res.* **1971**, *50*, 516. [CrossRef] [PubMed]

23. Fukada, E. Piezoelectricity of Natural Biomaterials. *Ferroelectrics* **1984**, *60*, 285. [CrossRef]

24. Maiti, S.; Kumar Karan, S.; Lee, J.; Kumar Mishra, A.; Bhusan Khatua, B.; Kon Kim, J. Bio-waste onion skin as an innovative nature-driven piezoelectric material with high energy conversion efficiency. *Nano Energy* **2017**, *42*, 282. [CrossRef]

25. Ghosh, S.K.; Mandal, D. High-performance bio-piezoelectric nanogenerator made with fish scale. *Appl. Phys. Lett.* **2016**, *109*, 103701. [CrossRef]

26. Praveen, E.; Murugan, S.; Jayakumar, K. Investigations on the existence of piezoelectric property of a bio-polymer-Chitosan and its application in vibration sensors. *RSC Adv.* **2017**, *7*, 35490. [CrossRef]

27. Karan, S.K.; Maiti, S.; Kwon, O.; Paria, S.; Maitra, A.; Si, S.K.; Kim, Y.; Kim, J.K.; Khatua, B.B. Nature driven spider silk as high energy conversion efficient bio-piezoelectric nanogenerator. *Nano Energy* **2018**, *49*, 655. [CrossRef]

28. Lee, B.Y.; Zhang, J.; Zueger, C.; Chung, W.J.; Yoo, S.Y.; Wang, E.; Meyer, J.; Ramesh, R.; Lee, S.W. Virus-based piezoelectric energy generation. *Nat. Nanotechnol.* **2012**, *7*, 351. [CrossRef]

29. Lee, J.H.; Lee, J.H.; Xiao, J.; Desai, M.S.; Zhang, X.; Lee, S.W. Vertical Self-Assembly of Polarized Phage Nanostructure for Energy Harvesting. *Nano Lett.* **2019**, *19*, 2661. [CrossRef]

30. Heo, K.; Jin, H.-E.; Kim, H.; Lee, J.H.; Wang, E.; Lee, S.-W. Transient self-templating assembly of M13 bacteriophage for enhanced biopiezoelectric devices. *Nano Energy* **2019**, *56*, 716. [CrossRef]

31. Shin, D.-M.; Han, H.J.; Kim, W.-G.; Kim, E.; Kim, C.; Hong, S.W.; Kim, H.K.; Oh, J.-W.; Hwang, Y.-H. Bioinspired piezoelectric nanogenerators based on vertically aligned phage nanopillars. *Energy Environ. Sci.* **2015**, *8*, 3198. [CrossRef]

32. Lee, J.H.; Heo, K.; Schulz-Schonhagen, K.; Lee, J.H.; Desai, M.S.; Jin, H.E.; Lee, S.W. Diphenylalanine Peptide Nanotube Energy Harvesters. *ACS Nano* **2018**, *12*, 8138. [CrossRef] [PubMed]

33. Kholkin, A.; Amdursky, N.; Bdikin, I.; Gazit, E.; Rosenman, G. Strong piezoelectricity in bioinspired peptide nanotubes. *ACS Nano* **2010**, *4*, 610. [CrossRef] [PubMed]
34. Fukada, E.; Yasuda, I. On the Piezoelectric Effect of Bone. *J. Phys. Soc. Jpn.* **1957**, *12*, 1158. [CrossRef]
35. Andrew, C.; Bassett, L. Electrical effects in bone. *Sci. Am.* **1965**, *213*, 18.
36. Shamos, M.H.; Lavine, L.S.; Shamos, M.I. Piezoelectric Effect in Bone. *Nature* **1963**, *197*, 81. [CrossRef]
37. Tian, J.J.; Shi, R.; Liu, Z.; Ouyang, H.; Yu, M.; Zhao, C.C.; Zou, Y.; Jiang, D.J.; Zhang, J.S.; Li, Z. Self-powered implantable electrical stimulator for osteoblasts' proliferation and differentiation. *Nano Energy* **2019**, *59*, 705. [CrossRef]
38. Yu, P.; Ning, C.; Zhang, Y.; Tan, G.; Lin, Z.; Liu, S.; Wang, X.; Yang, H.; Li, K.; Yi, X.; et al. Bone-Inspired Spatially Specific Piezoelectricity Induces Bone Regeneration. *Theranostics* **2017**, *7*, 3387. [CrossRef]
39. Ren, L.; Yang, P.; Wang, Z.; Zhang, J.; Ding, C.; Shang, P. Biomechanical and biophysical environment of bone from the macroscopic to the pericellular and molecular level. *J. Mech. Behav. Biomed. Mater.* **2015**, *50*, 104. [CrossRef] [PubMed]
40. Chorsi, M.T.; Curry, E.J.; Chorsi, H.T.; Das, R.; Baroody, J.; Purohit, P.K.; Ilies, H.; Nguyen, T.D. Piezoelectric Biomaterials for Sensors and Actuators. *Adv. Mater.* **2019**, *31*, 1802084. [CrossRef] [PubMed]
41. Yuan, H.; Lei, T.; Qin, Y.; He, J.-H.; Yang, R. Design and application of piezoelectric biomaterials. *J. Phys. D* **2019**, *52*, 194002. [CrossRef]
42. Ben Amar, A.; Kouki, A.B.; Cao, H. Power Approaches for Implantable Medical Devices. *Sensors* **2015**, *15*, 28889–28914. [CrossRef] [PubMed]
43. Chung, W.J.; Oh, J.W.; Kwak, K.; Lee, B.Y.; Meyer, J.; Wang, E.; Hexemer, A.; Lee, S.W. Biomimetic self-templating supramolecular structures. *Nature* **2011**, *478*, 364. [CrossRef] [PubMed]
44. Jin, H.E.; Jang, J.; Chung, J.; Lee, H.J.; Wang, E.; Lee, S.W.; Chung, W.J. Biomimetic Self-Templated Hierarchical Structures of Collagen-Like Peptide Amphiphiles. *Nano Lett.* **2015**, *15*, 7138. [CrossRef] [PubMed]
45. Rakonjac, J.; Bennett, N.J.; Spagnuolo, J.; Gagic, D.; Russel, M. Filamentous Bacteriophage: Biology, Phage Display and Nanotechnology Applications. *Curr. Issues Mol. Biol.* **2011**, *13*, 51. [PubMed]
46. Goldbourt, A. Structural characterization of bacteriophage viruses by NMR. *Prog. Nucl. Magn. Reson. Spectrosc.* **2019**, *114–115*, 192. [CrossRef]
47. Xu, J.; Dayan, N.; Goldbourt, A.; Xiang, Y. Cryo-electron microscopy structure of the filamentous bacteriophage IKe. *Proc. Natl. Acad. Sci. USA* **2019**, *116*, 5493. [CrossRef]
48. Oh, J.W.; Chung, W.J.; Heo, K.; Jin, H.E.; Lee, B.Y.; Wang, E.; Zueger, C.; Wong, W.; Meyer, J.; Kim, C.; et al. Biomimetic virus-based colourimetric sensors. *Nat. Commun.* **2014**, *5*, 3043. [CrossRef]
49. Wang, J.L.; Wang, L.; Yang, M.Y.; Zhu, Y.; Tomsia, A.; Mao, C.B. Untangling the Effects of Peptide Sequences and Nanotopographies in a Biomimetic Niche for Directed Differentiation of iPSCs by Assemblies of Genetically Engineered Viral Nanofibers. *Nano Lett.* **2014**, *14*, 6850. [CrossRef]
50. Wang, J.L.; Yang, M.Y.; Zhu, Y.; Wang, L.; Tomsia, A.P.; Mao, C.B. Phage Nanofibers Induce Vascularized Osteogenesis in 3D Printed Bone Scaffolds. *Adv. Mater.* **2014**, *26*, 4961. [CrossRef]
51. Frenkel, D.; Solomon, B. Filamentous phage as vector-mediated antibody delivery to the brain. *Proc. Natl. Acad. Sci. USA* **2002**, *99*, 5675. [CrossRef]
52. Yoo, S.Y.; Kobayashi, M.; Lee, P.P.; Lee, S.-W. Early osteogenic differentiation of mouse preosteoblasts induced by collagen-derived DGEA-peptide on nanofibrous phage tissue matrices. *Biomacromolecules* **2011**, *12*, 987. [CrossRef] [PubMed]
53. Zhu, H.; Cao, B.; Zhen, Z.; Laxmi, A.A.; Li, D.; Liu, S.; Mao, C. Controlled growth and differentiation of MSCs on grooved films assembled from monodisperse biological nanofibers with genetically tunable surface chemistries. *Biomaterials* **2011**, *32*, 4744. [CrossRef] [PubMed]
54. Tom, S.; Jin, H.E.; Heo, K.; Lee, S.W. Engineered phage films as scaffolds for $CaCO_3$ biomineralization. *Nanoscale* **2016**, *8*, 15696. [CrossRef] [PubMed]
55. Kilper, S.; Jahnke, T.; Aulich, M.; Burghard, Z.; Rothenstein, D.; Bill, J. Genetically Induced In Situ-Poling for Piezo-Active Biohybrid Nanowires. *Adv. Mater.* **2019**, *31*, e1805597. [CrossRef] [PubMed]
56. Jeong, C.K.; Kim, I.; Park, K.I.; Oh, M.H.; Paik, H.; Hwang, G.T.; No, K.; Nam, Y.S.; Lee, K.J. Virus-Directed Design of a Flexible $BaTiO_3$ Nanogenerator. *ACS Nano* **2013**, *7*, 11016. [CrossRef]
57. Slocik, J.M.; Crouse, C.A.; Spowart, J.E.; Naik, R.R. Biologically tunable reactivity of energetic nanomaterials using protein cages. *Nano Lett.* **2013**, *13*, 2535. [CrossRef]

58. Perriman, A.W.; Colfen, H.; Hughes, R.W.; Barrie, C.L.; Mann, S. Solvent-free protein liquids and liquid crystals. *Angew. Chem. Int. Ed.* **2009**, *48*, 6242. [CrossRef]
59. Ryadnov, M.G. A self-assembling peptide polynanoreactor. *Angew. Chem. Int. Ed.* **2007**, *46*, 969. [CrossRef]

Review

Recent Advances in Organic Piezoelectric Biomaterials for Energy and Biomedical Applications

Dong-Myeong Shin [1],*, Suck Won Hong [2] and Yoon-Hwae Hwang [3]

[1]   Department of Mechanical Engineering, The University of Hong Kong, Hong Kong 999077, China
[2]   Department of Cogno-Mechatronics Engineering, Department of Optics and Mechatronics Engineering, Pusan National University (PNU), Busan 46241, Korea; swhong@pusan.ac.kr
[3]   Department of Nanoenergy Engineering & BK21 PLUS Nanoconvergence Technology Division, Pusan National University (PNU), Busan 46241, Korea; yhwang@pusan.ac.kr
*   Correspondence: dmshin@hku.hk

Received: 30 November 2019; Accepted: 7 January 2020; Published: 9 January 2020

Abstract: The past decade has witnessed significant advances in medically implantable and wearable devices technologies as a promising personal healthcare platform. Organic piezoelectric biomaterials have attracted widespread attention as the functional materials in the biomedical devices due to their advantages of excellent biocompatibility and environmental friendliness. Biomedical devices featuring the biocompatible piezoelectric materials involve energy harvesting devices, sensors, and scaffolds for cell and tissue engineering. This paper offers a comprehensive review of the principles, properties, and applications of organic piezoelectric biomaterials. How to tackle issues relating to the better integration of the organic piezoelectric biomaterials into the biomedical devices is discussed. Further developments in biocompatible piezoelectric materials can spark a new age in the field of biomedical technologies.

Keywords: piezoelectric materials; organic materials; biomaterials; energy applications; biomedical applications

---

1. Introduction

Piezoelectric materials are a class of solid materials that can accumulate an electric charge in response to applied mechanical agitation, facilitating the conversion from mechanical energy to electrical energy and vice versa. Piezoelectricity has been found in both organic and inorganic materials, where the physical principles of piezoelectricity are varied upon material classification. In inorganic piezoelectric materials, the piezoelectric effect arises from the rearrangement of ions in the dielectric materials that possess a lack of inversion symmetry in crystalline structure [1]. In contrast, the reorientation of molecular dipole mainly induces polarization in organic piezoelectric materials under applied mechanical stress [2,3]. These materials have taken over the entire market of electromechanical devices, such as sensors [4–6], actuators [7], energy harvesting [8–10] and storage [11,12]. Recently, medically implantable and mountable devices have attracted considerable attention [13,14], and are the newly emerging applications for piezoelectric materials.

Organic piezoelectric biomaterials offer several benefits over inorganic piezoelectric materials, which include a high biocompatibility, excellent flexibility, environmental friendliness, and a high level of processability. Ever since the discovery of polarization in asymmetric biological tissue in 1941 [15], many researchers have looked not only to unveil the primary principle underlying the piezoelectricity of those materials, but also to enhance its physical and chemical properties by designing a molecular structure, nanostructuring, and adding dopants [2,16]. Although organic piezoelectric biomaterials exhibit weak piezoelectricity compared to inorganic counterparts, recent research suggests that biocompatible piezoelectric materials, which are interfaced with the biological system of human beings, can serve as the functional materials in the field of medically implantable and mountable applications when they are well-processed. Organic piezoelectric materials are applicable in broad range of devices, including nano- to millimeter-scaled devices, so there might be some challenges in the device fabrication due to local damage and nonlocal elasticity [17,18]. However, our manuscript will only detail views on organic piezoelectric materials.

The rapid development in organic piezoelectric biomaterials calls for a comprehensive review that can provide a useful reference for researchers in relevant fields. Herein, we provide a thorough review of organic piezoelectric biomaterials that are used in energy and biomedical applications. We review the working principle and properties of the different types of organic piezoelectric biomaterials. Efforts to improve the piezoelectric performance of each materials are discussed. The applications of these materials are introduced in terms of energy harvesting, sensor, and cell and tissue regeneration. Meanwhile, the challenges that need to be addressed for practical application are also presented.

2. Mechanism of Piezoelectricity in Biomaterials

Piezoelectricity in organic biomaterials mainly originated from the reorientation of the molecular dipole [2,3] while the breaking of structural symmetry in crystal lattices results in piezoelectricity in traditional inorganic materials [1]. As a piezoelectric biomaterial is deformed under stress, the molecular chains with a permanent dipole in the material are aligned along one direction, yielding or changing in net polarization so that it is able to represent the piezoelectric behavior regardless of the absence of non-centrosymmetry. Therefore, the piezoelectric biomaterials are required to possess the presence of permanent molecular dipoles, the ability to orient the molecular dipoles, and the ability to maintain the dipole alignment [19]. The following sections provide detail on piezoelectric mechanisms and the properties of these materials, including the proteins, peptides, and biopolymers, and the piezoelectric constants summarized in Table 1.

2.1. Piezoelectric Proteins

It is well known that the collagen, being a main structural protein in the extracellular matrix in the tissues, mainly causes the piezoelectric effect in bone, but its clear fundamental principle has not yet been discovered. Several hypotheses have been established to elucidate the origin of piezoelectricity in the collagen fibril [20], which involve the noncentrosymmetric structure, the existence of polar bonding at the molecular level, reorientation of the C=O–NH bond in the α-helix structure, and the polarization of hydrogen bonds in collagen [21–24]. A recent study reveals that the piezoelectric effect in collagen comes from the reorientation of, and a magnitude change, in the permanent dipoles of individual charged and polar residues towards the long axis of the collagen fibril [20], as shown in Figure 1a. The shear piezoelectric constant reported is varied from $d_{14} = 0.2$ pC/N to $d_{14} = 2.0$ pC/N for collagens impregnated by bone and tendon [25], respectively.

The M13 bacteriophage is recently emerging as the functional material for multiple applications in energy harvesting [26–28], chemical sensor [29–31], and tissue regeneration [32,33]. It is a filamentous bacterial virus with the well-defined dimension of 880 nm in length and 6.6 nm in width, consisting of single-stranded DNA which is wrapped up with 2700 copies of major proteins (pVIII) and lidded with five copies of minor proteins (pIII/pVI or pVIII/pIX) on the top or bottom ends, respectively. The major protein has an ~20° tilt angle with respect to the DNA axial direction, and is arranged with a combined five-fold rotational and two-fold screw symmetry. Each major protein holds a dipole moment directed toward the DNA axis, leading to permanent polarization in both the axial and radial direction of the phage (Figure 1b). The piezoelectric constant for radial direction is $d_{33} = \sim7.8$ pm/V [26], and the constant for axial direction is improved by up to around three times as high as that of radial direction [27]. Recently, Lee et al. improved the value for axial direction up to $d_{33} = \sim26.4$ pm/V by unidirectionally sticking the M13 bacteriophages [28].

2.2. Piezoelectric Peptides

Glycine (G) is a zwitterionic amino acid and a good model building block for investigating the process of polymorphic crystallization [34–36]. At ambient conditions, the crystalline glycines have three distinct structures, α-, β-, and γ-structures, where the β- and γ-structures have shear piezoelectricity due to their acentric structures [37–40]. Herein, the piezoelectricity arises from the displacement of ion in the crystal, as it does in the inorganic materials, and such displacement creates a dipole in local, and a net polarization in bulk, material, as shown in Figure 1c. The β-structured glycine has a high shear piezoelectric constant, d_{16} = ~190 pm/V [41], which is comparable to the normal piezoelectric constant of barium titanate ($BaTiO_3$) [42].

Diphenylalanine (FF) is composed of two phenylalanine (F) amino acids and can be self-assembled into semi-crystalline peptide nanotubes and microrods, exhibiting multiple advantages including morphological diversity, functional diversity, high biocompatibility, and a high Young's modulus [16,43]. The nanostructured diphenylalanines are widely studied piezoelectric materials that have a non-centrosymmetric hexagonal space group ($P6_1$) [44]. This crystalline class serves to demonstrate their diverse physical effects including piezoelectricity, second harmonic generation, optical activity, pyroelectricity, ferroelectricity, and enantiomorphism [45] (Figure 1d). The peptide nanotubes have achieved a high shear piezoelectric constant up to d_{15} = ~60 pm/V [46]. In order to improve the scalability as well as the uniformity of semi-crystalline film, the unidirectionally polarized and aligned diphenylalanine nanotubes films were fabricated using the meniscus-driven self-assembly process [47], representing a similar value of d_{15} = ~45 pm/V to that of the highly crystalline structure. In addition, Nguyen et al. have devoted to obtaining the normal piezoelectricity (reached up to d_{33} = ~17.9 pm/V) of the peptide nanostructures by the vertical alignment of individual microrods [48,49].

2.3. Other Piezoelectric Biopolymers

The poly(vinylidene fluoride) (PVDF) films have shown one of the highest piezoelectric performances among all piezoelectric polymers found to date [50]. The PVDF has five different crystalline structures, α-, β-, γ-, δ-, and ε-structures where the β-structured PVDF has a normal piezoelectricity of d_{33} = −33 pC/N [51]. A dipole moment is induced perpendicular to the polymer chain in each unit of PVDF due to the presence of a branched fluorine atom with a large van der Waals radius together with the electronegativity [52–54]. In the β-phase, the orthorhombic crystal structure, with aligned fluorine and hydrogen branches parallel to each other, contributes to the net dipole moment and piezoelectricity [55], whereas the net dipole moment is cancelled out by the anti-parallel alignment of the dipole in the α-phase hexagonal structure [56], as depicted in Figure 1e. The nature of the negative piezoelectric effect is related to the redistribution of the electron molecular orbitals and total charges under an electrical field applied [57]. The copolymer approaches have been applied to enhance the piezoelectric constant by conjugating with trifuoroethylene (TrFE) [58,59], hexafluoropropylene (HPF) [60,61], or chlorotrifluoroethylene (CTFE) [62].

The poly(*L*-lactic acid) (PLLA) is a polymorphic polymer with excellent biodegradability and biocompatibility. The thermodynamically stable conformation is the α-crystalline structure where the dipoles introduced by carbonyl groups (C=O) are not aligned along the main polymer chain. The external stimuli, such as electrospinning (Figure 1f), allows for the dipoles to be unidirectionally oriented along the stretched direction [63], which is termed the β-crystalline structure, resulting in the shear piezoelectricity of d_{14} = 12 pC/N [64,65]. It is worth to note that the β-crystalline PLLA, along with decent piezoelectricity, requires no polling process due to its helical structure [3], widening the application area in the biocompatible mobile devices [66].

Natural polymers are gaining more importance owing to their biocompatibility, as recent research is aiming at the investigation of the utility of piezoelectric materials for body implantable or mountable devices. The piezoelectricity of silk arises from the combined effects of a high degree of silk II, β-sheet crystallinity, and a crystalline orientation [67]. The reported piezoelectric constant is

$d_{14} = -1.5$ pC/N [67]. Cellulose, which is the most abundant natural polymer on earth, is also known to have a shear piezoelectricity of $d_{14} = 0.2$ pC/N [68].

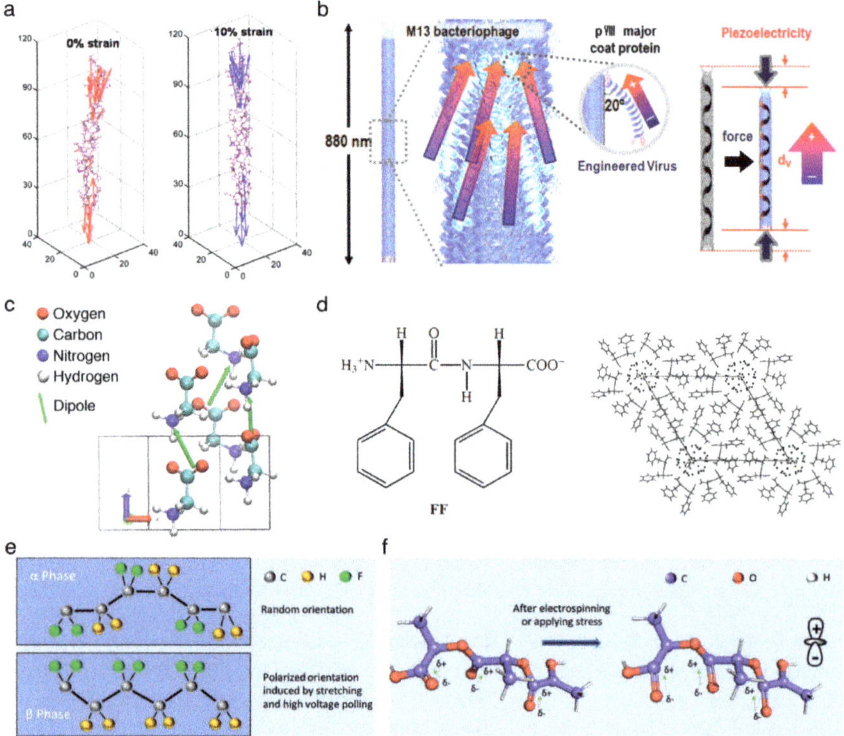

Figure 1. Piezoelectricity in the organic piezoelectric biomaterials. (**a**) Molecular origin of the piezoelectric effect in collagen. Reproduced with permission from [20]. Copyright American Chemical Society, 2016. (**b**) Schematic illustration of piezoelectric M13 bacteriophage. Reproduced with permission from [27]. Copyright The Royal Society of Chemistry, 2015. (**c**) Unit cell of β-glycine crystal has two molecules, where two molecular dipole moments form the net dipole moment along the *z*-axis. Reproduced with permission from [40]. Copyright Nature Publishing Group, 2019. (**d**) Unit cell and molecular packing of diphenylalanine. Reproduced with permission from [44]. Copyright Wiley–VCH, 2001. (**e**) Structures of non-piezoelectric (α-phase) and piezoelectric (β-phase) poly(vinylidene fluoride) (PVDF). Reproduced with permission from [2]. Copyright Wiley–VCH, 2018. (**f**) Molecular structure of poly(L-lactic acid) (PLLA) chain. Reproduced with permission from [2]. Copyright Wiley–VCH, 2018.

Table 1. Comparison of piezoelectric constants for various organic piezoelectric biomaterials.

Piezoelectric Organic Biomaterials	Piezoelectric Constant		References
	Normal Piezoelectric	**Shear Piezoelectric**	
Collagen	-	$d_{14} = 0.2\text{–}2.0$ pC/N	[25]
M13 bacteriophage	$d_{33} = 7.8\text{–}26.4$ pm/V		[26–28]
Glycine	-	$d_{16} = \sim190$ pm/V	[41]
Diphenylalanine	$d_{33} = \sim17.9$ pm/V	$d_{15} = 45\text{–}60$ pm/V	[46–49]
PVDF	$d_{33} = -3$ pC/N $d_{31} = 23$ pC/N	-	[51]
PVDF–TrFE	$d_{33} = -25$ to -40 pC/N $d_{31} = 12\text{–}25$ pC/N	-	[52,58,59]
PVDF–HPF	$d_{33} = -24$ pC/N $d_{31} = 30$ pC/N	-	[60,61]
PVDF–CTFE	$d_{33} = -140$ pC/N	-	[62]
PLLA	-	$d_{14} = 12$ pC/N	[64,65]
Silk	-	$d_{14} = -1.5$ pC/N	[67]
Cellulose	-	$d_{14} = 0.2$ pC/N	[68]

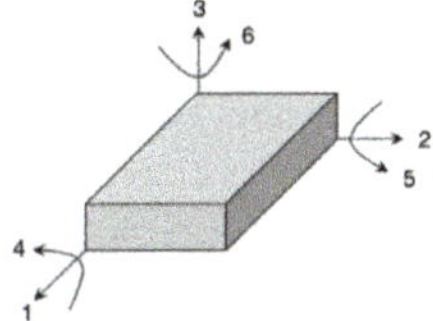

Note: 1. The piezoelectric constant d_{ij} is the ratio of the strain in the j-axis to the electric field applied along the i-axis. In other words, the i and j correspond to the response and excitation of materials, respectively. The inset image below depicts the coordination system. 2. The TrFE, HPF, and CTFE stand for trifuoroethylene, hexafluoropropylene, and chlorotrifluoroethylene, respectively.

3. Device Applications

3.1. Energy Harvesting

Self-powered electronics (SPEs) have gained an attraction as an alternative powering technology due to their independence, sustainability, and maintenance-free nature. SPEs are defined as electronics that can be operated themselves without feeding from external electrical power. In these electronics, electrical energy is provided from renewable resources, such as solar, thermal and mechanical energy. Nanogenerators that are converting mechanical to electrical energy have led the SPE field as the nanogenerators possess multiple advantages, including easy fabrication, portability, and high conversion efficiency, over conventional renewable energy technologies. Ever since the discovery of the piezoelectric and triboelectric nanogenerators (PENG and TENG, respectively) [69,70], their rapid development has shifted the paradigm for mechanical energy harvesting from fast and periodic energy resources to slow and random energy resources. A number of publications have tried over the last decade to develop the various nanogenerators possessing novel device structures and advanced materials [27,71–76]. Organic piezoelectric biomaterials have taken on an important role as the functional materials for applications in humans, being implantable and mountable nanogenerators due to their remarkable biodegradability and biocompatibility.

Vivekananthan et al. [77] reported a piezoelectric collagen nanofibril film that is capable of both converting mechanical energy to electrical energy and functioning as a humidity sensor. A schematic of the device is illustrated in Figure 2a. Collagen-based PENG produced electrical outputs of 250 nA and 45 V. In addition, it served as a humidity sensor that showed a linear response with a good sensitivity (0.1287 μA/% RH) in the range of 50–90% room humidity. These results demonstrated a field of eco-friendly multifunctional biomaterials, towards the development of noninvasive, implantable, smart bio-medical systems.

Lee et al. [26] firstly used a self-assembled M13 bacteriophage film for a piezoelectric energy harvester in 2012. The nanogenerator yields a current of up to 6 nA and a voltage of up to 400 mV. The M13 bacteriophages are expected to have a high piezoelectric response compared to laterally assembled phages, due to their high elasticity properties along the axial direction of the DNA [78]. Shin et al. [27] reported vertically aligned M13 bacteriophage nanopillars using enforced infiltration. The vertically aligned M13 bacteriophage-based nanogenerator exhibits electrical outputs that are up to about 2.6-fold greater than those of the laterally assembled, bacteriophages-based nanogenerator. There was still a limitation on the ability to control the directionality of individual M13 bacteriophage, but Lee et al. [28] addressed this issue using genetic engineering techniques (Figure 2b–d). The resulting structure-based PENG produced up to 2.8 V of potential, 120 nA of current, and 236 nW of power from 17 N of force. The apparent versatility of the M13 bacteriophage suggests that the piezoelectric M13 bacteriophages can serve as functional nanomaterials for numerous electronic and optoelectronic applications.

Piezoelectric peptide nanostructures have also been implemented into the PENGs. Nguyen et al. [49] fabricated the vertical FF microrod arrays by applying an electric field, and then the arrays were integrated into the PENGs, representing an open-circuit voltage of 1.4 V and a power density of 3.3 nW. The performance voltage of the FF-based PENG was improved up to 2.2 V in tandem with a TENG comprising the polyethylene terephthalate and Kapton films as triboelectrically active materials [79]. Recently, Lee et al. [47] developed large-scale, unidirectionally polarized, aligned FF nanotubes and fabricated peptide-based PENGs. They used the meniscus-driven self-assembly process to fabricate horizontally aligned FF nanotubes. The fabricated FF nanotubes-based PENGs can generate voltage, current, and power of up to 2.8 V, 37.4 nA, and 8.2 nW, respectively. Hence, the FF nanostructures will act as a compatible energy source for biomedical applications in the future.

The PVDF and its copolymers have been adopted for flexible PENGs due to their inherent flexibility, high processability, and mechanical rigidity [80–82]. Chang et al. [83] developed a method to directly fabricate the PVDF nanofibers with a β-crystalline structure using the near-field electrospinning process, which provides a peak current of 3 nA and a peak voltage of 30 mV after integration into the PENG. A hybrid nanogenerator, demonstrated by Hansen et al. [84], is made of a piezoelectric PVDF nanofibers-based PENG and a flexible biofuel cell, and was used for powering a single nanowire-based ultraviolet sensor to build an SPEs. As shown in Figure 2e,f, Ishida et al. [85] demonstrated a self-powered pedometer, which consisted of a PVDF sheet, a 2 V organic circuit, and a flexible printed circuit board. This work suggested that the PVDF sheet can not only harvest the mechanical energy from footsteps but also serve as a footstep sensor. Sun et al. [86] and Xue et al. [87] envisioned a PVDF nanostructures-based PENG to harvest energy from human respiration. Persano et al. [88] demonstrated the textile-based PENG, featuring the highly aligned electrospun fibers of the PVDF–TrFE, exhibiting superior flexibility and mechanical robustness.

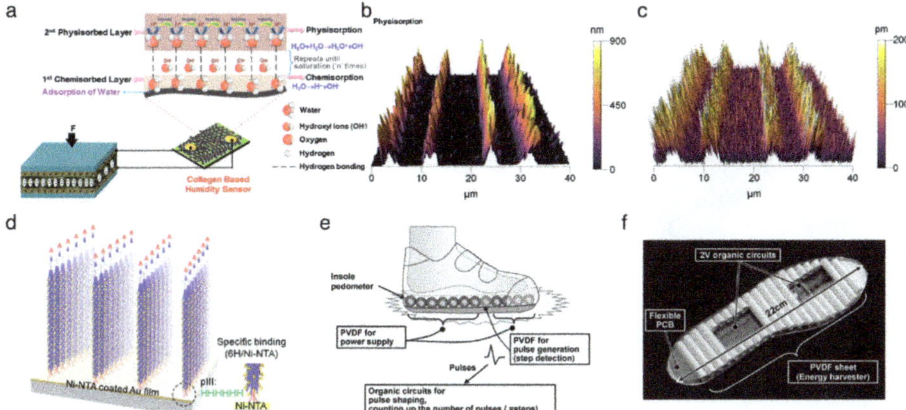

Figure 2. Applications of organic piezoelectric biomaterials in energy harvesting. (**a**) Schematic of sustainable energy harvesting and battery-free humidity sensor using biocompatible collagen nanofibrils. The collagen nanofibrils deposited on cotton cloth serve as a humidity sensor by measuring current signal at a fixed bias voltage. In order to demonstrate the self-powered sensing system, the energy harvester comprising collagen nanofibrils film sandwiched between Al electrodes is parallelly connected to the humidity sensor. Reproduced with permission from [77]. Copyright American Chemical Society, 2018. (**b–d**) Vertical self-assembly of polarized M13 bacteriophage nanostructure for energy harvesting. Reproduced with permission from [28]. Copyright American Chemical Society, 2019. (**b**) 3D-atomic force microscope (AFM) topography image of vertically aligned M13 bacteriophages. (**c**) Piezoresponse force microscope amplitude image corresponding to the 3D–AFM topography image. (**d**) The direction of polarization of the vertically aligned M13 bacteriophage with specific binding between the 6H tag on phage tail and the Ni-nitrilotriacetic acid (NTA) substrate. (**e,f**) Insole pedometer with piezoelectric energy harvester. Reproduced with permission from [85]. Copyright IEEE, 2012. (**e**) Schematics of the proposed insole pedometer. The pieces of PVDF sheet are used for the piezoelectric energy harvester as well as the pulse generator to detect steps in which each PVDF piece was rolled to increase the total area. The organic circuits are integrated with PVDF pieces to count the number of steps. (**f**) Photograph of the prototype insole pedometer.

3.2. Sensors

The organic biomaterials have been studied as the platform materials for biomedical pressure-sensing applications due to their high flexibility and high sensitivity to small force. The aligned PVDF–TrFE nanofibers on polyimide substrate were employed to build a flexible and lightweight pressure sensor [88]. This pressure sensor can measure small pressures down to ~0.1 Pa over the course of cyclic bending. The wearable piezoelectric PVDF sensor can serve as a healthcare monitoring device to monitor respiration signals, human gestures, and vocal cord vibrations, as demonstrated by Liu et al. [89] (Figure 3a–c). Bodkhe et al. [90] developed a pressure sensor comprised of 10% of barium titanate nanoparticle and β-crystalline phase PVDF ball mill nanocomposites using 3D printing techniques, which generated a voltage of 4 V upon gentle finger taps. The composite films of PVDF and graphene oxide developed by Park et al. [91] were used as multifunctional electronic skins to monitor multiple stimuli, including static/dynamic pressure and temperature, exhibiting a high sensitivity for monitoring simultaneous artery pulse pressures and temperature (Figure 3d,e). Recently, biodegradable and implantable sensors have gained great interest in medical applications where there is a demand for short-term functionality because biodegradable sensors are not required for the medical surgery of removal. Curry et al. [92] fabricated the piezoelectric pressure sensor featuring all the biodegradable materials of piezoelectric PLLA, molybdenum electrodes, and polylactic acid encapsulators. This device was capable of measuring a wide range of pressure, from 0 to 18 kPa. More interestingly, the sensor was completely degraded over a period of 56 days at an elevated temperature of 74 °C, indicating that this biodegradable sensor holds promise for clinical implementation. A number of publications has utilized the nature-driven piezoelectric materials to develop human physiological monitoring and electronic skins [93–96].

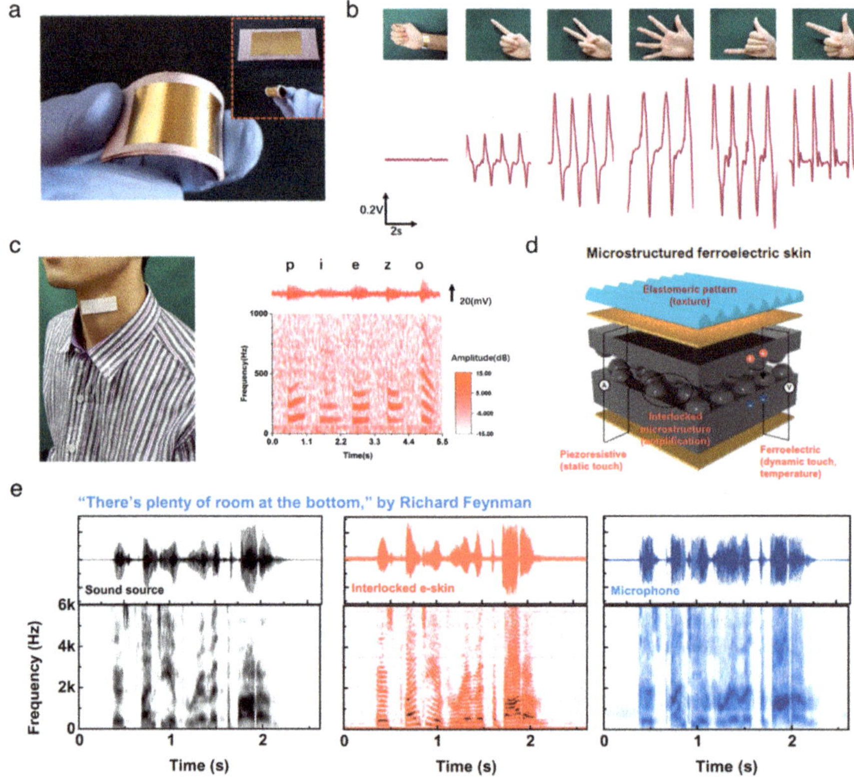

Figure 3. Applications of organic piezoelectric biomaterials in sensors. (**a–c**) Flexible piezoelectric nanogenerator in a wearable, self-powered active sensor for healthcare monitoring. Reproduced with permission from [89]. Copyright IOP Publishing, 2017. (**a**) Photographs of the flexible piezoelectric nanogenerator. (**b**) hand gesture sensing. (**c**) human voice recording. (**d,e**) Electronic skins for discriminating static/dynamic pressure stimuli. Reproduced with permission from [91]. Copyright The American Association for the Advancement of Science, 2015. (**d**) Schematic illustration of flexible and multimodal ferroelectric e-skin. (**e**) The waveform and short-time Fourier transform (STFT) signals of the sound source, readout signals from the interlocked e-skin, and microphone.

3.3. Cell and Tissue Regeneration

Organic piezoelectric biomaterials have been chosen as the functional materials for fabricating a scaffold to grow and differentiate cells in the field of tissue engineering [16]. Several studies have shown that the biocompatible piezoelectric materials can serve as tissue stimulators and scaffolds to promote tissue regeneration. Damaraju et al. [97] found that the cell growth on the β–phase PVDF nanofibers film exhibited higher alkaline phosphatase activity and earlier mineralization compared to the growth on random-phase PVDF film. Then, Damaraju et al. [98] showed that the 3D fibrous scaffolds decorated with electrospun PVDF–TrFE fibers stimulated the differentiation of human mesenchymal stem cells. The electromechanical actuation under high voltage helped osteogenic differentiation, whereas the actuation under low voltage aided chondrogenic differentiation (Figure 4a). Muscle cell adhesion and proliferation were improved due to the fact that the negatively charged β-phase PVDF fibers helped to elongate the muscle cells along the aligned fibers [99]. Hoop et al. [100] demonstrated that wireless stimulations helped to induce the potential in piezoelectric β-phase PVDF, improving the neurite generation in PC12 cells using the ultrasonic technique (Figure 4b–d). Similarly, the PVDF–TrFE fibril scaffolds promoted the differentiation of neural cells, neurite extension and neuronal differentiation due to the piezoelectric effect of the scaffolds [101]. The mechanical stimulation facilitates the enhanced bone cell culture on the piezoelectric PVDF substrate by applying a voltage of 5 V.

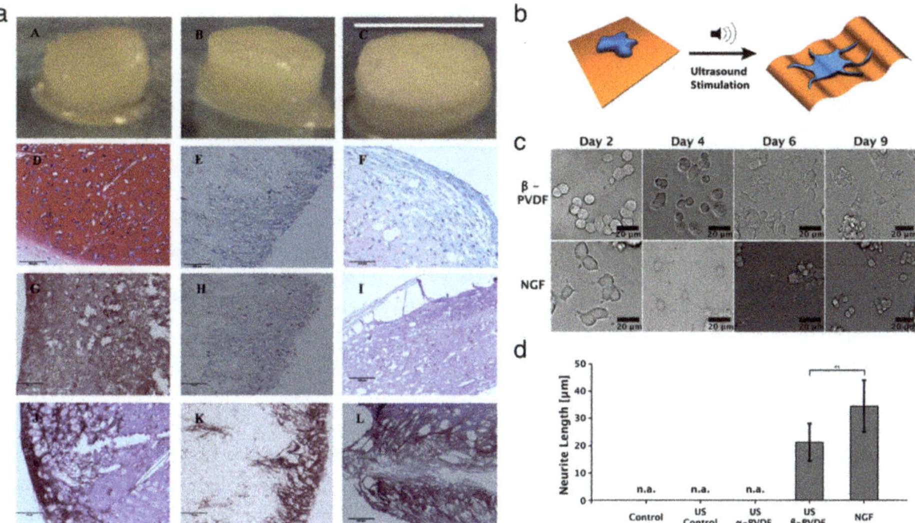

Figure 4. Applications of organic piezoelectric biomaterials in cell and tissue regenerations. (**a**) Representative gross images and histological images of scaffolds after 28 days undergoing chondrogenesis in dynamic conditions. As-spun PVDF–TrFE (left), annealed PVDF–TrFE (middle) and polycaprolactone (right) scaffolds. Reproduced with permission from [98]. Copyright Elsevier, 2017. (**b–d**) Ultrasound-mediated piezoelectric differentiation of neuron-like PC12 cells on PVDF membranes. Reproduced with permission from [100]. Copyright Nature Publishing Group, 2017. (**b**) Schematic of ultrasound stimulation of the piezoelectric β-PVDF membrane. (**c**) Comparison images of PC12 cells cultured under mechanical stimuli on PVDF substrate and neuronal growth factor stimuli. (**d**) Comparison of average neurite length of PC12 cells.

There have also been several attempts to utilize the biodegradable piezoelectric PLLA polymer as the tissue stimulator. Ikada et al. [102] intramedullary implanted PLLA rods in the cut tibiae of cats for internal fixation for up to eight weeks. The high aspect ratio of the PLLA rod enabled the enhanced fracture healing, indicated by improved callus formation, whereas the isotropic PLLA and a polyethylene control rod exhibited no effect on callus formation. Barroca et al. [103,104] observed that surface charges can change the orientation of the adsorbed proteins, resulting in the modulation of cell-binding domains. Indeed, the negatively charged PLLA improves the protein adsorption and cellular adhesion as well as proliferation.

4. Conclusions and Future Perspective

The present review has sought to offer insight into the importance of organic piezoelectric biomaterials in biomedical applications. We have reviewed the origin of piezoelectricity in organic piezoelectric biomaterials, including proteins, peptides, and biopolymers. The intrinsic piezoelectric property of those materials has been presented, and engineering and scientific endeavors to enhance these properties have also been reported. In summary, from current research, organic piezoelectric biomaterials have been likely to impact three major fields across many disciplines. First, they can serve as the functional materials for the power supply of implantable and mountable self-powered electronics because of their sensitivity to mechanical agitation and remarkable biocompatibility. Second, they have been utilized as platform materials for pressure sensing in biomedical applications, which is likely largely due to their high flexibility and high sensitivity to small forces in tandem with their biodegradability. Lastly, in cases where piezoelectric materials were integrated as the scaffolds for cell and tissue regeneration, those materials act as a tissue stimulator to promote the differentiation of the desired cells. An industry based on organic piezoelectric biomaterials is anticipated due to their variety of applications, but further improvements are required to smoothly implement them into practical biomedical devices.

We need to address several issues for the better integration of organic piezoelectric biomaterials into biomedical devices. Here are a few: (1) the fundamental physics of piezoelectricity in biomaterials. Even though researchers have focused on uncovering biological piezoelectricity, plenty of work remains to be done to exploit their electromechanical behavior in terms of unit cell properties. Such studies are in progress using not only single crystals of biomaterials but also calculations based on the first principle. (2) A relatively low piezoelectric constant compared to piezoelectric inorganic materials. The output performances in the applications of energy harvesting and sensors are related to the piezoelectric constant. The constant of biomaterials has been found to be much smaller than that of the state-of-the-art piezoelectric inorganic materials (d_{33} = 593 pC/N and d_{31} = −274 pC/N [21]), which has to be improved to achieve maximized performance. This is possible by creating proper nanostructures, aligning biomaterials, or fabricating a multilayer structure. (3) Biodegradability of the controlled manner. For biodegradable sensor and scaffold applications, the organic piezoelectric biomaterials must be decomposed within the desired time frame. The degradation rate of these materials can be engineered by different experimental treatments, such as temperature, stretching ratio, or poling electrical fields. Although researchers are still facing challenging issues, the promising physical properties of organic piezoelectric biomaterials have suggested feasible biomedical applications in energy harvesting, sensor, and tissue regeneration. We truly believe that organic piezoelectric biomaterials will continue their rapid growth in the next decade.

Author Contributions: D.-M.S., S.W.H. and Y.-H.H. formulated the project. D.-M.S. wrote the manuscript, and all authors contributed to revising the manuscript. All authors have read and agreed to the published version of the manuscript.

Funding: This research was supported by the Startup Fund, funded by Department of Mechanical Engineering and Faculty of Engineering in the University of Hong Kong. This work was also supported by National Research Foundation of Korea (NRF) grants funded by the Ministry of Science, ICT and Future Planning (MSIP) of Korea (No. NRF-2017R1A2B2006852).

Conflicts of Interest: The authors declare no conflict of interest.

References

1. Li, C.; Weng, G. Antiplane crack problem in functionally graded piezoelectric materials. *J. Appl. Mech.* **2002**, *69*, 481–488. [CrossRef]
2. Chorsi, M.T.; Curry, E.J.; Chorsi, H.T.; Das, R.; Baroody, J.; Purohit, P.K.; Ilies, H.; Nguyen, T.D. Piezoelectric biomaterials for sensors and actuators. *Adv. Mater.* **2019**, *31*, 1802084. [CrossRef] [PubMed]
3. Jacob, J.; More, N.; Kalia, K.; Kapusetti, G. Piezoelectric smart biomaterials for bone and cartilage tissue engineering. *Inflamm. Regen.* **2018**, *38*, 2. [CrossRef] [PubMed]
4. Sirohi, J.; Chopra, I. Fundamental understanding of piezoelectric strain sensors. *J. Intell. Mater. Syst. Struct.* **1999**, *11*, 246–257. [CrossRef]
5. Cannata, D.; Benetti, M.; Verona, E.; Varriale, A.; Staiano, M.; D'Auria, S.; Di Pietrantonio, F. Odorant detection via Solidly Mounted Resonator biosensor. In Proceedings of the 2012 IEEE International Ultrasonics Symposium, Dresden, Germany, 7–10 October 2012; IEEE: Piscataway, NJ, USA, 2012; p. 1537.

6. Chen, D.; Wang, J.; Xu, Y. Highly sensitive lateral field excited piezoelectric film acoustic enzyme biosensor. *IEEE Sens. J.* **2013**, *13*, 2217–2222. [CrossRef]

7. Gan, J.; Zhang, X. A review of nonlinear hysteresis modeling and control of piezoelectric actuators. *AIP Adv.* **2019**, *9*, 040702. [CrossRef]

8. Dagdeviren, C.; Yang, B.D.; Su, Y.; Tran, P.L.; Joe, P.; Anderson, E.; Xia, J.; Doraiswamy, V.; Dehdashti, B.; Feng, X.; et al. Conformal piezoelectric energy harvesting and storage from motions of the heart, lung, and diaphragm. *Proc. Natl. Acad. Sci. USA* **2014**, *111*, 1927–1932. [CrossRef]

9. Hwang, G.T.; Park, H.; Lee, J.H.; Oh, S.; Park, K.I.; Byun, M.; Park, H.; Ahn, G.; Jeong, C.K.; No, K.; et al. Self-powered cardiac pacemaker enabled by flexible single crystalline PMN-PT piezoelectric energy harvester. *Adv. Mater.* **2014**, *26*, 4880–4887. [CrossRef]

10. Zhu, G.; Wang, A.C.; Liu, Y.; Zhou, Y.; Wang, Z.L. Functional electrical stimulation by nanogenerator with 58 V output voltage. *Nano Lett.* **2012**, *12*, 3086–3090. [CrossRef] [PubMed]

11. He, Y.-B.; Li, G.-R.; Wang, Z.-L.; Su, C.-Y.; Tong, Y.-X. Single-crystal zno nanorod/amorphous and nanoporous metal oxide shell composites: Controllable electrochemical synthesis and enhanced supercapacitor performances. *Energy Environ. Sci.* **2011**, *4*, 1288–1292. [CrossRef]

12. Ryu, J.; Kim, S.W.; Kang, K.; Park, C.B. Synthesis of diphenylalanine/cobalt oxide hybrid nanowires and their application to energy storage. *ACS Nano* **2011**, *4*, 159–164. [CrossRef]

13. Kim, D.H.; Lu, N.; Ma, R.; Kim, Y.S.; Kim, R.H.; Wang, S.; Wu, J.; Won, S.M.; Tao, H.; Islam, A.; et al. Epidermal electronics. *Science* **2011**, *333*, 838–843. [CrossRef]

14. Choi, C.; Lee, Y.; Cho, K.W.; Koo, J.H.; Kim, D.H. Wearable and implantable soft bioelectronics using two-dimensional materials. *Acc. Chem. Res.* **2019**, *52*, 73–81. [CrossRef]

15. Martin, A.J.P. Tribo-electricity in wool and hair. *Proc. Phys. Soc.* **1941**, *53*, 186. [CrossRef]

16. Yuan, H.; Lei, T.; Qin, Y.; He, J.H.; Yang, R. Design and application of piezoelectric biomaterials. *J. Phys. D Appl. Phys.* **2019**, *52*, 194002. [CrossRef]

17. de Sciarra, F.M. A nonlocal model with strain-based damage. *Int. J. Solids Struct.* **2009**, *46*, 4107–4122. [CrossRef]

18. Barretta, R.; Fabbrocino, F.; Luciano, R.; de Sciarra, F.M. Closed-form solutions in stress-driven two-phase integral elasticity for bending of functionally graded nano-beams. *Phys. E Low Dimens. Syst. Nanostruct.* **2018**, *97*, 13–30. [CrossRef]

19. Fousek, J.; Cross, L.; Litvin, D. Possible piezoelectric composites based on the flexoelectric effect. *Mater. Lett.* **1999**, *39*, 287. [CrossRef]

20. Zhou, Z.; Qian, D.; Minary-Jolandan, M. Molecular mechanism of polarization and piezoelectric effect in super-twisted collagen. *ACS Biomater. Sci. Eng.* **2016**, *2*, 929–936. [CrossRef]

21. Bystrov, V.; Bdikin, I.; Heredia, A.; Pullar, R.; Mishina, E.; Sigov, A.; Kholkin, A. Piezoelectricity and ferroelectricity in biomaterials: From proteins to self-assembled peptide nanotubes. In *Piezoelectric Nanomaterials for Biomedical Applications, Nanomedicine and Nanotoxicology*; Ciofani, G., Menciassi, A., Eds.; Springer: Berlin, Germany, 2012; pp. 187–211.

22. Wojnar, R. Piezoelectric phenomena in biological tissues. In *Piezoelectric Nanomaterials for Biomedical Applications, Nanomedicine and Nanotoxicology*; Ciofani, G., Menciassi, A., Eds.; Springer: Berlin, Germany, 2012; pp. 173–2011.

23. Lemanov, V.; Popov, S.; Pankova, G. Piezoelectric properties of crystals of some protein aminoacids and their related compounds. *Phys. Solid State* **2002**, *44*, 1929–1935. [CrossRef]

24. Namiki, K.; Hayakawa, R.; Wada, Y. Molecular theory of piezoelectricity of α-helical polypeptide. *J. Polym. Sci. Polym. Phys. Ed.* **1980**, *18*, 993–1004. [CrossRef]

25. Fukada, E. History and recent progress in piezoelectric polymers. *IEEE Trans. Ultrason. Ferroelectr. Freq. Control.* **2000**, *47*, 1277–1290. [CrossRef]

26. Lee, B.Y.; Zhang, J.; Zueger, C.; Chung, W.-J.; Yoo, S.Y.; Wang, E.; Meyer, J.; Ramesh, R.; Lee, S.-W. Virus-based piezoelectric energy generation. *Nat. Nanotechnol.* **2012**, *7*, 351. [CrossRef]

27. Shin, D.-M.; Han, H.J.; Kim, W.-G.; Kim, E.; Kim, C.; Hong, S.W.; Kim, H.K.; Oh, J.-W.; Hwang, Y.-H. Bioinspired piezoelectric nanogenerators based on vertically aligned phage nanopillars. *Energy Environ. Sci.* **2015**, *8*, 3198–3203. [CrossRef]

28. Lee, J.-H.; Lee, J.H.; Xiao, J.; Desai, M.S.; Zhang, X.; Lee, S.-W. Vertical self-assembly of polarized phage nanostructure for energy harvesting. *Nano Lett.* **2019**, *19*, 2661–2667. [CrossRef]

29. Moon, J.-S.; Kim, W.-G.; Shin, D.-M.; Lee, S.-Y.; Kim, C.; Lee, Y.; Han, J.; Kim, K.; Yoo, S.Y.; Oh, J.-W. Bioinspired M-13 bacteriophage-based photonic nose for differential cell recognition. *Chem. Sci.* **2017**, *8*, 921–927. [CrossRef]

30. Moon, J.-S.; Lee, Y.; Shin, D.-M.; Kim, C.; Kim, W.-G.; Park, M.; Han, J.; Song, H.; Kim, K.; Oh, J.-W. Identification of endocrine disrupting chemicals using a virus-based colorimetric sensor. *Chem. Asian J.* **2016**, *11*, 3097–3101. [CrossRef]

31. Kim, W.-G.; Zueger, C.; Kim, C.; Wong, W.; Devaraj, V.; Yoo, H.-W.; Hwang, S.; Oh, J.-W.; Lee, S.-W. Experimental and numerical evaluation of a genetically engineered M13 bacteriophage with high sensitivity and selectivity for 2,4,6-trinitrotoluene. *Org. Biomol. Chem.* **2019**, *17*, 5666–5670. [CrossRef]

32. Shin, Y.C.; Kim, C.; Song, S.-J.; Jun, S.; Kim, C.-S.; Hong, S.W.; Hyon, S.-H.; Han, D.-W.; Oh, J.-W. Ternary aligned nanofibers of RGD peptide-displaying M13 bacteriophage/PLGA/graphene oxide for facilitated myogenesis. *Nanotheranostics* **2018**, *2*, 144. [CrossRef]

33. Raja, I.S.; Kim, C.; Song, S.-J.; Shin, Y.C.; Kang, M.S.; Hyon, S.-H.; Oh, J.-W.; Han, D.-W. Virus-incorporated biomimetic nanocomposites for tissue regeneration. *Nanomaterials* **2019**, *9*, 1014. [CrossRef]

34. Chew, J.W.; Black, S.N.; Chow, P.S.; Tan, R.B.H.; Carpenter, K.J. Stable polymorphs: Difficult to make and difficult to predict. *CrystEngComm* **2007**, *9*, 128–130. [CrossRef]

35. Poornachary, S.K.; Chow, P.S.; Tan, R.B.H. Influence of solution speciation of impurities on polymorphic nucleation in glycine. *Cryst. Growth Des.* **2008**, *8*, 179–185. [CrossRef]

36. Dowling, R.; Davey, R.J.; Curtis, R.A.; Han, G.; Poornachary, S.K.; Chow, P.S.; Tan, R.B.H. Acceleration of crystal growth rates: An unexpected effect of tailor-made additives. *Chem. Commun.* **2010**, *46*, 5924–5926. [CrossRef]

37. Iitaka, Y. A new form of glycine. *Proc. Jpn. Soc.* **1954**, *30*, 109–112. [CrossRef]

38. Kumar, R.A.; Vizhi, R.E.; Vijayan, N.; Babu, D.R. Structural, dielectric and piezoelectric properties of nonlinear optical γ-glycine single crystals. *Phys. Condens. B Matter* **2011**, *406*, 2594–2600. [CrossRef]

39. Iitaka, Y. The crystal structure of γ-glycine. *Acta Crystallogr.* **1958**, *11*, 225–226. [CrossRef]

40. Guerin, S.; Tofail, S.A.M.; Thompson, D. Organic piezoelectric materials: Milestones and potential. *NPG Asia Mater.* **2019**, *11*, 10. [CrossRef]

41. Guerin, S.; Stapleton, A.; Chovan, D.; Mouras, R.; Gleeson, M.; McKeown, C.; Noor, M.R.; Silien, C.; Rhen, F.M.F.; Kholkin, A.L.; et al. Control of piezoelectricity in amino acids by supramolecular packing. *Nat. Mater.* **2018**, *17*, 180. [CrossRef]

42. Newnham, R.E. *Properties of Materials: Anisotropy, Symmetry, Structure*, 1st ed.; Oxford University Press: New York, NY, USA, 2005.

43. Yan, X.; Zhu, P.; Fei, J.; Li, J. Self-assembly of peptide-inorganic hybrid spheres for adaptive encapsulation of guests. *Adv. Mater.* **2010**, *22*, 1283–1287. [CrossRef]

44. Görbitz, C.H. Nanotube formation by hydrophobic dipeptides. *Chem. Eur. J.* **2001**, *7*, 5153–5159. [CrossRef]

45. Amdursky, N.; Beker, P.; Rosenman, G. Physics of peptide nanostructures and their nanotechnology applications. In *Peptide Materials: From Nanostuctures to Applications*, 1st ed.; Aleman, C., Bianco, A., Venanzi, M., Eds.; Wiley-Blackwell: Hoboken, NJ, USA, 2013; pp. 1–37.

46. Kholkin, A.; Amdursky, N.; Bdikin, I.; Gazit, E.; Rosenman, G. Strong piezoelectricity in bioinspired peptide nanotubes. *ACS Nano* **2010**, *4*, 610–614. [CrossRef]

47. Lee, J.-H.; Heo, K.; Schulz-Schönhagen, K.; Lee, J.H.; Desai, M.S.; Jin, H.-E.; Lee, S.-W. Diphenylalanine peptide nanotube energy harvesters. *ACS Nano* **2018**, *12*, 8138–8144. [CrossRef]

48. Nguyen, V.; Jenkins, K.; Yang, R. Epitaxial growth of vertically aligned piezoelectric diphenylalanine peptide microrods with uniform polarization. *Nano Energy* **2015**, *17*, 323–329. [CrossRef]

49. Nguyen, V.; Zhu, R.; Jenkins, K.; Yang, R. Self-assembly of diphenylalanine peptide with controlled polarization for power generation. *Nat. Commun.* **2016**, *7*, 13566. [CrossRef]

50. Nalwa, H.S. *Ferroelectric Polymers: Chemistry: Physics, and Applications*, 1st ed.; CRC Press: New York, NY, USA, 1995.

51. Someya, T. *Stretchable Electronics*, 1st ed.; Wiley-VCH: Weinheim, Germany, 2013.

52. Martins, P.; Lopes, A.C.; Lanceros-Mendez, S. Electroactive phases of poly (vinylidene fluoride): Determination, processing and applications. *Prog. Polym. Sci.* **2014**, *39*, 683–706. [CrossRef]

53. Cui, Z.; Hassankiadeh, N.T.; Zhuang, Y.; Drioli, E.; Lee, Y.M. Crystalline polymorphism in poly(vinylidenefluoride) membranes. *Prog. Polym. Sci.* **2015**, *51*, 94–126. [CrossRef]

54. Wan, C.; Bowen, C.R. Multiscale-structuring of polyvinylidene fluoride for energy harvesting: The impact of molecular-, micro-and macro-structure. *J. Mater. Chem. A* **2017**, *5*, 3091–3128. [CrossRef]

55. Lovinger, A.J. Conformational defects and associated molecular motions in crystalline poly (vinylidene fluoride). *J. Appl. Phys.* **1981**, *52*, 5934–5938. [CrossRef]

56. Ameduri, B. From vinylidene fluoride (VDF) to the applications of VDF-containing polymers and copolymers: Recent developments and future trends. *Chem. Rev.* **2009**, *109*, 6632–6686. [CrossRef]

57. Bystrov, V.S.; Paramonova, E.V.; Bdikin, I.K.; Bystrova, A.V.; Pullar, R.C.; Kholkin, A.L. Molecular modeling of the piezoelectric effect in the ferroelectric polymer poly(vinylidene fluoride) (PVDF). *J. Mol. Model.* **2013**, *19*, 3591–3602. [CrossRef]

58. Lovinger, A.J. Ferroelectric polymers. *Science* **1983**, *220*, 1115–1121. [CrossRef]

59. Lutkenhaus, J.L.; McEnnis, K.; Serghei, A.; Russell, T.P. Confinement effects on crystallization and Curie transitions of poly(vinylidene fluoride-co-trifluoroethylene). *Macromolecules* **2010**, *43*, 3844–3850. [CrossRef]

60. Kunstler, W.; Wegener, M.; Seiss, M.; Gerhard-Multhaupt, R. Preparation and assessment of piezo- and pyroelectric poly(vinylidene fluoride-hexafluoropropylene) copolymer films. *Appl. Phys. A* **2002**, *73*, 641–645. [CrossRef]

61. Huan, Y.; Liu, Y.; Yang, Y. Simultaneous stretching and static electric field poling of poly(vinylidene fluoride-hexafluoropropylene) copolymer films. *Polym. Eng. Sci.* **2007**, *47*, 1630–1633. [CrossRef]

62. Li, Z.; Wang, Y.; Cheng, Z.-Y. Electromechanical properties of poly(vinylidene-fluoride-chlorotrifluoroethylene) copolymer. *Appl. Phys. Lett.* **2006**, *88*, 062904. [CrossRef]

63. Sultana, A.; Ghosh, S.K.; Sencadas, V.; Zheng, T.; Higgins, M.J.; Middya, T.R.; Mandal, D. Human skin interactive self-powered wearable piezoelectric bio-e-skin by electrospun poly-L-lactic acid nanofibers for non-invasive physiological signal monitoring. *J. Mater. Chem. B* **2017**, *5*, 7352–7359. [CrossRef]

64. Tajitsu, Y. Piezoelectricity of chiral polymeric fiber and its application in biomedical engineering. *IEEE Trans. Ultrason. Ferroelectr. Freq. Control* **2008**, *55*, 1000–1008. [CrossRef]

65. Fukada, E. New piezoelectric polymers. *Jpn. J. Appl. Phys.* **1998**, *37*, 2775. [CrossRef]

66. Yoshida, T.; Imoto, K.; Tahara, K.; Naka, K.; Uehara, Y.; Kataoka, S.; Date, M.; Fukada, E.; Tajitsu, Y. Piezoelectricity of poly(L-lactic acid) composite film with stereocomplex of poly(L-lactide) and poly(D-lactide). *Jpn. J. Appl. Phys.* **2010**, *49*, 09MC11. [CrossRef]

67. Yucel, T.; Cebe, P.; Kaplan, D.L. Structural origins of silk piezoelectricity. *Adv. Funct. Mater.* **2011**, *21*, 779–785. [CrossRef]

68. Kim, J.; Yun, S.; Ounaies, Z. Discovery of cellulose as a smart material. *Macromolecules* **2006**, *39*, 4202–4206. [CrossRef]

69. Wang, Z.L.; Song, J. Piezoelectric nanogenerators based on zinc oxide nanowire arrays. *Science* **2006**, *312*, 242–246. [CrossRef]

70. Fan, F.-R.; Tian, Z.-Q.; Wang, Z.L. Flexible triboelectric generator. *Nano Energy* **2012**, *1*, 328–334. [CrossRef]

71. Kim, G.H.; Shin, D.-M.; Kim, H.-K.; Hwang, Y.-H. Effect of the dielectric layer on the electrical output of a ZnO nanosheet-based nanogenerator. *J. Korean Phys. Soc.* **2015**, *67*, 1920–1924. [CrossRef]

72. Kim, T.; Jeon, S.; Lone, S.; Doh, S.J.; Shin, D.-M.; Kim, H.K.; Hwang, Y.-H.; Hong, S.W. Versatile nanodot-patterned Gore-Tex fabric for multiple energy harvesting in wearable and aerodynamic nanogenerators. *Nano Energy* **2018**, *54*, 209–217. [CrossRef]

73. Park, H.-Y.; Kim, H.K.; Hwang, Y.-H.; Shin, D.-M. Water-through triboelectric nanogenerator based on Ti-mesh for harvesting liquid flow. *J. Korean Phys. Soc.* **2018**, *72*, 449–503. [CrossRef]

74. Phan, H.; Shin, D.-M.; Jeon, S.H.; Kang, T.Y.; Han, P.; Kim, G.H.; Kim, H.K.; Kim, K.; Hwang, Y.-H.; Hong, S.W. Aerodynamic and aeroelastic flutters driven triboelectric nanogenerators for harvesting broadband airflow energy. *Nano Energy* **2017**, *33*, 476–484. [CrossRef]

75. Shin, D.-M.; Tsege, E.L.; Kang, S.H.; Seung, W.; Kim, S.-W.; Kim, H.-K.; Hong, S.W.; Hwang, Y.-H. Freestanding ZnO nanorod/graphene/ZnO nanorod epitaxial double heterostructure for improved piezoelectric nanogenerators. *Nano Energy* **2015**, *12*, 268–277. [CrossRef]

76. Tsege, E.L.; Shin, D.-M.; Lee, S.; Kim, H.-K.; Hwang, Y.-H. Highly Durable Ti-mesh based triboelectric nanogenerator for self-powered device applications. *J. Nanosci. Nanotechnol.* **2016**, *16*, 4864–4869. [CrossRef]

77. Vivekananthan, V.; Alluri, N.R.; Purusothaman, Y.; Chandrasekhar, A.; Selvarajan, S.; Kim, S.-J. Biocompatible collagen nanofibrils: An approach for sustainable energy harvesting and battery-free humidity sensor applications. *ACS Appl. Mater. Interfaces* **2018**, *10*, 18650–18656. [CrossRef]

78. Shi, Y.; He, S.; Hearst, J.E. Statistical mechanics of the extensible and shearable elastic rod and of DNA. *J. Chem. Phys.* **1996**, *105*, 714–731. [CrossRef]

79. Vu, N.; Kelly, S.; Yang, R. Piezoelectric peptide-based nanogenerator enhanced by single-electrode triboelectric nanogenerator. *APL Mater.* **2017**, *5*, 074108.

80. Fang, J.; Wang, X.; Lin, T. Electrical power generator from randomly oriented electrospun poly(vinylidene fluoride) nanofibre membranes. *J. Mater. Chem.* **2011**, *21*, 11088–11091. [CrossRef]

81. Pan, C.-T.; Yen, C.-K.; Wang, S.-Y.; Lai, Y.-C.; Lin, L.; Huang, J.C.; Kuo, S.-W. Near-field electrospinning enhances the energy harvesting of hollow PVDF piezoelectric fibers. *RSC Adv.* **2015**, *5*, 85073–85081. [CrossRef]

82. Liu, Z.H.; Pan, C.T.; Lin, L.W.; Huang, J.C.; Ou, Z.Y. Direct-write PVDF nonwoven fiber fabric energy harvesters via the hollow cylindrical near-field electrospinning process. *Smart Mater. Struct.* **2014**, *23*, 25003. [CrossRef]

83. Chang, C.; Tran, V.H.; Wang, J.; Fuh, Y.-K.; Lin, L. Direct-write piezoelectric polymeric nanogenerator with high energy conversion efficiency. *Nano Lett.* **2010**, *10*, 726–731. [CrossRef]

84. Hansen, B.J.; Liu, Y.; Yang, R.; Wang, Z.L. Hybrid nanogenerator for concurrently harvesting biomechanical and biochemical energy. *ACS Nano* **2010**, *4*, 3647–3652. [CrossRef]

85. Ishida, K.; Huang, T.-C.; Honda, K.; Shinozuka, Y.; Fuketa, H.; Yokota, T.; Zschieschang, U.; Klauk, H.; Tortissier, G.; Sekitani, T.; et al. Insole pedometer with piezoelectric energy harvester and 2 V organic circuits. *IEEE J. Solid State Circuits* **2013**, *48*, 255–264. [CrossRef]

86. Sun, C.; Shi, J.; Bayerl, D.J.; Wang, X. PVDF microbelts for harvesting energy from respiration. *Energy Environ. Sci.* **2011**, *4*, 4508–4512. [CrossRef]

87. Xue, H.; Yang, Q.; Wang, D.; Luo, W.; Wang, W.; Lin, M.; Liang, D.; Luo, Q. A wearable pyroelectric nanogenerator and self-powered breathing sensor. *Nano Energy* **2017**, *38*, 147–154. [CrossRef]

88. Persano, L.; Dagdeviren, C.; Su, Y.; Zhang, Y.; Girardo, S.; Pisignano, D.; Huang, Y.; Rogers, J.A. High performance piezoelectric devices based on aligned arrays of nanofibers of poly (vinylidenefluoride-co-trifluoroethylene). *Nat. Commun.* **2013**, *4*, 1633. [CrossRef]

89. Liu, Z.; Zhang, S.; Jin, Y.M.; Ouyang, H.; Zou, Y.; Wang, X.X.; Xie, L.X.; Li, Z. Flexible piezoelectric nanogenerator in wearable self-powered active sensor for respiration and healthcare monitoring. *Semicond. Sci. Technol.* **2017**, *32*, 064004. [CrossRef]

90. Bodkhe, S.; Turcot, G.; Gosselin, F.P.; Therriault, D. One-step solvent evaporation-assisted 3D printing of piezoelectric PVDF nanocomposite structures. *ACS Appl. Mater. Interfaces* **2017**, *9*, 20833–20842. [CrossRef]

91. Park, J.; Kim, M.; Lee, Y.; Lee, H.S.; Ko, H. Fingertip skin-inspired microstructured ferroelectric skins discriminate static/dynamic pressure and temperature stimuli. *Sci. Adv.* **2015**, *1*, e1500661. [CrossRef]

92. Curry, E.J.; Ke, K.; Chorsi, M.T.; Wrobel, K.S.; Miller, A.N.; Patel, A.; Kim, I.; Feng, J.; Yue, L.; Wu, Q.; et al. Biodegradable piezoelectric force sensor. *Proc. Natl. Acad. Sci. USA* **2018**, *115*, 909–914. [CrossRef]

93. Joseph, J.; Singh, S.G.; Vanjari, S.R.K. Leveraging innate piezoelectricity of ultra-smooth silk thin films for flexible and wearable sensor applications. *IEEE Sens. J.* **2017**, *17*, 8306–8313. [CrossRef]

94. Wang, X.; Gu, Y.; Xiong, Z.; Cui, Z.; Zhang, T. Silk-molded flexible, ultrasensitive, and highly stable electronic skin for monitoring human physiological signals. *Adv. Mater.* **2014**, *26*, 1336–1342. [CrossRef]

95. Ghosh, S.K.; Mandal, D. Sustainable energy generation from piezoelectric biomaterial for noninvasive physiological signal monitoring. *ACS Sustain. Chem. Eng.* **2017**, *5*, 8836–8843. [CrossRef]

96. Moreno, S.; Baniasadi, M.; Mohammed, S.; Mejia, I.; Chen, Y.; Quevedo-Lopez, M.A.; Kumar, N.; Dimitrijevich, S.; Minary-Jolandan, M. Flexible electronics: Biocompatible collagen films as substrates for flexible implantable electronics. *Adv. Electron. Mater.* **2015**, *1*, 1500154. [CrossRef]

97. Damaraju, S.M.; Wu, S.; Jaffe, M.; Arinzeh, T.L. Structural changes in pvdf fibers due to electrospinning and its effect on biological function. *Biomed. Mater.* **2013**, *8*, 045007. [CrossRef]

98. Damaraju, S.M.; Shen, Y.; Elele, E.; Khusid, B.; Eshghinejad, A.; Li, J.; Jaffe, M.; Arinzeh, T.L. Three-dimensional piezoelectric fibrous scaffolds selectively promote mesenchymal stem cell differentiation. *Biomaterials* **2017**, *149*, 51–62. [CrossRef] [PubMed]

99. Martins, P.M.; Ribeiro, S.; Ribeiro, C.; Sencadas, V.; Gomes, A.C.; Gama, F.M.; Lancerosmendez, S. Effect of poling state and morphology of piezoelectric poly(vinylidene fluoride) membranes for skeletal muscle tissue engineering. *RCS Adv.* **2013**, *3*, 17938–17944. [CrossRef]

100. Hoop, M.; Chen, X.Z.; Ferrari, A.; Mushtaq, F.; Ghazaryan, G.; Tervoort, T.; Poulikakos, D.; Nelson, B.; Pane, S. Ultrasound-mediated piezoelectric differentiation of neuron-like PC12 cells on PVDF membranes. *Sci. Rep.* **2017**, *7*, 4028. [CrossRef]
101. Lee, Y.S.; Arinzeh, T.L. The influence of piezoelectric scaffolds on neural differentiation of human neural stem/progenitor cells. *Tissue Eng. Part A* **2012**, *18*, 2063–2072. [CrossRef]
102. Ikada, Y.; Shikinami, Y.; Hara, Y.; Tagawa, M.; Fukada, E. Enhancement of bone formation by drawn poly(L-lactide). *J. Biomed. Mater. Res. Part A* **1996**, *30*, 553–558. [CrossRef]
103. Barroca, N.; Daniel-da-Silva, A.; Gomes, P.; Fernandes, M.; Lanceros-Mendez, S.; Sharma, P.; Gruverman, A.; Fernandes, M.; Vilarinho, P. Suitability of PLLA as piezoelectric substrates for tissue engineering evidenced by microscopy techniques. *Microsc. Microanal.* **2012**, *18*, 63–64. [CrossRef]
104. Barroca, N.; Vilarinho, P.M.; Daniel-da-Silva, A.L.; Wu, A.; Fernandes, M.H.; Gruverman, A. Stability of electrically induced-polarization in poly (L-lactic) acid for bone regeneration. *Appl. Phys. Lett.* **2011**, *98*, 133705. [CrossRef]

MDPI
St. Alban-Anlage 66
4052 Basel
Switzerland
Tel. +41 61 683 77 34
Fax +41 61 302 89 18
www.mdpi.com

Nanomaterials Editorial Office
E-mail: nanomaterials@mdpi.com
www.mdpi.com/journal/nanomaterials